Molecular Biology of HLA Class II Antigens

Editor

Jack Silver, Ph.D.

Chief
Division of Molecular Medicine
North Shore University Hospital
Cornell University Medical College
Manhasset, New York

CRC Press, Inc.
Boca Raton, Florida

Library of Congress Cataloging-in-Publication Data

Molecular biology of HLA class II antigens / editor, Jack Silver.
    p. cm.
    Includes index.
    ISBN 0-8493-4785-8
    1. HLA class II antigens. I. Silver, Jack, 1946-
    [DNLM: 1. Histocompatibility Antigens Class II. 2. Molecular
Biology. WO 680 M718]
QR184.32.M65 1990
616.07′92—dc20
DNLM/DLC
for Library of Congress
                                     89-15776
                                           CIP

Direct all inquiries to CRC Press, Inc., 2000 Corporate Blvd., N.W., Boca Raton, Florida 33431.

© 1990 by CRC Press, Inc.

International Standard Book Number 0-8493-4785-8

Library of Congress Card Number 89-15776
Printed in the United States

PREFACE

The HLA field has made remarkable progress over the past 20 years. From its humble beginnings as a region known for its polymorphism and where it was used primarily for tissue typing, it has developed to the point where it now represents one of the major areas of research in immunology. The twelve chapters in this book show how the HLA field, when viewed from several different perspectives, is not only a fascinating area of research but also highly informative with respect to many basic concepts in biology and medicine.

The first chapter of the book by Inoko et al. deals with the genetic organization of the class II region in a number of different haplotypes as determined by pulsed field gel electrophoresis. It demonstrates how remarkably dynamic and diverse this region is even when viewed from the macromolecular level. This is followed by a chapter by Bach et al. dealing with an analysis of class II diversity using cellular reagents (homozygous typing cells and T-cell clones) and demonstrates the exquisite specificity of these reagents.

The next two chapters by Charron et al. and Long et al. deal with two additional types of diversity, aside from allelic diversity, displayed by class II molecules — diversity generated by the *trans*-complementation of DQα and DQβ chains and, mixed isotype diversity generated by the association of α and β chains of different isotype (e.g., DRαDQβ). The possible role of the latter in the etiology of autoimmune diseases is also discussed.

The chapter by Wilkinson et al. combines T-cell clones and L cells transfected with class II cDNA to demonstrate how the two together can effectively be used to study the antigen presentation function of class II genes. This is followed by four chapters, all of which analyze class II diversity in greater detail at the molecular level (DNA sequencing, and allele specific oligonucleotide typing of PCR amplified DNA). These chapters also discuss the relationship of allelic diversity to human autoimmune diseases such as rheumatoid arthritis (Gregersen and Silver), insulin-dependent diabetes mellitus (Erlich et al., Sinha et al., Nepom and Nepom), and *pemphigus vulgaris* (Erlich et al., Sinha et al.).

The next chapter by Abastado and Kourilsky deals with gene conversion, a genetic mechanism frequently ascribed to the class II system as a means of generating diversity. This is followed by two chapters by Mach et al. and Calman and Peterlin which describe in detail the analysis of several enhancers and promoters found in class II genes and provide the first glimpses of how the expression of class II genes is regulated. This is an area of research that is rapidly growing and in which additional major discoveries are yet to be made.

This book represents only the tip of the iceberg as far as our understanding of the function of the HLA system is concerned. Future developments such as the X-ray crystallographic structure of HLA class II molecules, their role in determining the T-cell repertoire and finally an understanding of how they determine susceptibility to autoimmune diseases will lead to an even more comprehensive picture of the HLA system.

HLA CLASS II NOMENCLATURE

It is safe to say that one of the main obstacles for researchers entering the HLA field is its nomenclature and that this alone represents a greater barrier to entry than the molecular complexity of the system itself. I feel obligated, therefore, to at least attempt to provide some introduction to this arcane area with the hope that the novice reader will then still have sufficient patience to peruse through the text of the book.

The terminology used to define the HLA class II system has evolved over a period of 15 years in parallel with, and as a consequence of our increasing knowledge of the function, structure, and genetic organization of the HLA class II region. Indeed, this region, which was initially thought to represent a single genetic locus designated HLA-D, is now known to encompass 1.1 megabases of genomic DNA and to contain multiple genes encoding the α and β chains that make up class II molecules. As with other multigene families the HLA class II genes can be clustered into groups and placed within subregions (DP, DQ, and DR) on the basis of chromosomal position and DNA sequence similarity; α(β) genes in the same subregion are approximately 90 to 95% identical while those in different subregions differ by at least 30%. The positions and names currently applied to these genes (as well as their previous designations) are shown in the figure below (Figure 1). As can be seen, there are at least 6 α and 8 β chain genes. However, despite this large number of genes, only four class II complexes are actually produced, DP, DQ, and 2 DR. This is due to the fact that many of the genes are either defective (pseudogenes) or expressed at low levels (e.g., DZα, DOβ) and, because, in general, only α and β chains encoded within the same subregion can form functional heterodimers.

One of the more striking characteristics of both class I (HLA-A, -B, and -C) and class II genes is their high degree of polymorphism. Indeed, the HLA complex is the most polymorphic genetic region known in man. One of the earliest means of detecting and defining this polymorphism was serologic typing. Sera from multiparous women who had become naturally immunized to paternal class II antigens could be used to distinguish various alleles. As additional sera were obtained and screened, additional class II specificities were

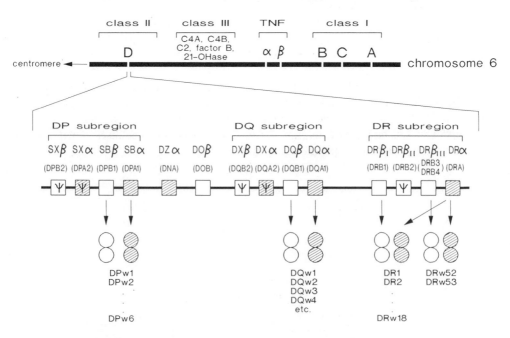

FIGURE 1.   Genetic organization and nomenclature of the HLA class II system.

identified resulting often in what are referred to as "splits" of previously designated spec-ificities. One of the drawbacks to this method was that different laboratories used different sera to perform their typing. As a consequence, different designations were often applied to the same allele by different laboratories. Many of these differences began to be resolved with the convening of International Histocompatibility workshops whose primary mission was the standardization of reagents and nomenclature. At the conclusion of the last workshop the nomenclature shown in the figure above in which the prefixes DQ and DR are used for serologically identifiable specificities associated with these two subregions was adopted. It should be noted that some of these specificities are based on detection with monoclonal antibodies (e.g., DQw7 and DQw8 formerly known as DQw3.1 and DQw3.2, respectively) which are a rapidly evolving and increasingly powerful tool in defining the polymorphism of the HLA class II system.

In addition to serologic typing it is also possible to define the polymorphism of the class II system with cellular reagents. Two methods have been employed. One is known as the mixed lymphocyte reaction (MLR) and is based on the observation that lymphocytes of disparate HLA class II type stimulate each other and proliferate in culture due to the rec-ognition of "foreign" class II molecules on B cells by the T cell receptor. This property, which distinguishes the class II genes from the class I genes, is the basis for some of the Dw nomenclature that was originally used to indicate differences within the class II region. This form of typing became standardized with the availability of HLA homozygous B cell lines (HTC or homozygous typing cells) that could serve as stimulators in the MLR and thus represented a common panel of cells available to everyone (Figure 2A). Typing by this

A. Mixed lymphocyte typing

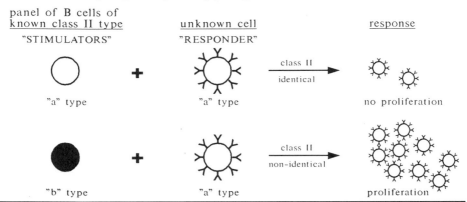

B. Primed lymphocyte typing

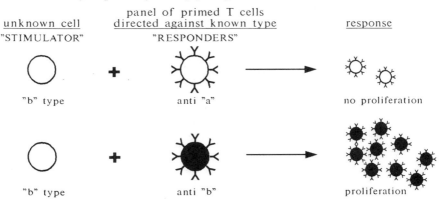

FIGURE 2.   HLA class II typing with cellular reagents.

method was rapidly followed by another cellular method which used panels of T cells primed in an MLR as reagents to detect differences. In this procedure, lymphocytes from individuals of unknown type are tested against a panel of primed T cells directed against known class II specificities (Figure 2B); it differs from the traditional MLR (Dw typing) where T cells are tested against a panel of B-cell lines of known type (HTC). This method of typing, known as PLT or primed lymphocyte typing, has proven to be especially useful for distinguishing alleles at loci which are not detectable by Dw (MLR) typing; variants detected by primed lymphocyte typing (PLT) were originally given the designations SB1, SB2, etc. and are now denoted, in most cases, as DP antigens since most of them appear to be encoded by genes at this locus (Figure 1). More recently, just as with serologic typing, cloned T cells have become powerful reagents in the further delineation of class II specificities and are responsible, for example, for the subdivision of the DRw52 specificity into the Dw24, Dw25, and Dw26 specificities. Continued ''splitting'' of the Dw specificities will undoubtedly occur as additional cellular reagents become available.

With the advent of molecular biology and its application to the HLA system it has now become possible to precisely define class II allelic variation. Methods that have been employed include restriction fragment length polymorphism (RFLP), oligonucleotide typing, and direct DNA sequencing of alleles (see Appendix). These studies have revealed even greater complexity and allelic variation in the class II system than was previously recognized and are the subject of several of the chapters in this book.

THE EDITOR

Jack Silver, Ph.D., is Chief of the Division of Molecular Medicine in the Department of Medicine at North Shore University Hospital and Professor of Medicine at Cornell University Medical College. Dr. Silver received his B.S. degree from Brooklyn College, Brooklyn, New York in 1966 and his Ph.D. degree in 1971 from the Department of Biochemistry, Boston University School of Medicine. After doing postdoctoral work in the Division of Biology, California Institute of Technology, Pasadena, he was appointed Assistant Member at Scripps Clinic and Research Foundation, La Jolla, in 1976. From 1981-1984, Dr. Silver was an Associate Professor in the Department of Microbiology, Michigan State University, E. Lansing. In 1985, he was appointed Director of the Cellular and Molecular Biology Unit at the Hospital for Joint Diseases, New York University Medical Center. He assumed his current position in 1989. Dr. Silver is a member of the American Society for Histocompatibility and Immunogenetics, American Association of Immunologists, and an Associate Editor for the Journal of Immunology. He was a recipient of the Established Investigatorship Award from the American Heart Association (1974 to 1979).

He has been the recipient of research grants from the National Institutes of Health, the National Science Foundation, the American Cancer Society, the National Multiple Sclerosis Society, the Kroc Foundation, and the American Heart Association. He has published more than 100 papers on the structure and function of the major histocompatibility complex and on the regulation of gene expression. His current research interests relate to the structure and function of HLA Class II molecules and their role in determining susceptibility to autoimmune diseases.

CONTRIBUTORS

Jean-Pierre Abastado, Ph.D.
Head of Research at C.N.R.S.
Pasteur Institute
Paris, France

Hans Acha-Orbea, Ph.D.
Research Fellow
The Ludwig Institute for Cancer Research
Lausanne, Switzerland

Daniel M. Altmann
Department of Human Immunogenetics
Imperial Cancer Research Fund
Holborn, London, United Kingdom

Asako Ando, M.D.
Department of Transplantation
School of Medicine
Tokai University
Kanagawa, Japan

Fritz H. Bach, M.D.
Harry Kay Chair in Immunobiology
Department of Laboratory Medicine and
  Pathology
University of Minnesota
Minneapolis, Minnesota

C. Berte
Department of Microbiology
University of Geneva Medical School
Geneva, Switzerland

Teodorica Bugawan
Research Associate
Department of Human Genetics
Cetus Corporation
Emeryville, California

David Burroughs, M.D.
Resident
Laboratoire d'Immunogénétique
  Moléculaire
Institut Biomedical des Cordeliers
Paris, France

Andrew F. Calman, M.D., Ph.D.
Resident Physician
Department of Ophthalmology
University of California
San Francisco, California

D. Charron, M.D.
Professor and Director
Department of Hematology and
  Laboratoire d'Immunogenetique
  Moleculaire
Institut Biomedical des
  Cordeliers
Chu Pitie Salpetriere
Paris, France

Henry Erlich, Ph. D.
Director
Department of Human Genetics
Cetus Corporation
Emeryville, California

Annick Faille, M.D.
Laboratoire d'Immunogénétique
  Moléculaire
Institut Biomedical des Cordeliers
Paris, France

Peter K. Gregerson, M.D.
Director
Laboratory of Molecular Immunology
Division of Molecular Medicine
North Shore University Hospital
Manhasset, New York

Pierre Hermans
Research Assistant
Laboratoire d'Immunogénétique
  Moléculaire
Institut Biomedical des Cordeliers
Paris, France

Glenn Horn, Ph.D.
Integrated Genetics
Framingham, Massachusetts

Hitoshi Ikeda
Department of Human Genetics
Imperial Cancer Research Fund
Holborn, London, United Kingdom

Hidetoshi Inoko, M.D., Ph.D.
Department of Transplantation
School of Medicine
Tokai University
Kanagawa, Japan

Steven Jacobson, Ph.D.
Senior Staff Fellow
Neuroimmunology Branch
National Institute of Neurological and
  Communicative
Disorders and Stroke
National Institutes of Health
Bethesda, Maryland

Li Ya Ju
Laboratoire d'Immunogénétique
  Moléculaire
Institut Biomedical des Cordeliers
Paris, France

Jun Kawai
Department of Transplantation
Tokai University School of Medicine
Kanagawa, Japan

Philippe Kourilsky, M.D.
Director of Research at C.N.R.S.
Pasteur Institute
Paris, France

Eric O. Long, Ph.D.
Visiting Scientist
Laboratory of Immunogenetics
National Institute of Allergy and
  Infectious Diseases
National Institutes of Health
Bethesda, Maryland

Vincent Lotteau, Ph.D.
Post-doctoral Fellow
Laboratoire d'Immunogénétique
  Moléculaire
Institut Biomedical des Cordeliers
Paris, France

Bernard Mach, Ph.D.
Professor and Chairman
Department of Microbiology
University of Geneva Medical School
Geneva, Switzerland

Hugh O. McDevitt, M.D.
Professor and Chairman
Department of Microbiology and
  Immunology
Stanford University School of Medicine
Stanford, California

Barbara S. Nepom, M.D.
Affiliate Member
Department of Immunology
Virginia Mason Research Center
Seattle, Washington

Gerald T. Nepom, M.D., Ph.D.
Director
Immunology Program
Virginia Mason Research Center
Seattle, Washington

B. Matija Peterlin, M.D.
Assistant Professor
Departments of Medicine and
  Microbiology and Immunology
University of California
San Francisco, California

Nancy L. Reinsmoen, Ph.D.
Scientist
Immunology Laboratory
University of Minnesota Hospital and
  Clinic
Minneapolis, Minnesota

W. Reith
Department of Microbiology
University of Geneva Medical School
Geneva, Switzerland

Sandra Rosen-Bronson, Ph.D.
Research Associate
Department of Microbiology
Georgetown University
Washington, D.C.

Stephen Scharf
Research Associate
Department of Human Genetics
Cetus Corporation
Emeryville, California

Rafick P. Sekaly, Ph.D.
Laboratory Director
Department of Molecular Immunology
Clinical Research Institute of Montreal
Montreal, Quebec, Canada

Jack Silver, Ph.D.
Chief
Division of Molecular Medicine
North Shore University Hospital
Cornell University Medical College
Manhasset, New York

Animesh A. Sinha, M.D., Ph.D.
Post-doctoral Research Scholar
Department of Microbiology and
  Immunology
Stanford University
Stanford, California

Luc Teyton, M.D., Ph.D.
Post-doctoral Fellow
Laboratoire d'Immunogénétique
  Moléculaire
Institut Biomedical des Cordeliers
Paris, France

J. -M. Tiercy
Department of Microbiology
University of Geneva Medical School
Geneva, Switzerland

Luika A. Timmerman, M.S.
Life Science Research Associate
Department of Medical Microbiology
Stanford Medical School
Stanford, California

John Todd, Ph.D.
Research Fellow
Nuffield Department of Surgery
Institute of Molecular Medicine
John Radcliffe Hospital
Oxford, England

John Trowsdale, Ph.D.
Department of Human Immunogenetics
Imperial Cancer Research Fund
Holborn, London, United Kingdom

Kimiyoshi Tsuji, M.D.
Department of Transplantation
Tokai University School of Medicine
Kanagawa, Japan

Pascale Turmel
Technical Assistant
Laboratoire d'Immunogénétique
  Moléculaire
Institut Biomedical des Cordeliers
Paris, France

David Wilkinson
Department of Human Immunogenetics
Imperial Cancer Research Fund
Holborn, London, United Kingdom

Toshio Yabe, Ph.D.
Research Associate
Immunobiology Research Center
Department of Laboratory Medicine and
  Pathology
University of Minnesota
Minneapolis, Minnesota

TABLE OF CONTENTS

Chapter 1

MAPPING OF THE HLA-D REGION BY PULSED-FIELD GEL ELECTROPHORESIS: SIZE VARIATION IN SUBREGION INTERVALS

Hidetoshi Inoko, Asako Ando, Jun Kawai, John Trowsdale, and Kimiyoshi Tsuji

TABLE OF CONTENTS

I. ABSTRACT

Pulsed field gel electrophoresis (PFGE) and cosmid walking analysis were used to construct genomic maps of the human leucocyte antigen (HLA)-D region. The total size of the HLA class II gene region varies from 1000 to 1100 kb depending on the HLA haplotype. In the course of these experiments a new HLA class II β chain sequence, DVβ, was located close to DXα between DXα and DQβ. All other markers studied from the DP, DZ/DO, DQ, and DR subregions were consistent in their positions in the haplotypes studied: centromere-DPβ2-DPα2-DPβ1-DPα1-DO(DZ)α-DOβ-DXβ-DXα-DVβ-DQβ-DQα-DRβI-DRβII-DRβIII-β1 exon-DRα-telomere. However, the distance between DQα and DRα or DVβ and DQβ differed in various haplotypes. For example, in DR7 (MANN) the distance between DQα and DRα was approximately 380 kb, but in the DR2 haplotype (AKIBA) it was 270 kb. This may be due to variation, either in the number of duplicated DRβ genes or in the length of other uncharted DNA, in the HLA-D region.

II. INTRODUCTION

The human HLA-D region of the major histocompatibility complex (MHC) constitutes a multigene family encoding polymorphic cell-surface glycoproteins called class II antigens, which are involved in regulation of the immune response. Each class II antigen is a heterodimer composed of a 34 kb α chain and a 28 kb β chain. Knowledge of the organization of the HLA-D region has developed through a combination of techniques, including class immunology and biochemistry, molecular cloning, and, more recently, PFGE.[1-5] The region may be divided into several subregions, DP, DZ/DO, DQ, and DR, although each contains a complex array of genes and pseudogenes. The PFGE data obtained by Hardy et al.[2] and Lawrance et al.[4] indicate that the HLA-D region spans over 1000 kb, but the precise genomic organization of the class II gene loci, including the gene order and orientation or the distances between the subregions, has not been determined. We used the PFGE technique to establish detailed maps of the HLA-D region from serologically different HLA-homozygous B-cell lines with the aid of the restriction maps obtained by digestion of cosmid clone DNAs. α- and β-chain probes from all of the subregions so far available were used, including DVβ, a new HLA class II pseudogene isolated recently,[6] and a DX region-specific probe. During the course of these experiments, size variation in the subregion intervals was found among different HLA haplotypes.

III. RESULTS

A. RESTRICTION ENZYMES USEFUL FOR PFGE ANALYSIS OF THE HLA CLASS II GENE REGION

The PFGE technique permits analysis of DNA fragments up to 2000 kb using infrequently cutting restriction enzymes which generally recognize 8 bp cleavage sites and/or contain CpG.[7,8] High molecular weight genomic DNA was prepared in agarose blocks, initially from the B cell line MANN. The DNA was cut with ten enzymes singly and in all pairwise combinations: *Sfi*I, *Cla*I, *Sal*I, *Sac*II, *Bss*HII, *Nar*I, *Mlu*I, *Not*I, *Nae*I, and *Nru*I. Digested DNAs were separated on PFGE gels, transferred to nylon membranes and hybridized sequentially with eleven probes from the HLA class II region (Table 1). Figure 1 shows the results of a typical experiment, using the same filter hybridized with DRα (a), DRβ (b), and DQα (c). *Sfi*I digestion was informative. Two bands, of 260 kb and 180 kb, were obtained when this enzyme was used with the DQα probe. This probe detects both DQα and DXα due to their close sequence homology.[15] Since the smaller, 180 kb *Sfi*I fragment hybridized to the DX region-specific probe (0.4 kb *Eco*RI genomic fragment 20 kb upstream

TABLE 1
HLA Probes Used for PFGE Analysis

Probe	Description	Ref.
DPα	1.2 kb *Eco*RI cDNA fragment	9
DPβ	1.0 kb *Eco*RI-*Hind*III cDNA fragment	10
DO(DZ)α	1.1 kb cDNA clone	40
DQβ	1.4 kb *Bam*HI cDNA fragment	11
DXβ upstream	0.4 kb *Eco*RI genomic fragment: 20 kb upstream from DXβ	38
DVβ	2 kb *Pst*I genomic fragment	6
DQβ	0.4 and 0.8 kb *Eco*RI cDNA fragments	41
DQα	1.6 kb *Bam*HI-*Hind*III cDNA fragment	42
DRβ	0.8 kb *Hind*III-*Sal*I cDNA fragment	12
β1 exon	0.95 kb *Hind*III-*Pst*I genomic fragment	13
DRα	2 × 0.7 kb *Pst*I cDNA fragment	14

of DXβ), the 260 kb and 180 kb fragments represent DQα and DXα, respectively. Similarly, the DOβ probe hybridized with the 180 kb fragment and the 260 kb DQα band was shared with one of the DRβ hybridizing bands, establishing the order of the four subregions as DO-DX-DQ-DR. *Nar*I and *Sac*II digestion produced 150 kb bands which hybridized with both the DOβ and DXβ region specific probes, but not with DQα or DQβ, permitting the orientation of the DXβ and DXα loci to be determined, DXβ being proximal to DXα (Figure 2).

*Sfi*I digestion and hybridization with the DQβ probe revealed another distinctive band at 160 kb which, by elimination of the fragments described above, must represent the DQβ gene. Since it was established by gene cloning that DQα and DQβ are 15 kb apart,[17] the gene order is clearly DXβ-DXα-DQβ-DQα, and the distance between DXα and DQβ is more than 145 kb. These data are consistent with previously published results.[2]

B. A NEW β CHAIN PSEUDOGENE, DVβ, BETWEEN DXα AND DQβ

A phage genomic clone, λHMCβ19, specifying a new HLA class II antigen β chain pseudogene named DVβ, was isolated from a human genomic library[18] (not typed for HLA) with the DQβ cDNA probe under low stringency conditions. Southern hybridization analysis identified the β2 domain exon (exon 3) and the TM-CY exon (transmembrane-cytoplasmic exon, exon 4), but the first, second, and fifth exon encoding the 5'UT-leader, the β domain and the 3' UT domain of normal β chains, respectively, were entirely missing. Determination of the nucleotide sequence of DVβ showed several deleterious mutations including deletions and premature terminations (Figure 3). Two codons are deleted at positions 109 and 147, and two premature termination signals were detected at positions 153 and 188 (the amino acid numbers and exon designation are modeled on the DQβ chain gene). Amino acid replacement of a cysteine residue to tyrosine was found at position 173. This cysteine is necessary to form the disulfide bridge to preserve functional conformation of the class II antigen α chain. From these results, it can be concluded that the DVβ gene is a truncated pseudogene and cannot express a class II light chain. Comparison of the nucleotide and amino acid sequences of the β2 domain exon from the DVβ gene with those of other class II antigen β chains shows the highest degree of similarity with DQβ and DXβ (Table 2). With respect to the transmembrane-cytoplasmic domain exon, a similar level of nucleotide sequence matching (71 to 73%) was observed between DVβ versus DQβ, DXβ, DRβ, DRw53β and DRβ ψ (pseudo) genes.

Southern hybridization analysis using genomic DNAs from somatic human-mouse hybrids containing the X/6 translocation chromosome clearly showed that the DVβ sequences are encoded around the 6p2I region on the short arm of chromosome 6, where the HLA

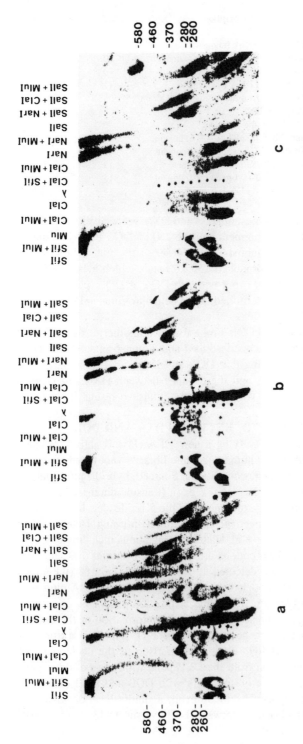

FIGURE 1.   PFGE analysis of genomic DNA from the human B cell line MANN (DR7). The DNA in each track was cleaved with the enzymes shown and subjected to PFGE separation. After Southern transfer, the filter was hybridized sequentially with three probes: (a) DRα, (b) DRβ and (c) DQα. The black dots show the positions of multimers of phage CI857S7, which has a unit size of 48.5 kb. In a and b, visualization of the markers was by hybridization to ^{32}P-lambda DNA. Yeast chromosomes separated in the tracks at both ends of the gel provided additional markers (sizes in kb and shown). Some *Cla*I and *Nar*I sites around the DR, DQ, and DQ subregions which are sensitive to partial cleavage were not taken into account for restriction mapping (Figure 2). *Sal*I digestion also produced some faint bands due to partial cleavage.

5

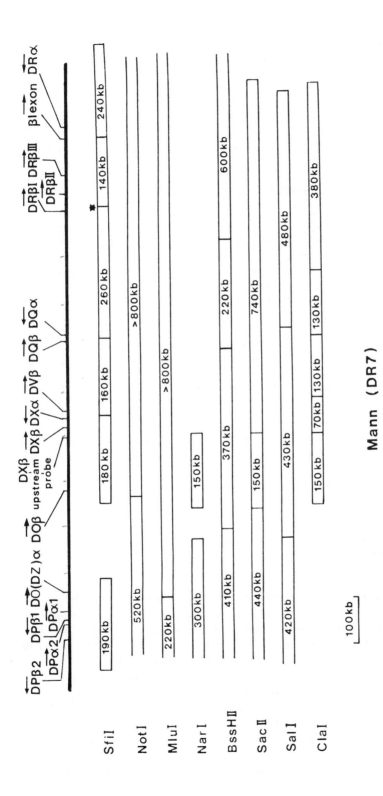

Mann (DR7)

FIGURE 2. Long range restriction maps of the HLA class II regions of DR7 (MANN) haplotype. Fragments obtained with each enzyme are illustrated to scale opposite the class II genes with which hybridization has been demonstrated. Open spaces mean that no fragment could be identified due to lack of probes available or to complex band patterns from persistent partial digestion. Restriction sites were mapped in relation to each other by double digestion, and ambiguous restriction sites due to partial digestion were not included (e.g., ClaI sites within the DR subregion as shown in Figure 1). The location of the DPα2 and DPβ2 genes and the gene order of the DRβIII and βI exon locus are assigned on the basis of gene cloning studies.[6,13,16] Arrows indicate gene orientation (5' → 3'). For details see text. The star indicates restriction sites sensitive to partial cleavage.

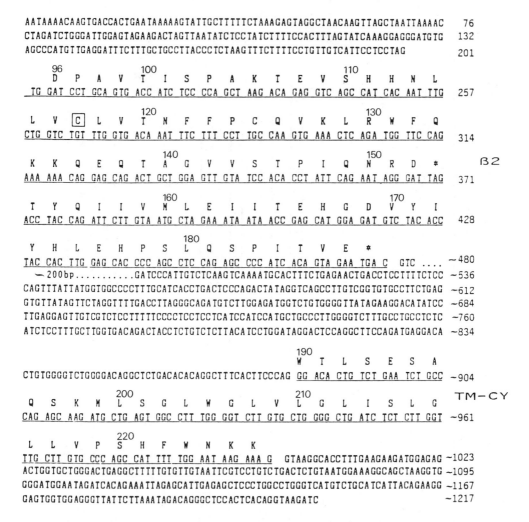

```
AATAAAACAAGTGACCACTGAATAAAAAGTATTGCTTTTTCTAAAGAGTAGGCTAACAAGTTAGCTAATTAAAAC    76
CTAGATCTGGGATTGGAGTAGAAGACTAGTTAATATCTCCTATCTTTTCCACTTTAGTATCAAAGGAGGGATGTG   132
AGCCCATGTTGAGGATTTCTTTGCTGCCTTACCCTCTAAGTTTCTTTTCCTGTTGTCATTCCTCCTAG          201
```

```
       96                    100                            110
        D   P   A   V   T   I   S   P   A   K   T   E   V   S   H   H   N   L
    TG GAT CCT GCA GTG ACC ATC TCC CCA GCT AAG ACA GAG GTC AGC CAT CAC AAT TTG   257
```

```
              120                            130
    L   V  [C]  L   V   T   N   F   F   P   C   Q   V   K   L   R   W   F   Q
    CTG GTC TGT TTG GTG ACA AAT TTC TTT CCT TGC CAA GTG AAA CTC AGA TGG TTC CAG   314
```

```
                             140                            150
    K   K   Q   E   Q   T   A   G   V   V   S   T   P   I   Q   N   R   D   *          β2
    AAA AAA CAG GAG CAG ACT GCT GGA GTT GTA TCC ACA CCT ATT CAG AAT AGG GAT TAG   371
```

```
                             160                            170
    T   Y   Q   I   I   V   M   L   E   I   I   T   E   H   G   D   V   Y   I
    ACC TAC CAG ATT CTT GTA ATG CTA GAA ATA ATA ACC GAG CAT GGA GAT GTC TAC ACC   428
```

```
                             180
    Y   H   L   E   H   P   S   L   Q   S   P   I   T   V   E   *
    TAC CAC TTG GAG CAC CCC AGC CTC CAG AGC CCC ATC ACA GTA GAA TGA C  GTC ....  ~480
    — 200bp..........GATCCCATTGTCTCAAGTCAAAATGCACTTTCTGAGAACTGACCTCCTTTTCTCC    ~536
    CAGTTTATTATGGTGGCCCCTTTGCATCACCTGACTCCCAGACTATAGGTCAGCCTTGTCGGTGTGCCTTCTGAG   ~612
    GTGTTATAGTTCTAGGTTTTGACCTTAGGGCAGATGTCTTGGAGATGGTCTGTGGGGTTATAGAAGGACATATCC   ~684
    TTGAGGAGTTGTCGTCTCCTTTTTCCCCTCCTCCTCATCCATCCATGCTGCCCTTGGGGTCTTTGCCTGCCTCTC   ~760
    ATCTCCTTTGCTTGGTGACAGACTACCTCTGTCTCTTACATCCTGGATAGGACTCCAGGCTTCCAGATGAGGACA   ~834
```

```
                                             190
                                              W   T   L   S   E   S   A
    CTGTGGGGTCTGGGGACAGGCTCTGACACACAGGCTTTCACTTCCCAG GG ACA CTG TCT GAA TCT GCC  ~904
```

```
                 200                            210                          TM—CY
    Q   S   K   M   L   S   G   L   W   G   L   V   L   G   L   I   S   L   G
    CAG AGC AAG ATG CTG AGT GGC CTT TGG GGT CTT GTG CTG GGG CTG ATC TCT CTT GGT   ~961
```

```
                 220
    L   L   V   P   S   H   F   W   N   K   K
    TTG CTT GTG CCC AGC CAT TTT TGG AAT AAG AAA G  GTAAGGCACCTTTGAAGAAGATGGAGAG   ~1023
    ACTGGTGCTGGGACTGAGGCTTTTTGTGTTGTAATTCGTCCTGTCTGACTCTGTAATGGAAAGGCAGCTAAGGTG   ~1095
    GGGATGGAATAGATCACAGAAATTAGAGCATTGAGAGCTCCCTGGCCTGGGTCATGTCTGCATCATTACAGAAGG   ~1167
    GAGTGGTGGAGGGTTATTCTTAAATAGACAGGGCTCCACTCACAGGTAAGATC                         ~1217
```

FIGURE 3. Nucleotide and predicted amino acid sequences of the DVβ gene. A 2.0 kb *Pst*I fragment from lambda HMCβ19 isolated from the human genomic library constructed by Lawn et al.[18] was subjected to sequence analysis by the M13 chain termination method. The translated portion of the putative third and fourth exons of the DVβ gene was identified by comparison to the DQβ cDNA clone.[19] The conserved cysteine residue is boxed. The amino acid numbers and exon designations are modeled on the DQβ chain gene. Stars indicate two premature termination codons.

genes are located. Further, in long-range physical mapping with the PFGE technique, the DVβ probe detected similar bands to DXα and DXβ in most cases, suggesting close linkage. However, the 160 kb *Sfi*I fragment positive with DVβ was shared only with DQβ (Figure 2). Based on these results, the gene order is established as DOβ-DXβ-DXα-DVβ-DQβ-DQα. The linkage of DVβ and DXα was confirmed by cosmid cloning.

One DVβ genomic clone, pA412, was isolated from the HLA-homozygous DR2 cell line, AKIBA. Southern hybridization studies with two single-copy genomic probes isolated from this clone revealed that it overlapped one cosmid clone (pAX523, encoding the DX subregion) among our class II cosmid collection (Figure 4). About 100 kb of continuous genomic DNA around DX was mapped, placing the DVβ gene 15 kb upstream from the DXα gene.

This location of DVβ in close proximity to DXα was confirmed by RFLP studies. The 2.0 kb *Pst*I DVβ genomic fragment was hybridized to a Southern blot of *Taq*I- and *Msp*I-

TABLE 2
Percent Nucleotide and Amino Acid Identities between DVβ
and the Human or Mouse Class II Antigen β Chain Genes

	DVβ			
	β exon (282 bp)		CP-TM-CY exon (111 bp)	
	Nucleotide	Amino acid	Nucleotide	Amino acid
DQβ	75	65	73	57
DXβ	75	62	73	60
DRβ(DR4)	66	54	72	57
DRβ(DRw53)	64	52	73	57
DRβψ(DR4)	63	51	71	57
DPβ	66	56	63	49
DPβψ	63	51	64	46
DOβ	62	47	61	35
I-Aβ^b	66	55	67	51
I-Eβ^d	62	47	67	51

Note: Numbers represent the percentage of identical position.

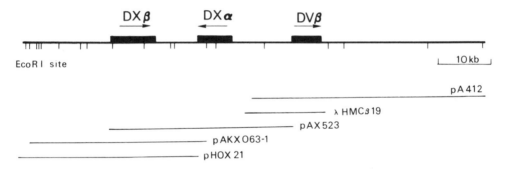

FIGURE 4.   Molecular maps of the DX-DV subregion. The location and orientation of the DXα, DXβ, and DVβ genes, shown as black boxes, were determined by Southern blot analyses using the separated 5′ and 3′ end fragments from the DQα or DQβ cDNA clones as hybridization probes, and by nucleotide sequence analysis. Five overlapping phage and cosmid clones contain one DQα gene, one DXβ gene, and one DVβ gene. Lambda HMCβ19 was isolated from the human phage genomic library[18] (not typed for HLA). The other four clones were isolated from the AKIBA cosmid genomic library (DR2).

digested cellular DNAs from 36 HLA-homozygous human B cell lines with different DR specificities, and two alleles were defined with each enzyme, 3.4 kb vs. 2.3 kb (*Taq*I) or 23 kb vs. 10 kb (*Msp*I). Although these polymorphisms around DVβ were not obviously associated with any of the DR, DQ, DP, and Dw serological or cellular types nor with any variation around the DR, DQ, DP, or DO (DZ) subregions as detected by RFLP studies, they showed complete association with alleles of the DXα gene detected with the *Taq*I enzyme. Sequence analysis[5] and restriction fragment length polymorphism (RFLP) analysis[20] defined 1.9 kb and 2.1 kb *Taq*I fragments as alleles of DXα using a DQα probe. As summarized in Table 3, all of the HLA-homozygous cell lines showing the DVβ *Taq*I 2.3 fragments gave the DXα *Taq*I 1.9 kb fragments, whereas all other cell lines with the DVβ *Taq*I 3.4 kb fragments gave the DXα *Taq*I 2.1 kb fragments. Similarly, with *Msp*I polymorphism of DVβ there was virtually complete association of the DVβ and DXα. *Bam*HI digestion detected more extensive polymorphism (four alleles) around DVβ, which also correlated with DXα polymorphism but not obviously with any polymorphisms in other HLA class II subregions.

TABLE 3
Relationship between TaqI and MspI RFLPs of the DVβ and Those of DXα genes

Cell	Dw	DR	DXα TaqI		DVβ TaqI		DVβ MspI	
			2.1 kb	1.9 kb	3.4 kb	2.3 kb	23 kb	10 kb
CRB	1	1	−	+	−	+	−	+
Sa	1	1	−	+	−	+	−	+
CAH	1	1	−	+	−	+	−	+
Holm-1	1	1	−	+	−	+	−	+
MAM	1	1	−	+	−	+	−	+
WK-1	1	1	−	+	−	+	−	+
CI	2	2	+	−	+	−	+	−
PGF	2	2	+	−	+	−	+	−
AKIBA	12	2	−	+	−	+	−	+
TOK	12	2	−	+	−	+	−	+
FJO	FJO	2	−	+	−	+	−	+
SUD	4	4	+	−	+	−	+	−
JAH	4	4	+	−	+	−	+	−
PEA	13	4	−	+	−	+	−	+
Has15	15	4	−	+	−	+	−	+
Wa	15	4	−	+	−	+	−	+
KT3	15	4	−	+	−	+	−	+
KT2	KT2	4	+	−	+	−	+	−
KT17	KT2	4	+	−	+	−	+	−
HVB5B	5	5	−	+	−	+	−	+
MD	6	6	−	+	−	+	−	+
HOR	EN	6	+	−	+	−	+	−
0934ZUK	18	13	−	+	−	+	−	−
Car-6	19	13	+	−	+	−	+	−
SLE	19	13	+	−	+	−	+	−
KT11	19	13	+	−	+	−	+	−
MANN	7	7	−	+	−	+	−	+
OH7	7	7	−	+	−	+	−	+
BOR	7	7	−	+	−	+	−	+
PITOUT	17	7	−	+	−	+	−	+
BTB	8	8	−	+	−	+	−	+
Riek100	8	8	−	+	−	+	−	+
BH8B	8	8	−	+	−	+	−	+
KT12	DB5	9	−	+	−	+	−	+
KT14	DB5	9	−	+	−	+	−	−
HID	Ky	9	−	+	−	+	−	−

Probe: DXα: pDCα 107 (DQα cDNA) PstI 1.1 kb fragment; DVβ: λHMCβ 19 PstI 2.0 kb fragment.

		DXα TaqI				DXα TaqI	
		1.9 kb	2.1 kb			1.9 kb	2.1 kb
DVβ TaqI				DVβ MspI			
	2.3 kb	26	0		10 kb	26	0
	3.4 kb	0	10		23 kb	0	10

C. LARGE-SCALE GENOMIC RESTRICTION MAPS OF DR7 AND DR2 HAPLOTYPES

We adopted eight or nine of the ten infrequently cutting restriction enzymes described above for the construction of genomic restriction maps of the DR7 (MANN, Figure 2) and DR2 (AKIBA, Figure 5) haplotypes. These maps are justified by self-consistency of the band patterns obtained after double enzyme digestion of each pairwise combination. Further, with respect to the DR2 haplotype (AKIBA) we have a large collection of cosmid genomic clones encompassing a total of 600 kb of the HLA class II region isolated from the AKIBA gene library (indicated by black bars in Figure 5[38]), and all the restriction sites which are inferred to be located in the region encoded by our cosmid clones (from our PFGE analysis) are confirmed by restriction enzyme digestion of each cosmid clone as expected (marked arrowheads in Figure 3). The other restriction sites are located in the gap region between the genes and are thought not to be cloned in our cosmid clones so far. The low frequency of fragments containing both DOβ and DOα (DZα) is consistent with the placement of DOα close to DPα. However, the 340 kb ClaI and 330 kb MluI fragments hybridizing to DOβ also encompassed DOα, but not DPα, when DNA from AKIBA was used, establishing the gene order as DPα-DOα-DOβ. Four restriction enzyme sites (SfiI, SalI, SacII, and NarI) were in the 3' flanking region of the DO(DZ)α gene cloned by the cosmid clones (see the arrowheads in Figure 5), determining the orientation of DOα with its 5' end toward the centromere, which is the same as the orientation of the DPα1 and DPα2 genes in the DP subregion, although DOα has not been linked to the DP subregion by cosmid clones. In a similar way, the 3' end of the DOβ gene was determined to be toward the telomere, which is the same as the orientation of the β genes in the DX-DQ-DR subregions including DVβ, by restriction enzyme mapping of the cloned cosmids. DOβ does not appear to be a partner with DO(DZ)α as the loci are not coordinately expressed.[2,9]

In the DR subregion there are three or four DRβ genes, named DRβI, DRβII, DRβIII, (DRβIV) from the centromere side. One or two of the DRβ genes are pseudogenes.[16,21-24] In the DR3 and DR4 haplotypes, the DRβI gene encodes the polymorphic serologic specificity (DR3 or DR4) and the most telomeric DRβIII or IV gene comprises the supertypic or nonpolymorphic one (DRw52 or DRw53).[16,21] Similarly, in the DR2 haplotype two DRβ genes are known to be expressed, leading to formation of two DR products on the basis of biochemical and cDNA sequencing studies.[25-28] One of these expressed genes encodes a nonpolymorphic protein which is believed to correspond to the supertypic DR allospecificity in other haplotypes controlled by the DRβIII locus, such as DRw52 and DRw53 (in DR3 and DR4, respectively), although in the DR2 serotype such a supertypic antigen has not yet been serologically defined. The other encodes a protein polymorphic among the Dw2, Dw12, and Dw21 (MN2) subtypes of DR2, which is believed to specify the polymorphic DR2 allodeterminants and to correspond to the DRβI locus-encoding molecule. From the DR2 AKIBA genomic library, three DRβ and one DRα genes could be isolated and characterized (Figure 6). The DRβII gene is a pseudogene lacking the first exon which encodes the leader peptide, whereas the DRβI and DRβIII genes were expressed independently together with the DRα chain on mouse L cells by the method of gene-mediated transfection. Unexpectedly, restriction enzyme mapping clones and reactivity of their products expressed on the L-cell transfectants against monoclonal antibodies such as a nonpolymorphic anti DR1, DR2 mAb (Hu30)[29] clearly showed that the DRβI and DRβIII genes encoded the nonpolymorphic and polymorphic DRβ chains, respectively.[39] This gene arrangement is the reverse of those observed in haplotypes DR3, -4 and -6. The alignment of the HLA class II genes (including the DRβ genes) on chromosome 6 of the DR2 haplotype, however, was consistent with other haplotypes, namely DXβ-DXα-DVβ-DQβ-DQα-DRβI-DRβII-DRβIII-DRα from the centromere. These results suggest that the susceptibility to mutations or gene conversion responsible for genetic polymorphisms depends on the gene, not on its location on the

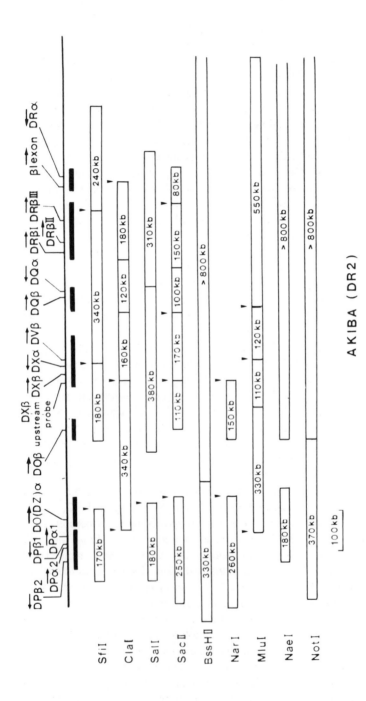

FIGURE 5.   Long range restriction maps of the HLA class II regions of DR2 (AKIBA) haplotype. Black bars represent a total of 600 kb of the HLA-D region defined by the cosmid clones isolated from the AKIBA gene library.[38] Their analysis showed that the DR subregion also contained three DRβ genes (DRβI, βII and βIII) and that DRβII was a pseudogene. Arrowheads indicate the restriction enzyme sites confirmed by enzyme digestion of these cosmid clones. For others, see the legend to Figure 2. Ambiguous restriction sites, due to partial digestion, were not included (e.g., *ClaI* and *BssHII* sites present in cloned DNA encoding the DR subregion).

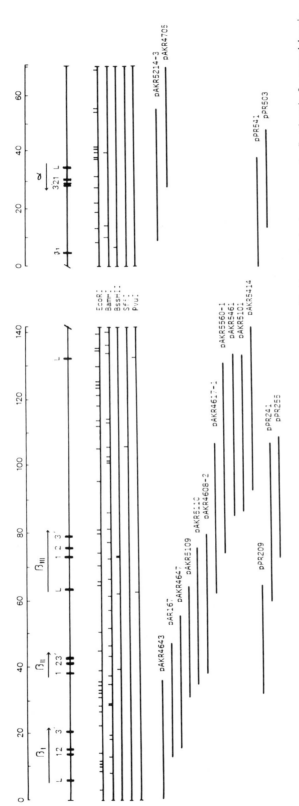

FIGURE 6.   Molecular map of the DR subregion. The top line gives the scale in kb. The location of exons is shown by filled boxes. L = leader sequence, 1 = first domain, 2 = second domain, 3' = 3' untranslated region, β₁ = an exon homologous to the first domain of the DRβ chain. The exons for the transmembrane region and the cytoplasmic tail are not indicated. Arrows above the map indicate the direction of transcription of each gene. The 140- and 70- kb stretches of DNA are defined by a number of overlapping cosmid clones shown below the cleavage sites of the restriction enzymes EcoRI, BamHI, BssHII, SfiI, and PvuI. Cosmid clones designated pAKR or pAR and pPR are derived from the libraries of AKIBA (DR2Dw12) and PFG (DR2Dw2), respectively. BssHII sites within the DRβ_{II} and DRβ_{III} genes present in cloned DNA were not found when genomic DNA was digested, probably due to partial digestion caused by methylated CpG dinucleotides.

chromosome. Furthermore, absorption experiments of anti-DR2 allosera using the DRα/β transfectants revealed that the so-called DR2 specificities were determined by multiple epitopes and that both the DRβI and DRβIII genes could specify these polymorphic DR2 epitopes.

D. SIZE POLYMORPHISMS IN SUBREGION INTERVALS

PFGE analyses demonstrate that the distances between adjacent subregions are fairly large, namely, about 200 kb between DOα and DOβ, 150 kb between DOβ and DXβ, and 130 kb between DVβ and DQβ (Figures 2 and 5). These distances were consistent for both cell lines, MANN (DR7) and AKIBA (DR2). In contrast, the distance between DQα and DRα differed in the two haplotypes, being greater in DR7. This is exemplified by the following three points. (1) The 340 kb *Sfi*I fragment in the DR2 haplotype contained the DV (but not DX), DQβ and DQα, as well as all of the DRβ-related (but not DRα) genes. In DR7 the total length of *Sfi*I fragments covering the same genes amounted to 560 kb. (2) The DQβ gene is located close to one end of the 220 kb *Bss*HII fragment in DR7 but it contained no DRβ sequences, indicating a space of about 200 kb between DQβ and DRβ. (3) Both DR7 and DR2 haplotypes appear to contain conserved *Sal*I sites, one in close proximity to the 5′ end of the DQα gene and the other about 100 kb away from the 5′ end of the DRα gene (Figures 2 and 5). This fragment was sized to 310 kb in DR2 and 480 kb in DR7, reflecting the size difference of the region between DQα and DRβ in the two haplotypes.

In order to measure the distance between the DQα and DRα genes in other haplotypes, PFGE analysis was extended to samples from other B cell lines with different DR specificities from DR1 to DRw8, and *Sal*I fragments derived from DNA of nine B cell lines with different HLA specificities hybridizing with the DRβ probe were examined. 1BW4 (DR1), PGF (DR2, Dw2), and OLL (DRw8) gave a *Sal*I fragment with a similar size (310 kb) to AKIBA (DR2, Dw12), whereas WT51 (DR4) was similar (480 kb) to MANN (DR7). A series of experiments using other restriction enzymes and other class II sequence probes demonstrated that two *Sal*I sites referred to above are generally conserved among these cell lines. One site is in close proximity to the 5′ end of the DQα gene and the other is located 100 kb away from the 5′ end of the DRα gene, as described above. These facts indicate that DR1, DR2, and DRw8 haplotypes most likely possess the shorter genomic sequence between DQα and DRβ with the same size (80 kb) as AKIBA (DR2, Dw12 haplotype), whereas DR4 haplotypes probably contain the longer one with the same size (240 kb) as MANN (DR7 haplotype). On the other hand, the DRβ probe recognized 550 kb and 500 kb *Sal*I fragments in WT49 (DR3) or WT46(DR6) and JVM (DR5) cell lines, respectively. The DQα probe also gave the same *Sal*I fragments in all these three cell lines, suggesting that the *Sal*I site detected just close to the DQα gene in other haplotypes is missing in these three cell lines. These facts prompted us to construct restriction maps of the DR, DQ, and DX subregions of JVM (DR5) and WT46 (DR6) by more extensive PFGE analysis, as shown in Figure 7 (DR3 and DR6 haplotypes are genetically related and might be expected to show the same restriction map).[30,31] This distance between DQα and DRα was found to be similar to that of AKIBA in these two additional cell lines. During the course of these experiments, we have also found that the distance between the DVβ and DQβ genes varies among different HLA haplotypes. DR5 and DR6 haplotypes possess longer genomic sequences between DVβ and DQβ with sizes of 260 kb and 170 kb, respectively, whereas DR2 and DR7 haplotypes contain the shorter one (130 kb) as described above. Size polymorphisms in the subregion interval are summarized in Table 4.

13

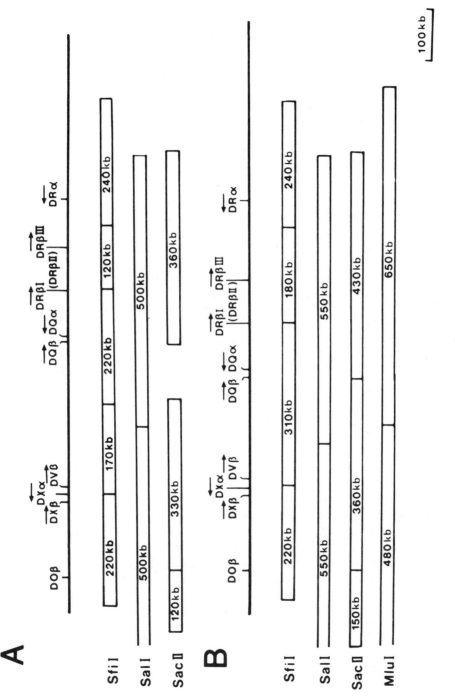

FIGURE 7. Long range restriction map of the HLA class II region of (A) DR5 (JVM) and (B) DR6 (WT46) haplotypes. See the legend to Figure 2 for details.

TABLE 4
Distances (kb) between the DQ-DR and
DV-DQ Regions in Different HLA
Haplotypes

HLA haplotype	Distance (kb)	
	DQα-DRβ	DVβ-DQα
IBW4(DR1)	80	N.D.
PGF(DR2Dw2)	80	N.D.
AKIBA(DR2Dw12)	80	130
WT49(DR3)	80	N.D.
WT51(DR4)	240	N.D.
JVM(DR5)	80	260
WT46(DRw6)	80	170
MANN(DR7)	240	130
OLL(DRw8)	80	N.D.

Note: N.D. = not determined.

IV. DISCUSSION

The PFGE analysis presented here establish the gene order in the HLA class II gene region as: centromere-DPβ-DPα-DO(DZ)α-DOβ-DXβ-DXα-DVβ-DQβ-DQα-(DRβI-DRβII-DRβIII)-DRα-telomere. The gene order and the rare-cutting restriction enzyme map constructed from DR2 (AKIBA) are quite consistent with our cosmid clone analysis as shown in Figure 5. As described above, there are gaps or open spaces between adjacent subregions which are large enough to encode new HLA class II genes or other expressed genes which may be candidates for susceptibility to HLA-associated diseases. In fact, the restriction sites for *Cla*I, *Nar*I, and *Sac*II enzymes are clustered away from the 5′ end of DXβ (Figures 2 and 3) and conserved among different HLA haplotypes. Such a clustering of sites is characteristic of the short unmethylated CpG-rich regions known as HpaII tiny fragment (HTF) islands, which are usually associated with the 5′ ends of expressed genes.[32]

The DVβ gene mapped very close to the 5′ end of DXα, which was confirmed by cosmid clone analysis. These data are consistent with the fact that *Bam*HI, *Taq*I, and *Msp*I polymorphisms in DVβ correlate with variations at the DXα and DXβ loci, but not at the DQ and DR loci by RFLP analysis. The DQ and DR loci are in strong linkage disequilibrium,[1,5] indicating that hot spots for recombination may be present between DVβ and DQβ.

We have cloned three DRβ genes by overlapping cosmid clones encompassing a 140-kb length of continuous genomic DNA, and also one DRα gene from the DR2 genomic libraries (Figure 6). Bohme et al.[30] suggested that three DRβ genes existed in the DR2 haplotype. Indeed, digestion with *Eco*RI, *Bam*HI, and *Pst*I followed by hybridization with the DRβ 3′UT specific probe revealed the presence of three bands in genomic DNAs isolated from AKIBA (DR2, Dw12) and PGF (DR2, Dw2), corresponding to each of three DRβ genes cloned here, suggesting that these cosmid clones span the entire DRβ subregion. Restriction enzyme digestion and transfection experiments show that the DRβI and DRβIII genes encode the nonpolymorphic and polymorphic DRβ chains, respectively. Namely, mRNA transcribed from the DRβI gene corresponds to the nonpolymorphic cDNA, and its gene product reacts with the supertypic mAb Hu30[29] after association with the DRα chain. On the other hand, mRNA from the DRβIII gene corresponds to the polymorphic cDNA containing the *Bss*HII sites in the βI exon, and its product does not react with Hu30. Although in other haplotypes such as DR4, DR3, and DR6,[16,21,24] the nonpolymorphic or supertypic DRβ chain was demonstrated to be encoded by one of the DRβ genes most telomeric to the DQ subregion (DRβIII in DR3 and DR6, and the DRβIV gene in DR4), it must be noted that in DR2 the DRβI gene represents the less polymorphic DRβ chain. However, the organization and orientation of the class II genes, including the three DRβ genes, in DR2 is similar to those in other haplotypes (DR3, DR4, DR6 and DR7). These facts indicate that the susceptibility to mutations or gene conversions responsible for genetic polymorphism or diversity may not depend on the location of the gene on the chromosome, but on the gene itself.

Size polymorphisms in the distance between DQα and DRα or DVβ and DQβ were detected. Thus, the total size of the HLA class II gene region varies from 1100 kb (DR4) to 1000 kb (DR2) (Figure 8). The distance between the DQ and DR subregions was estimated to be longer in DR4 (and DR7) than in other HLA haplotypes. It must be noted that DR4 and DR7 haplotypes are evolutionarily related by virtue of sharing the DRw53 supertypic specificity.[33] Complexity and variation in the numbers of DRβ genes and pseudogenes are not surprising in view of the fact that members of different supertypic groups have been found to contain β chain genes that are not truly allelic.[34] In this context it is worth noting that DR4 and DR7 belong to the same supertypic group, characterized by a common DRw53 gene, which shares a number of features, including RFLP patterns for DQα.[35] Thus, in view of all of these data, it seems reasonable to propose that supertypic DR groups, for example DR4 and DR7, belong to the same ancestral groups that have remained intact at the gross genomic level for a considerable time. This is underlined by the fact that the DRw53 sequences from DR7 and DR4 haplotypes are identical at the nucleotide level.[36] In spite of this, marked variation may take place at key locations: for example, the βI domains of DR4 and DR7 β chains are as different from each other as they are from equivalent domains of other DRβ specificities.[25] Gene conversion may be responsible for this localized polymorphism. The data are also consistent with the fact that productive recombination in the DQ-DR interval may be very rare since the supertypic groups tend to stay together. The region between DQα and DRβ could represent junk or selfish DNA which has accumulated as a result of gene duplication.[3,5,37] It cannot be ruled out that the additional DNA is due, in part at least, to extra duplications of DRβ genes in DR4 and DR7 haplotypes, which are known to have occurred, although our data indicate that some sections of DNA do not contain DRβ-related sequences.

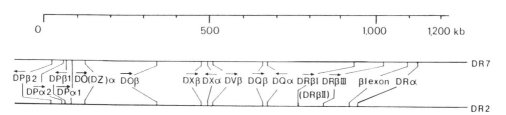

FIGURE 8.   Comparison of the total length of the HLA class II gene region between the DR2 (AKIBA) and DR7 (MANN) haplotypes.

NOTE ADDED IN PROOF

The genes in the HLA-D region have been given official designations as follows: DPβ2 → DPB2, DPα2 → DPA2, DPβ1 → DPB1, DPα1 → DPA1, DO(DZ)α → DNA, DOβ → DOB, DXβ → DQB2, DXα → DQA2, DVβ → DQB3, DQβ → DQB1, DQα → DQA1, DRβI → DRB1, DRβII → DRB2, DRβIII → DRB3, DRβIV → DRB4, DRα → DRA.

REFERENCES

1. **Bodmer, W. F.,** The HLA system, in *Histocompatibility Testing 1984,* Albert, E. D., Baur, M. P., and Mayr, W. R., Eds., Springer-Verlag, Heidelberg, 1984, 11.
2. **Hardy, D. A., Bell, J. I., Long, E. D., Lindsten, T., and McDevitt, H. O.,** Mapping of the class Ii region of the human major histocompatibility complex by pulsed-field gel electrophoresis, *Nature (London),* 323, 453, 1986.
3. **Korman, A. J., Boss, J. M., Spies, T., Sorrentino, R., Okada, K., and Strominger, J. L.,** Genetic complexity and expression of human class II histocompatibility antigens, *Immunol. Rev.,* 85, 45, 1985.
4. **Lawrance, S. K., Smith, C. L., Srivastava, R., Cantor, C. R., and Weissman, S. M.,** Megabase-scale mapping of the HLA gene complex by pulsed field gel electrophoresis, *Science,* 235, 1387, 1987.
5. **Trowsdale, J., Young, J. A. T., Kelly, A. P., Austin, P. J., Carson, S., Meunier, H., So, A., Erlich, H., Spielman, R., Bodmer, J., and Bodmer, W.,** Structure, sequence and polymorphism in the HLA-D region, *Immunol. Rev.,* 85, 4, 1985.
6. **Ando, A., Inoko, H., Kimura, M., Ogata, S., and Tsuji, K.,** Isolation and characterization of genomic clones encoding new HLA class II antigen heavy and light chains, in *HLA in Asia-Oceania 1986,* Aizawa, M., Ed., Hokkaido University Press, Sapporo, 1986, 859.
7. **Schwartz, D. C. and Cantor, M. V.,** Separation of yeast chromosome-sized DNAs by pulsed field gradient gel electrophoresis, *Cell,* 37, 67, 1984.
8. **Carle, G. F. and Olson, M. V.,** Separation of chromosomal DNA molecules from yeast by orthogonal-field-alteration gel electrophoresis, *Nucleic Acids Res.,* 12, 5647, 1984.
9. **Erlich, H., Stetler, D., Sheng-Dong, R., and Saiki, R.,** Analysis by molecular cloning of the human class II genes, *Fed. Proc.,* 43, 3025, 1984.
10. **Roux-Dosseto, M., Auffray, C., Lillie, J. W., Boss, J. M., Cohen, D., DeMars, R., Mawas, C., Seidman, J. G., and Strominger, J. L.,** Genetic mapping of a human class II antigen β chain cDNA clone to the SB region of the HLA complex, *Proc. Natl. Acad. Sci. U.S.A.,* 80, 6036, 1983.
11. **Tonnelle, C., DeMars, R., and Long, E. O.,** DOβ a new β chain gene in HLA-D with a distinct regulation of expression, *EMBO J.,* 2, 389, 1982.
12. **Long, E. O., Wake, C. T., Gorski, J., and Mach, B.,** Complete sequence of an HLA DRβ chain deduced from a cDNA clone and identification of multiple non-allelic DRβ chain genes, *EMBO J.,* 2, 389, 1982.
13. **Meunier, H. F., Carson, S., Bodmer, W. F., and Trowsdale, J.,** An isolated βI exon next to the DRα gene in the HLA-D region, *Immunogenetics,* 23, 172, 1986.
14. **Gustafsson, K., Wiman, K., Larhammar, D., Rask, L., and Peterson, P. A.,** Signal sequences distinguish class II histocompatibility antigen β chains of different loci, *Scand. J. Immunol.,* 19, 91, 1984.
15. **Jonsson, A. K., Hyldig-Nielsen, J. J., Servenius, B., Larhammar, D., Andersson, G., Jorgensen, F., Peterson, P. A., and Rask, L.,** Class II genes of the human major histocompatibility complex, *J. Biol. Chem.,* 262, 8767, 1987.
16. **Spies, T., Sorrentino, R., Boss, J. M., Okada, K., and Strominger, J. L.,** Structural organization of the DR subregion of the human major histocompatibility complex, *Proc. Natl. Acad. Sci. U.S.A.,* 82, 5156, 1985.
17. **Okada, K., Boss, J., Prentice, H., Spies, T., Mengler, R., Auffray, C., Lillie, J., Grossberger, D., and Strominger, J. L.,** Gene organization of the DC and DX subregions of the human major histocompatibility complex, *J. Biol. Chem.,* 262, 8767, 1987.
18. **Lawn, R. M., Fritsch, E. F., Parker, R. C., Blake, G., and Maniatis, T.,** The isolation and characterization of linked δ- and β-globin genes from a cloned library of human DNA, *Cell,* 15, 1157, 1978.
19. **Schenning, L., Larhammar, D., Bill, D., Wiman, K., Jonsson, A. K., Rask, L., and Peterson, P. A.,** Both α and β chains of HLA-DC class II histocompatibility antigens display extensive polymorphism in their amino-terminal domains, *EMBO J.,* 3, 447, 1984.
20. **Trucco, M., Rwenshine, S., Cascino, I., and Duquesnoy, R. J.,** HLA-DR polymorphism analyzed by sequential restriction endonuclease DNA digestion, *Immunogenetics,* 24, 184, 1986.
21. **Rollini, P., Mach, B., and Gorski, J.,** Linkage map of three HLA-DRβ-chain genes: Evidence for a recent duplication event, *Proc. Natl. Acad. Sci. U.S.A.,* 82, 7197, 1985.
22. **Rollini, P., Mach, B., and Gorski, J.,** Characterization of an HLA-DRβ pseudogene in the DRw52 supertypic group, *Immunogenetics,* 25, 336, 1987.
23. **Larhammar, D., Servenius, B., Rask, L., and Peterson, P. A.,** Characterization of an HLA DRβ pseudogene, *Proc. Natl. Acad. Sci. U.S.A.,* 82, 1475, 1985.
24. **Andersson, G., Larhammar, D., Widmark, E., Servenius, B., Peterson, P. A., and Rask, L.,** Class II genes of the human major histocompatibility complex, *J. Biol. Chem.,* 262, 8748, 1987.
25. **Lee, B. S. M., Rust, N. A., McMichael, A. J., and McDevitt, H. O.,** HLA-DR2 subtypes from an additional supertypic family of DRβ alleles, *Proc. Natl. Acad. Sci. U.S.A.,* 84, 4591, 1987.
26. **Nakatsuji, T., Moriya, K., Ando, A., Inoko, H., and Tsuji, K.,** HLA-DQ structural polymorphism in HLA-DR2 associated HLA-D clusters, *Hum. Immunol.,* 16, 157, 1986.

27. **Nepom, G. T., Nepom, B. S., Wilson, M. E., Mickelson, E., Antonelli, P., and Hansen, J. A.,** Multiple Ia-like molecules characterize HLA-DR2-associated haplotypes which differ in HLA-D, *Hum. Immunol.,* 10, 143, 1984.

28. **Wu, S., Yabe, T., Madden, M., Saunders, T. L., and Bach, F. H.,** cDNA cloning and sequencing reveals that the electrophoretically constant DRβ2 molecules, as well as the variable DRβI molecules, from HLA-DR2 subtypes have different amino acid sequences including a hypervariable region for a functionally important epitope, *J. Immunol.,* 138, 2953, 1987.

29. **Nishimura, Y., Tsukamoto, K., Sone, T., Hirayama, K., Takenouchi, T., Ogasawara, K., Kasahara, M., Wakisaka, A., Aizawa, M., and Sasazuki, T.,** Molecular and functional analysis of class II molecules characteristic to HLA-Dw2 and Dw12, in *Histocompatibility Testing 1984,* Albert, E. D., Baur, M. P., and Mayr, W. R., Eds., Springer-Verlag, Heidelberg, 1984, 504.

30. **Bohme, J., Andersson, M., Andersson, G., Moller, E., Peterson, P. A., and Rask, L.,** HLA-DRβ genes vary in number between different DR specificities, whereas the number of DQβ genes is constant, *J. Immunol.,* 35, 2149, 1985.

31. **Gorski, J. and Mach, B.,** Polymorphism of human Ia antigens: gene conversion between two DRβ loci results in a new HLA-DR specificity, *Nature (London),* 322, 67, 1986.

32. **Lindsay, S. and Bird, A. P.,** Use of restriction enzymes to detect potential gene sequences in mammalian DNA, *Nature (London),* 327, 336, 1987.

33. **Gregersen, P. K., Moriuchi, T., Karr, R. W., Obata, F., Moriuchi, J., Maccari, J., Goldberg, D., Winchester, R. J., and Silver, J.,** Molecular diversity of HLA-DR4 haplotypes, *Proc. Natl. Acad. Sci. U.S.A.,* 83, 9149, 1986.

34. **Gorski, J., Rollins, P., and Mach, B.,** Characterization of an HLA-DRβ pseudogene in the DRw52 supertypic group, *Immunogenetics,* 25, 336, 1987.

35. **Trowsdale, J., Lee, J., Carey, J., Grosveld, F., and Bodmer, W. F.,** Sequences related to HLA-DRα chain on human chromosome 6: restriction enzyme polymorphism detected with DCα chain probes, *Proc. Natl. Acad. Sci. U.S.A.,* 80, 1972, 1983.

36. **Young, J. A. T., Wilkinson, D., Bodmer, W. F., and Trowsdale, J.,** Sequences of DRβ cDNA clones derived from the HLA-DR7 haplotype: implications for the evolution of DRβ genes within the DRw53 supertypic subgroup, *Proc. Natl. Acad. Sci. U.S.A.,* 84, 4929, 1987.

37. **Figueroa, F. and Klein, J.,** The evolution of class II MHC genes, *Immunol. Today,* 7, 78, 1986.

38. **Inoko, H., et al.,** unpublished.

39. **Kawai, J., et al.,** submitted.

40. **Young, J. and Trowsdale, J.,** unpublished.

41. **Erlich, H., et al.,** unpublished.

42. **Auffray, C., et al.,** unpublished.

Chapter 2

THE T-CELL PERSPECTIVE OF THE HLA-D REGION*

Fritz H. Bach, Toshio Yabe, and Nancy L. Reinsmoen

TABLE OF CONTENTS

* IRC paper #500 is supported in part by NIH grants AI 17687, 18326, 22682, and the Harry Kay Charitable Foundation.

I. INTRODUCTION

It is approximately a half-century ago that the pioneering studies of Snell, Gorer, and colleagues led to recognition of the major histocompatibility complex (MHC) in mouse (H-2). A panoply of different approaches to the study of the MHC in several species including man has focused on questions as fundamental as the DNA sequence basis of the extensive polymorphism of MHC genes and the three-dimensional structures of MHC-encoded mol-.ecules to clinical studies on the role of the MHC in transplantation and in association with a number of diseases. Invitation by the editor, if not frank encouragement, to focus on work published from one's own laboratory, and thus on the particular perspective that has provided the basis for our fascination with the MHC, is a wonderful opportunity. The realization that many of the major investigators in this field will contribute to this volume makes it easier to succumb to the temptation of such a singular focus.

A. THE MIXED LEUKOCYTE CULTURE TEST AND HLA

The perspective that this laboratory, first at the University of Wisconsin and more recently at the University of Minnesota, has pursued is based largely on our interest in recognition of MHC-encoded molecules by T lymphocytes rather than through the use of serological reagents. Our work originated in man in the 1960s, with the description of the mixed leukocyte culture (MLC) test[1,2] combined with the establishment of a one-way method of stimulation in MLC.[3] These methods allowed the correlation of the presence or absence of a proliferative response in MLC between siblings with results of serological studies[4] and the conclusion that a single genetic locus (in the transmissional sense) controlled reactivity in MLC as well as encoding the antigens defined serologically; we proposed that this was the major histocompatibility complex in man, now known as human leucocyte antigens (HLA).[5]

That the chromosomal region of HLA which controls the strongest stimulatory determinants for MLC (now known as the HLA-D region) was different from the class I encoding genes was already suggested in two of our early studies.[4,6] Definitive evidence in this regard came from the elegant study of Yunis and Amos.[7]

B. THE DIFFERENTIAL FUNCTION OF CLASS II AND CLASS I ANTIGENS: THE LD-SD DICHOTOMY

In an attempt to obtain a further dissection of the regions of the MHC that are responsible for stimulating the proliferative response in MLC, we turned, at the beginning of the 1970s, to studies in mouse. Our work, largely with Jan Klein, demonstrated that it was H-2 I region-encoded molecules (i.e., class II antigens) that were the primary stimulus of proliferation in MLC,[8,9] a finding confirmed independently by Meo and colleagues.[10]

Work from our laboratory and two others demonstrated in 1970 that antigen-specific cytotoxic T cells (Tc) could be generated in an MLC.[11-13] We thus performed an analysis of both the proliferative and cytotoxic responses generated in MLC and found in mouse,[14,15] as did Eijsvoogel and colleagues in man,[16] that class I antigens (H-2K and H-2D) serve as the primary target structures for Tc, even though the class II antigens stimulated the majority of the proliferative response.

1. Collaboration between Th and Tc Responsive to Class II and Class I Antigens

Our work also showed that it was collaboration between class II-responsive Th and class I-responsive Tc (which we referred to as LD-SD collaboration) that resulted in the generation of maximal cytotoxic responses to class I antigens.[15,17] Although exceptions have been described to this dichotomy of antigen function, i.e., the stimulation of Tc by class I antigens and the stimulation of helper Th by class II antigens, the general model evolving from those studies for both allo-recognition and later for recognition of foreign antigens restricted by the two classes of antigens, continues to serve as a basic tenet of our understanding of the function of MHC-encoded molecules.

We attempted to encapsulate certain concepts regarding these phenomena by use of the metaphor "the LD-SD dichotomy", with LD referring to lymphocyte defined and SD to serologically defined, to differentiate between what are now known as the class II and class I antigens, respectively.[15,18] Our basis at that time for distinguishing between LD (class II) and SD (class I) antigens was the difference in the ability of these two classes of antigens to stimulate proliferation vs. acting as targets for Tc, although we, and others, demonstrated that class II antigens can function as targets for Tc[19,20] and allo-class I antigens can stimulate a proliferative response.[21-23]

2. Different Determinants Recognized on an MHC Molecule by T Cells vs. Antibody

A later application of the concept of the LD-SD dichotomy, which was to form the theoretical basis for attempts to define the T lymphocyte-defined polymorphism as separate from that defined serologically, rested on the premise that antigenic determinants recognized by T lymphocytes are different, in some significant measure, from those recognized serologically. Thus, we began to speak of LD and SD determinants on single class I or II molecules. A part of this extension of the LD-SD concept suggested that the determinants recognized by T lymphocytes on class II molecules may well be different from those recognized serologically and that a similar situation might exist for class I molecules in that Tc may recognize one determinant(s) on the class I molecule, whereas antisera would react to a different determinant(s).

It was on the basis of this conceptualization of immunological recognition of MHC-encoded molecules that we performed, over the next 15 years, a series of analyses of HLA-D region encoded molecules and their attendant polymorphism. Rather than evaluating the overall HLA polymorphism, we focused our attention on what turned out to be two families of haplotypes: those encoding the serologically defined specificity DR4 and those encoding DR2. In each case, we referred to the polymorphisms that exist within the population of individuals expressing DR4 or those expressing DR2 as Dw/LD subtype polymorphisms. Dw is the designation used for specificities recognized by T cells; the Dw numbers are assigned by a WHO nomenclature committee. LD is the term introduced by us, to differentiate conceptually between the serologically defined specificities (SD) and those recognized by T lymphocytes (LD), and was used by the International Histocompatibility Workshops for the designation of the T lymphocyte-defined specificities prior to the time that they were given official Dw designations.

Can the LD determinants (epitopes) recognized by T lymphocytes be defined serologically? This is a difficult, if not impossible, question to answer given present knowledge and techniques; nor, at least in the opinion of some, is it the most important question. Given currently available reagents, the determinants most frequently recognized by T cells are often different from the determinants most readily recognized serologically. If it seems surprising that an epitope is not recognized readily by both T and B cells, one must remember that certain allogeneic differences recognized strongly by T lymphocytes (such as certain H-2K locus mutants[8,24]) have been extremely difficult to define serologically, despite very extensive efforts to do so.

II. ASSIGNMENT OF HLA CLASS II SPECIFICITIES BASED ON T LYMPHOCYTE RESPONSES

Recognition in the 1970s that the polymorphism recognized by T lymphocytes was, at least in part, different from and more extensive than the polymorphism defined serologically, based on studies with homozygous typing cells,[25,26] persuaded us to attempt to develop additional methods that would allow definition of T lymphocyte-recognized determinants (referred to as LD determinants) by more discriminatory tests than the use of HTCs in MLC.[27-31] With this goal in mind, we developed the primed lymphocyte (LD) typing (PLT) test[32,33] in which, by priming against more restricted allo-differences than would be recognized on an entire HLA haplotype, one could develop reagents that had greater discriminatory ability than HTC testing. As what must be regarded as the ultimate cellular reagent, we followed the lead of investigators who cloned T lymphocytes in mouse,[34] by achieving the cloning of T lymphocytes in man and using such cloned reagents to help define the HLA class II LD polymorphism.[35] These procedures using reagents from HTC, PLT, and cloned T cells form the basis of defining the Dw/LD polymorphism.

A. THE HOMOZYGOUS TYPING CELL (HTC) TEST

Definition of the HLA-Dw1 through Dw23 specificities is based on proliferative responses of T lymphocytes in a primary MLC to stimulating cells that are "homozygous" (HTC), at least for the DR and DQ products. (Dw24-26, representing DRw52 subgroups, are defined with T cell clones, not HTCs). In some cases the HTCs are autozygous, i.e., the donors are offspring of consanguinous marriage, in which case the cells are almost always homozygous for all HLA-region products; in other cases, HTCs are phenotypically homozygous (allozygous) for DR and DQ but can be heterozygous for DP. Allozygous HTC may also not be identical in the two haplotypes for the DR and DQ products.

HTCs are chosen as typing reagents if (1) the cells do not stimulate a significant response in the appropriate combinations within the family from which they were derived, (2) they do not stimulate (or are weak in stimulating) cells of other HTCs used to "define" the same Dw specificity, and (3) they can be used successfully to "type" an unrelated panel, i.e., to distinguish between unrelated individuals whose cells respond positively in MLC to the HTC and those who show no response or a weak response, i.e., carry the specificity defined by the HTC.

With some HTCs, responding cells of the panel form a biphasic response; those responding weakly are assigned the Dw specificity of the HTC, while those responding strongly are not. More often, however, there is no clear biphasic response and thus an arbitrary "cut-off" is chosen. It must be emphasized that this arbitrary threshold may represent as much as 30% of the T-cell proliferative response, by those same responding cells, to antigenic differences associated with a full HLA-D disparity. Assignment of Dw specificities, therefore, frequently not only does not guarantee HLA-D region (including DP) identity between responding and stimulating cells but leaves a large likelihood of possible nonidentity.

HTC testing has been most useful in defining the series of Dw specificities which represent clusters of antigenic determinants associated with the various class II products. (Class I antigens can, when disparate, also stimulate T lymphocytes to proliferate, albeit relatively weakly). The response to an HTC represents the aggregate reactions of clones that can recognize determinants associated chiefly with DR, DQ, and DP. It depends not only on the number of determinants carried by the HTC, which the responding cell does not have, but also on the "strength" of those determinants, i.e., the frequency of clones that respond to those disparate determinants and the extent of proliferation by cells of those clones.

The ability to define Dw specificities with HTCs, despite the ability of all HLA products to stimulate T cell proliferation, rests to a great extent on three factors. First, the DR product(s) appear to stimulate most of the proliferating cells in an MLC,[36] a finding that appears to apply to restricted responses as well. Second, loose criteria, i.e., a 40% to 50% relative response, have been used to designate a typing response; more stringent criteria, such as a 10 to 20% relative response, would result in fewer "typing responses". Third,

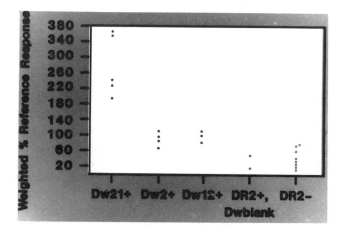

FIGURE 1.   Stimulation of the PLT reagent primed against a Dw21 haplotype are indicated as weighted percent reference response, which averages
tritiated thymidine incorporation results at four concentrations of primed
cells tested. The Dw21 positive cells significantly restimulate this reagent
with clear discrimination from other DR2 positive and DR2 negative cells.

linkage disequilibrium holds together on given haplotypes certain alleles encoding DR, DQ,
and other specificities; this linkage disequilibrium in turn leads to sufficient gene, and thus
antigen, sharing between the HTCs and the cells of individuals being tested to provide typing
responses. To the extent that HLA alleles, such as those of DP, are not in linkage disequilibrium with the DR-DQ combination of a given haplotype, stimulation of proliferation by
products of those DP genes creates "noise," which makes more difficult the assignment of
a Dw specificity. There is ample evidence that products are held together in linkage disequilibrium with different products in different populations. Related but different Dw specificities may therefore be disclosed in different populations by a single HTC.

B. PRIMED LYMPHOCYTE (LD) TYPING (PLT)

The PLT test is an alternative method of defining LD determinants. HTCs are difficult
to find and, as discussed above, carry LD determinants associated with class I and class II
products that stimulate a proliferative response. In the PLT test there is selective activation
of T lymphocytes to determinants associated with single or multiple class I or class II
products. In order to develop a PLT reagent, responding and stimulating cells are chosen
which differ for various class I and/or class II products. The cells are incubated in a primary
MLC for a period of 10 d, providing time for the responding cells to revert to nondividing,
but now "primed" cells. These make an accelerated, and very strong, secondary response
to the determinants recognized in the priming combination. When third-party cells carry the
LD specificities recognized in the sensitizing MLC, they will stimulate a proliferative response similar to that evoked by cells of the original sensitizing cell donor; third party cells
not carrying those specificities will not evoke such a response.

PLT reagents can be used to define determinants closely associated with HLA-Dw as
defined by HTCs. These reagents are especially useful in defining new HLA-D specificities
for which an HTC has not been defined. Dw21 (MN2) an HLA-D specificity associated
with DR2, was initially defined by PLT (Figure 1) not HTC reagents.[37] The cloning of PLT
reactive cells (see below) has refined the PLT test. Clones presumably define single determinants; cloned reagents, therefore, can define determinants associated with a single product
of the class II region. The alleles of DP, for instance, have to date largely been defined
with PLT reagents.[38,39] Whereas it is extremely difficult to derive bulk PLT reagents to

define the DP antigens, cloning of anti-DP reagents (generated in DR/DQ "identical" combinations) allows the ready preparation of PLT cells defining the DPw1-DPw3 and DPw5 and DPw6 antigens.[40] It is thus possible to obtain PLT reagents which correspond very closely to the Dw specificities defined with HTCs or to define individual LD determinants that make up a part of a Dw specificity.

C. STUDIES WITH CLONED T CELLS

It was not initially clear whether there are LD determinants associated with both DR and DQ products that contribute to the Dw haplotype assignment, i.e., have a population distribution corresponding to the subtypic Dw specificity. We thus used cloned T cells to study panels of phenotypically well-characterized cells, as well as monoclonal antibody blocking studies to ascertain with which class II dimer cells of a given clone react. We wished to evaluate whether determinants associated with the DR as well as the DQ products are recognized and can contribute to the definition of a Dw specificity.[41] We found that cloned T cells respond to what are probably both DR αβ dimers in haplotypes expressing two DR products as well as to DQ. Further, we were prompted to investigate individual allodeterminants associated with DR and DQ with cloned T cells to study further the more extensive polymorphisms recognized by T lymphocytes as compared with the serologically defined DR and DQ polymorphisms. We have performed/participated in several studies that demonstrate that restricting determinants for several different nominal antigens are subtypic to the serologically defined DR and DQ specificities and appear to be closely related to the Dw associated allodeterminants.[42-44]

Bulk primed cell populations were generated using cells matched for class I determinants and disparate for class II determinants, cloned and assayed for proliferative and cytolytic function with panels of cells well-characterized for HLA. The results suggest that there are multiple stimulatory determinants on the DR and DQ molecules; many of these determinants are commonly shared with cells which type for the same HLA-Dw specificity as the original sensitizing cell. In addition, these determinants may be occasionally shared by, or highly cross-reactive with, cells which do not express the same Dw, DR, or DQ specificities as the sensitizing cell. We give some examples of this approach below.

Cytolytic clones generated from the anti-DR4/Dw4 priming combination (Table 1) or the anti-DR2/Dw2 priming combination (Table 2), in general lysed targets sharing the same Dw type as the sensitizing cell, with some clones lysing additional target cells. (The numbers given in the tables represent percent cytotoxicity values; 10% or more lysis is regarded as a positive and thus indicates that the cell lysed carries a determinant recognized by that clone.) Whether the clones which only lyse targets bearing the same Dw specificity as the sensitizing cell are all directed against a single determinant or whether several determinants are frequently found on cells of that Dw type remains to be elucidated.

Monoclonal antibody inhibition studies of those clones identified most clones derived from the anti-DR4/Dw4/DQw3 priming combination as being directed at DR in that they were blocked by the mAb directed at DR monomorphic determinants but not by the anti-DRw53 mAb (DRαβIII), PL3 (Table 3). One clone (#21) appeared to be directed at a determinant associated or identical with DRw53 in that it lysed 8 of 8 DRw53 positive targets and was inhibited by the mAb PL3.[45]

Some clones from the anti-DR2/Dw2/DQw1 priming combination appeared to be DR-directed while other appeared to be DQ-directed based on blocking of cytolytic reactivity with mAb (Table 4). Certain of the DR-directed as well as the DQ-directed groups of clones lysed only target cells bearing the same Dw specificity as the sensitizing cell (Table 2). These results demonstrate that some clones directed against the DR or DQ molecules recognize determinants that are associated with the same Dw specificity as the sensitizing cell. Thus, it appears that determinants on both the DR and DQ molecules could contribute to

TABLE 1
Cytolytic Clones — Anti DR/Dw4 Priming[a]

Targets

	S1	1	2	3	4	5	6	7	8	9	10
DR	3,4	4,4	4,4	4,4	4,4	4,4	4,4	4,4	4,4	4,4	4,4
DRw	53	(53)	(53)	53	(53)	(53)	53	53	53	53	53
Dw	3,4	4,4	4,4	13,4	13,13	13,13	14,14	14,14	15,15	10,10	10,10
DQ	3	3	3	(3)	(3)	3	3	3	—	3	3
DP	1	4	3,4	NT	NT	5	2,3	3,6	5	NT	4
clone											
15	18[b]	34	25	15	3	1	-2	-19	2	-2	3
33	43	62	46	34	1	8	8	-10	2	-4	1
46	16	62	37	28	9	3	5	-7	8	10	7
57	37	57	25	27	2	5	2	-13	3	1	3
9	13	56	17	-2	-4	-8	1	NT	-4	-3	4
44	9	56	16	-1	-2	-5	-1	NT	4	-1	5
45	11	56	39	1	-1	-3	12	NT	21	-2	2
12	55	65	60	37	6	0	3	-18	6	-4	5
67	35	59	44	46	-1	43	-1	-19	3	1	68
37	21	86	58	16	-2	5	16	NT	3	-6	1
21	14	46	24	22	17	14	26	17	9	17	71
56	31	64	35	32	62	24	55	NT	25	-4	3
48	14	46	25	27	34	5	62	64	19	10	3
63	34	70	42	11	50	53	32	NT	16	22	67

Targets

	11	12	13	14	15	16	17	18	19	20	21	22
DR	5,5	5,5	3,3	2,2	1,9	2,6	2,2	1,1	2,3	3,3	6,6	7,7
	52	NT	52	NT	53	52	—	—	52	NT	52	53
Dw	5,5	5,5	3,3	2,2	1,23	6,21	2,2	1,1	2,3	3,3	6,6	7,7
DQ	3	NT	2	NT	1	1	1	1	1	NT	1	2
DP	4	4	3,4	4,5	NT	NT	2,5	3,4	4,6	1,3	2,4	4
clone												
15	-5	-4	6	1	3	-4	-2	-5	-5	-18	-9	-7
33	-6	-7	2	-2	4	-6	-6	3	-4	-2	-3	-4
46	-7	-7	4	-1	3	-5	-4	-2	4	-8	5	-8
57	-5	-4	3	-1	5	-5	-4	-1	3	-10	1	1
9	NT	NT	NT	NT	NT	NT	NT	NT	-3	NT	NT	NT
44	-6	-5	17	18	36	-5	-1	16	-4	-20	-7	-6
45	-5	-4	29	24	21	-3	-3	-3	2	-17	-8	-9
12	-4	-5	16	10	8	-4	-1	-7	1	-16	-6	-9
67	2	6	10	NT	NT	-2	16	2	7	-4	-2	-1
37	-4	4	17	6	13	-4	-1	-5	-4	-21	-9	-11
21	-3	-5	-5	4	51	-2	-4	-2	2	-20	-7	29
56	-4	-5	12	8	11	-4	-4	-5	-7	-20	-8	-10
48	-7	-6	4	9	49	-5	-6	55	4	-12	-8	-5
63	-7	-4	15	7	27	-3	1	-4	-7	-21	-9	3

Note: Values greater than or equal to 10 are underscored. Specificities indicated within parenthesis have not been tested but are as indicated based on the phenotyping data.

[a] Clones 56, 63, and 46 also demonstrate proliferative reactivity.
[b] Results expressed as % cytotoxicity.

TABLE 2
Cytolytic Clones — Anti DR2/Dw2 Priming[a]

										Targets								
	S1	S2	1	2	3	4	5	6	7	8	9	10	11	12	13	14	15	16
DR	2,3	2,3	2,2	2,2	2,2	2,2	2,2	2,2	6,6	1,1	4,4	4,4	4,4	4,4	3,3	3,3	7,7	5,5
Dw	2,3	2,3	2,2	2,2	2,2	21	21	22	6,6	1,1	4,4	10,10	14,14	13,13	3,3	3,3	7,7	5,5
DQ	1,2	1,2	1	1	NT	NT	NT	NT	1	1	3	3	3	NT	NT	2	2	3
DP	4,6	4	4,5	2,5	NT	NT	NT	NT	2,4	3,4	6,2	4	6,3	NT	1,3	3,4	4	4
clone #																		
3-89(DQ)	19[b]	38	5	23	41	8	1	-3	-9	-8	2	-.1	-.7	.1	6	-17	-1	.2
3-17(DQ)	24	26	19	29	29	4	2	-3	-4	-8	4	-.6	-.9	-.7	4	-19	2	3
3-19(DR)	15	28	14	40	55	4	.2	-2	-2	-5	7	-.1	-.5	-2	5	-17	.4	-.1
1-12	28	44	35	54	40	20	-.5	-.6	-6	2	-.9	-1	-2	-3	-.6	-17	-4	-.7
3-84	11	22	14	44	40	30	4	-3	-7	-5	3	-2	-5	-2	1	-16	-1	-.7
3-91	11	6	14	14	19	18	5	5	-8	-7	7	-1	-1	-2	2	-19	-.6	5
3-29(DR)	27	27	22	41	31	29	47	43	51	26	8	-.7	-3	-4	13	-16	-2	-1
1-17(DQ)	11	20	22	20	24	-.5	5	-3	47	-7	.7	-.7	-2	-3	-.6	-17	-1	-.6
3-27(DQ)	27	45	18	53	50	36	18	-1	51	-5	4	51	-2	-1	2	-15	2	4
1-109	15	57	34	55	44	9	16	45	13	-9	14	37	-3	-3	6	-18	23	4

a   Clone 3-19, 1-84, 3-91, 3-29, 3-27, 1-109 also demonstrated strong proliferative reactivity (>10,000 cpm) and clones 3-89, 3-17, and 1-17 demonstrated weaker proliferative reactivity.

b   Results expressed as percent of cytotoxicity. Values greater than or equal to 10 are underscored.

TABLE 3
Dissection of Determinants Associated with Different DR Dimers Anti-DR4/Dw4/ DQw3 Priming Combination

	Anti-DR mAb						Anti-DRw53 mAb			Anti-DQ mAb		
	L243			Hu4			PL3			Tu22		
Clones	80[a]	400	4000	80	400	4000	80	400	4000	80	400	4000
15	113[b]	104	60	121	98	54		NT		14	25	31
37	121	129	121	111	79	57		NT		43	32	67
48	116	106	66	91	81	6	29	−29	−43	−3	9	3
67	111	89	38	117	109	57	24	17	24	−13	−11	2
63	97	32	38	145	104	31	−25	−150	0	38	45	62
33	71	20	21	71	29	17	−8	−10	12	8	29	32
56	100	97	33		NT			NT			NT	
45	100	100	100		NT			NT			NT	
44	97	100	88		NT			NT			NT	
21	7	8	14	20	2	10	101	104	76	14	5	29

Note: NT = not tested.

[a] Reciprocal mAb dilutions.

[b] Results expressed as percent inhibition; values > 50% are underscored.

the definition of a Dw subtype specificity assignment for a given cell in that both products stimulate T cells.

Other clones appeared to recognize determinants on DR molecules from cells which typed for the same DR, but different Dw, specificities; occasionally a target cell which did not bear the same DR type as the sensitizing cell was lysed to as great a degree as the sensitizing cell. These results indicate that clones are recognizng several different determinants on the DR molecules, some of which may be shared or highly cross-reactive with determinants on DR molecules from cells of a different DR type.

In summary, panel-cell analysis of clones directed against DR4 or DR2 haplotypes demonstrated several patterns of reactivity for both DR- and DQ-directed clones. The results of clones directed against DR4 haplotypes demonstrate the presumed detection of both DR molecules, one polymorphic and one nonpolymorphic (DRw53); however, no DQ-directed clones were observed. The results suggest the detection of several determinants on DR molecules, some Dw-type specific and others shared by cells of different DR2 or DR4-associated Dw subtypes. The Dw subtype-specific DR- and DQ-directed clones suggest both molecules play a role in the allogeneic response in DR2 haplotypes. However, DR is probably immunodominant in most responder-stimulator cell combinations[36] and, as in the DR4 haplotypes, one of the DR molecules may be immunodominant.

III. THE Dw/LD SUBTYPES: POLYMORPHISM WITHIN SEROLOGICALLY DEFINED DR AND DQ SPECIFICITIES

There is polymorphism within single serologically defined specificities of DR and DQ. Two major lines of evidence have led to this conclusion. First, individuals expressing a given DR specificity, with the associated DQ specificity (such as DR4-DQw3), can be divided into several groups based on assignment of Dw specificities.[25,26] Second, protein studies of the β chains of DR and DQ have demonstrated a polymorphism in isoelectric focusing that correlates with the Dw types;[46,47] this finding, mentioned in the introduction, will not be discussed extensively in this paper although it represents an important link in the overall story, in that it showed a protein polymorphism correlating with the Dw poly-

TABLE 4
Dissection of Cloned Cytolytic T Cell Reactivity with Monoclonal Antibodies Anti-DR2/Dw2/DQw1 Priming Combination

Clones	Anti-DR mAb									Anti-DQw1 mAb						Anti-DP			Anti-class I		
	L243			Hu4			Hu30			S 3/4			Genox 3.53			FA			w632		
	80[a]	400	4000	80	400	4000	800	4000	40000	80	400	4000	80	400	4000	80	400	4000	80	400	4000
3-19	<u>97</u>[b]	<u>67</u>	24		NT			NT		12	6	6	15	12	21		NT			NT	
3-29	<u>108</u>	<u>62</u>	0		NT			NT		15	23	8	31	8	46		NT			NT	
3-27	<u>89</u>	<u>53</u>	42	<u>92</u>	<u>97</u>	47	12	15	15	47	37	<u>53</u>	47	26	32	27	<u>64</u>	18	25	35	20
3-89	-8	10	6	<u>-52</u>	<u>-52</u>	-34	0	3	3	<u>115</u>	<u>100</u>	<u>64</u>	49	26	21		NT		0	7	14
3-17	-69	-50	<u>50</u>		NT			NT		<u>138</u>	<u>106</u>	<u>106</u>	<u>113</u>	<u>81</u>	<u>50</u>		NT			NT	
1-17	-57	-86	-29		NT			NT		<u>157</u>	<u>200</u>	<u>200</u>	<u>143</u>	<u>143</u>	<u>114</u>		NT			NT	

Note: NT = not tested.

a Reciprocal mAb dilution.

b Results expressed as % inhibition; values greater than or equal to 50% are underscored.

morphism and thus strengthened the view that it was polymorphism of the DR molecules that led to Dw designations.

Certain relationships exist between the class II serologically defined (Ia) and T lymphocyte defined (Dw/LD) specificities. One speaks of one specificity as being "supertypic" to another (the latter being called "subtypic" to the first) based on population studies. For instance, DQw1 is generally supertypic to DR1, DR2, and DRw6; individuals positive for DQw1 can be divided into those that are positive for DR1, DR2, etc. Any individual typing positively for DR1, DR2, or DRw6 is also likely to carry the specificity DQw1.

The Dw specificities defined with HTCs were first thought by some to be the equivalent of the serologically defined DR antigens but there is now conclusive evidence that one serologically defined DR specificity such as DR2, DR4, or others can be associated with several Dw subtypes as defined with HTC or PLT reagents. Although the Dw subtypes are usually referred to as being related to a given DR specificity, they also relate to the supertypic DQ specificities. Thus, DQ and DR product-associated LD determinants contribute to the definition of the Dw specificity. One can appropriately speak of DR4-DQw3 or DR2-DQw1 Dw subtypes in the Caucasian population in which DR4 is generally in linkage disequilibrium with DQw3 and DR2 with DQw1.

A. DEFINITION OF Dw/LD SPECIFICITIES WITH HTCs AND PLT REAGENTS

At least four different Dw specificities can be defined within DR4-DQw3: Dw4, Dw10, Dw13, and Dw14; Dw15 is a subtype of DR4 but is not associated with DQw3.[26] Results of HTC testing defining these subtypes is shown in Figure 2. Similarly, DR2 can be split into four Dw subtypes; Dw2, Dw12, Dw21 (formerly referred to as MN2, FJO, AZH) and Dw22 (LD-5a).[48,37] We have studied these DR-associated clusters in detail.

B. MOLECULAR BASIS FOR Dw/LD SUBTYPES

In an attempt to detect a molecular polymorphism at the protein level that correlated with the Dw subtypes, we studied DR4β chains.[46] Work from Goyert and Silver had already demonstrated polymorphism of DQβ within DR5;[49] it was our interest to evaluate whether protein polymorphism correlated with Dw subtype. HTCs from DR4+ individuals whose cells expressed different Dw subtypes were used in these studies; several HTCs were included for each subtype where possible. The radiolabeled DR product was precipitated by a monoclonal antibody directed at DR and analyzed in a two-dimensional SDS-IEF system. We found a very strong correlation between IEF positions of DR4β chains and their Dw subtype.[46] DRβ chains focused differently dependent on the Dw subtype; all β chains from cells of different individuals expressing the same subtype, however, focused identically (Figure 3). Four different focusing patterns were found for the five different subtypes (β chains of Dw4 and Dw14 cells focused identically). Similar results were obtained after neuraminidase treatment removing charge contribution differences attributable to sialic acid.[50] Nepom, Hanson, and colleagues[47] obtained very similar results with DR4β chains and extended their findings to the Dw subtypes of DR2.

Digestion of DRβ chains with α-chymotrypsin, which cleaves the β chain between the two disulfide domains to yield two fragments that correspond quite closely to the protein products of the first and second exons for DRβ,[51] revealed that the protein differences resulting in differential positional focusing in IEF of the different β chains resides in the N-terminal domain of the molecule.[50] Thus, Dw4 and Dw14β chains do differ in their peptide maps following trypsin digestion, by analysis on high performance liquid chromatography (HPLC)[52] and by sequence.[53] Although the majority of such studies performed to date have examined the DR product, work from several laboratories has revealed differences in isoelectric focusing between DQβ chains associated with a single DQ Ia specificity. The molecular polymorphism of DR and DQβ chains, which is more extensive than the serol-

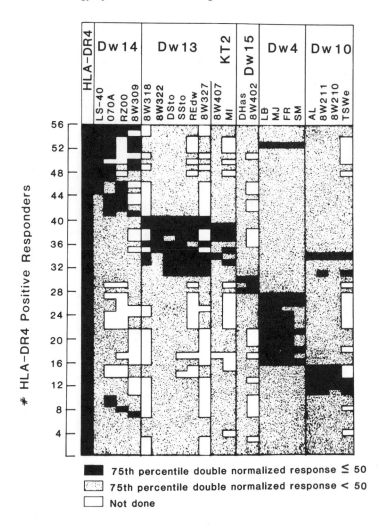

FIGURE 2.   HLA-DR4 associated HLA-Dw clusters. The figure represents sum-
marized compilation of double normalized responses from two to six testings of 56
DR4 positive responder cells tested with 22 HTCs defining the five HLA-Dw clusters:
Dw14, Dw13, Dw15, Dw4 and Dw10. (From Reinsmoen, N. L. and Bach, F. H.,
Hum. Immunol., 4, 249, 1982. With permission.)

ogically defined polymorphism, is thus probably related, in significant measure, to class II
product associated LD polymorphism.

C. STUDIES OF RESTRICTION FRAGMENT LENGTH POLYMORPHISM
(RFLP) ASSOCIATED WITH Dw/LD SUBTYPES

While our primary interest has always focused on the functional polymorphism of class
II products, with the overall organization of the HLA-D region genomic material being of
secondary interest, we nevertheless felt that a study of RFLP for the DR4 and DR2 subtypes
might be of interest. In fact, those studies provided evidence strongly supporting the very
recent evolution of the DR4 subtype polymorphism; further, the DR2 subtype polymorphism
appeared to be of more recent evolution than DR polymorphism in general as defined
serologically.

Our studies of RFLPs in the several haplotypes that represented the DR4 subtype polymorphism demonstrated that there was essentially no RFLP among the several haplotypes

ALL LCLs DR-4

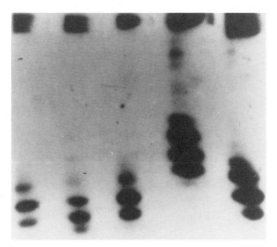

FIGURE 3. The results of isoelectric focusing of DRβ chains cut from an SDS gel are shown. These proteins have not been treated with neuriminidase; thus, several bands are seen for each β chain. β chains of Dw4 and Dw14 focus identically despite the now known differences in sequence; β chains of other subtypes focus differently reflecting their charge. β chains of cells expressing the same Dw subtype of DR4 focused identically (data in part not shown), whereas β chains of Dw10, Dw13, and Dw15 focus differently from each other as well as from the focusing pattern of Dw4 and Dw14.

that correlated with the Dw polymorphism; further, that the rare RFLP seen that distinguished one haplotype from one or more other haplotypes did not correlate with the Dw/LD specificities. These findings suggested to us that the DR4-associated Dw/LD subtype polymorphism was indeed of very recent evolutionary origin; to explain certain findings we speculated that this subtype polymorphism may have evolved after an evolutionary crisis or "bottleneck".[54,55]

Studies by Segall and co-workers of the DR2 subtypes demonstrated that clear RFLP patterns could be defined that correlated with each of the subtypes.[56] On the other hand, the degree of RFLP among the DR2 subtypes was again less than that seen among the different DR haplotypes as defined serologically.[57] Our studies, thus, suggested that these subtype polymorphisms evolved more recently than the polymorphism of class II genes defined serologically when comparing haplotypes that encoded different DR specificities. Since a majority of restriction sites are presumably in noncoding sequences, the studies gave little information about functional polymorphism. It is for this reason that, as did several different laboratories, we moved to DNA sequencing of class II polymorphic genes.

IV. SEQUENCING OF HLA CLASS II POLYMORPHIC GENES

As with all of our other studies, we focused our attention on sequences of the polymorphic genes associated with the Dw/LD subtypes of DR4 and DR2. At the time that we undertook these studies, there were already several class II sequences published from haplotypes expressing different DR specificities. From a comparison of those published sequences, it was clear that the polymorphism among class II polymorphic genes was primarily restricted to the exon encoding the N-terminal domain correlating with results of studies of proteins;

TABLE 5
Summary of Amino Acid Differences between DRβ1
Alleles

	Clone	Cell	DR	Dw	Amino acid position		
					71	74	86
a.	LS5.8.1	LS40	β1	14	Arg	Ala	Val
b.	S3.4	SSTO	β1	13	Arg	Glu	Val
c.		"DR4,6"[a]	β1	4	Lys	Ala	Gly
d.		Preiss[b]	β1	4	Lys	Ala	Gly
e.		Preiss[b]	β2[c]	4	Arg	Glu	Val

[a]   Taken from the unpublished sequence of a DRβ cDNA clone obtained by
      E. Long, B. Mach and co-workers.
[b]   Taken from T. Speiss et al.[5]
[c]   In new terminology this is encoded by the B4 locus.

further, that there were three (or by some accounts, four) hypervariable regions (HVR) in the exon encoding the N-terminal domain.

A. DR4 BI (DRβ1) SEQUENCES

We sequenced the DRB1 (β1) genes of DR4-Dw13 and DR4-Dw14 and compared those sequences with a sequence obtained from Long and Mach which was taken from a DR4[+] cell for which the Dw/LD subtype was not known. We identified the cell from which Long and co-workers had obtained their sequence as a DR4-Dw4 cell by using cloned T cell reagents that distinguished among the Dw/LD subtypes of DR4. We thus had available for comparison the DRB1 sequences of Dw4, Dw13, and Dw14 (see Table 5).[53]

Several findings were of interest. First, although DRβ sequences had, prior to our study, not been assignable as alleles or as isotypes, the minimal differences among the β genes for which we had sequences argued strongly that we were studying a series of three alleles. Second, differences among these three alleles were restricted to the third HVR (assuming that polymorphism around amino acid position 86 is considered a part of the third HVR). In one combination, the comparison of Dw14 and Dw13, the alleles differed by only a single amino acid at position 74. Third, based on the fact that it was extremely likely that these clones represent transcripts of alleles of a single locus, we could identify the clones as presumably representing transcripts of the DR B1 locus. The conclusion that these sequences represent DRB1 transcripts was also consistent with the observation that there was an approximately tenfold higher level of DRβI as compared with DRβIII expressed on the surface of EBV-transformed B cells; that finding may well be reflected in levels in the mRNA pool.

It was apparent from these data that a few amino acid differences may profoundly affect how the molecule is recognized by T lymphocytes. This situation was not without precedent; relatively minor amino acid substitutions in the heavy chain of the HLA-A2 molecule[58,59] and in mutants of the K[b] molecule[60] drastically alter T-cell recognition of those molecules.

These results were confirmed and extended by the work of Gregersen and colleagues.[61] Those workers obtained sequences for Dw10 and Dw15 as well as the ones we had compared.

The minimal polymorphism among the DRB1 alleles of the DR4 subtypes is consistent with the very recent evolution of the subtype polymorphism, as suggested from our RFLP studies. Since the Dw/LD subtype polymorphism of the DR4 haplotypes is also recognized in restricted recognition, it is clear that the differences described in the third HVR can underlie both allo- and restricted recognition. Whether the changes in DNA sequence noted here are directly responsible for encoding T-cell-recognized determinants, or whether they are responsible for conformational changes involving other parts of the molecule which are recognized by T cells, remains unclear.

B. DR2 B1 AND B3 SEQUENCES

Our studies of the DNA sequences of polymorphic genes of both DR and DQ relating to the Dw/LD subtypes of DR2 provided insights from several regards.[62-64] First, it was again possible to group the sequences into two presumed allelic series, the one relating to the more polymorphic (as judged by protein polymorphism) DR B1 locus and the other to the less polymorphic (on the same criteria) DR B3 locus. Second, gene conversion mechanisms had been proposed to explain several examples of clustered sequence variation in MHC genes; in all of these examples, however, the proposed gene conversion event was unidirectional, i.e., one of the two interacting genes acts as a sequence donor and the other as a sequence recipient. No examples of potential reciprocal genetic exchange in which the two interacting genes act as both donor and recipient of gene fragments, had been found in the MHC system or in other multigene families of higher organisms. The sequences of DRB genes from cells, all expressing the same serologically defined determinant (DR2) but different T cell-recognized (Dw/LD) specificities (Dw2 and Dw21), suggested that new coding sequences for DRB genes in the DR2 haplotypes are potentially generated by reciprocal intergenic exchange (reciprocal gene conversion). Third, and presumably related to the area of gene interconversion, we found a chi-like sequence upstream from the proposed area of gene conversion; such sequences have been proposed as promoting recombination.[62]

Shown in Table 6 are six amino acid sequences for DRβ genes from three different DR2 subtypes (Dw2, 12, and 21). The sequences can be clearly grouped. The chi-like sequence, which we have previously discussed, is shown in Table 7, which also gives the sequences for the potentially interconverted region.

C. DR2-DQw1 B1 SEQUENCES AND THE POSSIBLE ORIGIN OF THE DQw3 B1 GENE

In order to evaluate DQ polymorphism, we obtained and compared sequences from cDNA clones encoding DR2-Dw2, -Dw12, and -Dw21 DQβ chains.[68] Sequence analysis revealed that there is considerable polymorphism among the DQβ chains of these haplotypes, as there is for other haplotypes. Comparison with other DQβ sequences available in the literature shows that the DQw1β chain of FJO (Dw21) is very similar to DQw1β expressed by DR1-DQw1 cells. The polymorphism of DR2-associated DQw1β chains is in sharp contrast to the similar DQw2β chains expressed by DR3 and DR7 cells, and the essentially identical DQw3β chains described in DR4 subtypes.

One intriguing finding is the homology between the first domain exons of DQw3 and DQw1 vs DQw2β chains. The exon II sequence similarity suggests that the DQw3β gene may have been generated from the insertion of exon II from Dw2-DQw1 into a DQw2 gene. Two possible mechanisms for this are through gene conversion or intraallelic recombination; however, there is no strong evidence favoring one mechanism over the other (see Table 8).

V. STRUCTURE-FUNCTION RELATIONSHIP OF DR AND DQ MOLECULES

A major goal of the studies reviewed above was to reach a point when structure-function correlations could be made. Function would be the recognition of determinants on the DR and DQ products by T cells. The establishment of the crystal structure of an HLA class I molecule[65] and the conceptualization of the three-dimensional structure for class II molecules,[66] based on the known class I structure, as well as the accumulation of the sequences of the class II molecules has allowed us to approach this problem.

TABLE 6
Grouping and Comparison of Amino Acid Sequences of 6 DRβ Chains from HLA-DR2 Individuals

Group	β Chain	Dw	First domain															Second domain											CP 191	TM 203	CY 222	CY 231
			6	9	11	13	28	30	31	37	38	47	67	70	71	85	86	96	104	105	108	110	120	133	135	142	150	157	191	203	222	231
I	Dw2-A	Dw2	R	Q	D	Y	H	D	I	D	L	Y	F	D	R	V	G	E	A	R	T	E	–	R	S	V	–	–	–	V	–	–
I	DHO-7	Dw12	–	|	|	|	|	|	|	|	|	Y	F	D	R	V	G	E	A	R	T	Q	N	R	S	V	N	T	Q	V	K	H
I	MN2-2	MN2	C	Q	D	Y	H	G	I	N	V	Y	I	Q	A	A	V	E	A	R	T	Q	N	R	G	V	N	T	Q	I	K	H
II	Dw2-B	Dw2	R	W	P	R	D	Y	F	S	V	F	I	Q	A	V	V	Q	S	K	P	Q	S	L	G	M	D	T	R	V	R	Q
II	DHO-8	Dw12	R	W	P	R	D	Y	F	S	V	F	I	Q	A	V	G	Q	S	K	P	Q	S	L	G	M	N	T	R	V	R	Q
II	MN2-61	MN2	R	W	P	R	D	Y	F	S	V	Y	|	D	K	V	G	Q	S	K	P	Q	S	M	G	M	N	F	R	V	R	Q

TABLE 7
Nucleotide Sequences of cDNA Clones FJO-13, DHO-8, FJO-6, and DHO-7

```
            47                  50
(I) DHO-7   TAC CGG GCG GTG GTG ACG GA G CTG GGG   CGG CCT GAC GCT GAG TAC
(II) FJO-13 --- --- --- --- --- --- --- --- ---   --- --- --- --- --- ---
(II) DHO-8  -T- --- --- --- --- --- --- --- ---   --- --- --- --- --- ---
(I) FJO-6   --- --- --- --- --- --- --- --- ---   --- --- --- --- --- ---

                                        60                      70
            Phe Leu Glu Asp Arg Ala
(I) DHO-7   TTC CTG GAA GAC AGG GCC
(II) FJO-13 --- --- --- --- --- ---
            Ile Leu Glu Gln Ala Arg
(II) DHO-8  A-- --- --G C-G GC- CGG
(I) FJO-6   A-- --- --G C-G GC- CGG

                    80                          90          94
(I) DHO-7   GCC GCG GAC ACC TAC TGC AGA   GGT GAG AGC TTC   ACA CAG AAG GTG CAG CGG   CGA
(II) FJO-13 --- --- --- --- --- --- ---   --- --- --- ---   --- --- --- --- --- ---   ---
(II) DHO-8  --- --- --- --- --- -C- ---   --- --- --- ---   --- --- --- --- --- ---   ---
(I) FJO-6   --- --- --- --- --- --- -TG   --- --- --- ---   --- --- --- --- --- ---   ---
```

Note: Numbers correspond to the amino acid sequence of the DRβ molecules. Lines indicate nucleotide identity with DHO-7.

TABLE 8
Sequence Differences in the DQβ Genes: Relationship of DQw3 Allele to DQw1 (DR2-Dw2) and DQw2 (DR3) Alleles

Summary of Amino Acid Differences in the DQβ Genes

Haplotype	Leader: -6	-5	-4	First domain: 9	28	30	37	38	46	47	52	53	55	57	66	67	70	71	74	75	77	84	85	86	87	89	90	2D, TM, CY: 125	140	182	185	203	220	221
Dqw1.2 (DR2-Dw2)	S	L	F	T	Y	Y	A	V		Y	P	Q	R	D	E	V	G	T	E	L	T	E	V	A	F	G	I	G	A	S	T	G	R	Q
DQw3.1 (DR5-Swei)	T	P		Y	—	—	—	—	—	—	L	P	—	—	—	—	—	—	—	—	—	Q	L	E	L	T	T	A	T	—	I	H	—	H
DQw3.2 (DR4)	T	P	V	Y	—	—	—	—	—	—	L	P	A	—	—	R	—	—	—	—	—	Q	L	E	L	T	T	A	T	N	I	H	I	H
DQw2 (DR3-WT49)	T	P	V	S	S	I	V	E	F	L	L	A	D	I	R	K	A	V	R	—	—	Q	L	E	L	T	T	A	—	—	I	—	—	H

Comparison of Nucleotide Differences from Amino Acid Residue 9 to Residue 84

Haplotype	9	<u>21</u>	28	30	<u>35</u>	37	38	46	47	<u>48</u>	52	53	55	57	66	67	<u>68</u>	70	71	74	75	77	84
Dqw1.2 (DR2-Dw2)	TTC	ACC	TAC	GAG	TAC	GCG	GTG	TAC	CGC	CCG	CAG	CGG	GAT	GAA	GTC	CTG	GGG	ACC	GAG	TTG	ACG		GAG
DQw3.1 (DR5-Swei)	-A-	---	---	---	---	---	--G	---	-A-	---	-T-	-C-	--G	---	---	--A	---	---	---	---	---		C--
DQw3.2 (DR4)	-A-	---	---	---	---	---	--G	---	-A-	---	-T-	-C-	-CC	---	---	A--	---	---	---	---	---		C--
DQw2 (DR3-WT49)	-A-	-G-	-AG	--A	AT-	-T-	-A-	-T-	--G	-T-	--CC	-C	A--	---	---	A--	--AA	-C-	-G-	---	---		C--

Cloned Cytolytic T Cells Recognizing HLA-DR Associated Determinants

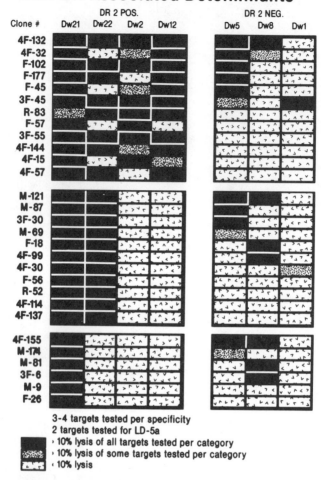

3-4 targets tested per specificity
2 targets tested for LD-5a
> 10% lysis of all targets tested per category
> 10% lysis of some targets tested per category
< 10% lysis

FIGURE 4.   Illustrates the results obtained with clones generated against Dw21 haplotypes and tested with DR2 positive and DR2 negative target cells. Three to four target cells were tested per category with the schematic illustration indicated as black when lysis was greater than 10% cytotoxicity for all target cells tested; stippled when results were greater than 10% for some but not all target cells per category; and predominantly white when all target cells demonstrated less than 10% lysis. The clones were tested with cells representative of all defined Dw specificities.

To approach the structure-function issue, we derived T cell clones sensitized to DR or DQ products associated with determinants of DR2 haplotypes. Specifically, we tested clones sensitized to either the Dw2 or Dw21 haplotypes and then (1) determined whether those clones were DR- or DQ-directed by mAb blocking studies and (2) performed panel studies to evaluate which products encoded by different haplotypes appeared to share determinants with the sensitizing cell. Clones were derived that reacted only with the sensitizing Dw specificity; with one or more additional Dw subtypes of DR2 but not with DR2 negative cells; or with both DR2 positive as well as DR2 negative cells. Data shown in Figure 4 show the patterns of reactivities of DR-directed clones tested on a panel of target cells. We then compared the known sequences of those class II products to determine whether we

TABLE 9
DQ-Directed T Cell Recognized Epitopes

DQw	DR	Dw	T cell clone reactivity				
			A	B	C	D	E
1	2	Dw2	+	+	−	+	−
1	6	Dw18	−	+	−	+	−
1	6	Dw9	−	−	+	+	−
1	2	Dw21	−	−	+	+	+
1	1	Dw1	−	−	+	+	−
3(7)	5 (Swei)	Dw5	−	−	−	+	−
3(7)	5	Dw5	−	−	−	−	−
1	2	Dw12	−	−	−	−	−
Shared/unique				53—57	66—71	70	57
sequence			None	66—77	(71)		

could correlate determinant sharing with sequence sharing. Shared sequences might indicate which part(s) of the three-dimensional structure of the class II molecule was involved in T-cell recognition. We have performed these studies for both DQ and DR.

The data presented in Table 9 are one example of the approach that we have taken. All of these clones react with the DQ product; five patterns of clone reactivity are shown. In pattern A are clones reactive with only the sensitizing haplotype, Dw2; in pattern B, clones reactive with Dw2 and Dw18; in pattern C, clones reactive with Dw21, Dw9, and Dw1, etc. Also shown in the Table are the positions of polymorphic amino acids of the N-terminal domain that are of interest in attempting to establish a correlation. To explain pattern B, for instance, one must invoke sharing of both the amino acids from 53 to 57 as well as amino acids 67 to 77. Invoking sharing of only the HVR comprising amino acids 66 to 77 would not explain the pattern since the DQw3 positive, DR5 positive, cell shares all amino acids in this HVR but is not lysed by the cytotoxic T lymphocyte clones. Pattern C, on the other hand, could be explained either by sharing of amino acids in the third HVR, or by amino acids 30 to 38. Pattern D can be explained by sharing of amino acid 70. Pattern E, the Dw21-specific clone, can be explained by sharing of amino acid 57.

The analyses of DR were both complicated, and made more interesting, by the expression of the two DR dimers and the shared sequences in the DRβI and DRβIII chains consequent to the presumed reciprocal intergenic conversion event[62] reviewed above. Several conclusions are possible.

First, in many instances, relating both to DR and DQ, sequence sharing in the third hypervariable region (HVR defined as amino acids 67 to 72), can explain LD determinant sharing. This is not surprising given the several lines of evidence implicating the third HVR as important in T-cell recognition; our own findings in the Dw4, Dw13, and Dw14 haplotypes showed the importance of this HVR for allo- and restricted recognition.

Second, and perhaps more interesting, is the finding that in certain comparisons of determinant and sequence sharing, it is amino acid 67, 70 to 71, and 74 that appear to be of importance. These are the amino acids on one face of the α helix of the first domain of the β chain and thus provide some evidence in man that it is this face of the helix that may be of import to allo-determinant recognition. (The only difference between Dw13 and Dw14 DRβI chains is in amino acid 74, for instance.)

Third, in some instances, one must invoke sharing of both the third and the second HVR to explain shared determinant recognition. Sharing of only one or the other HVR would not correlate with determinant sharing. Whereas there are several explanations for this finding of the apparent need for sharing of two HVRs to explain determinant sharing, one that we find interesting is based on the following. It has been suggested that recognition of allodeterminants is actually the restricted recognition of self-peptides[67] bound in the groove of the class II molecule, i.e., the β-pleated sheets. Polymorphism of the second hypervariable region, which is located on a β-pleated sheet immediately adjacent to the portion of the α helix encoding the third hypervariable region, may determine whether a given self-peptide binds or not. Binding would lead to allo-determinant expression. This model could explain the need for sequence identity in both the third HVR, which would be recognized by the T cell, and the second HVR, which would account for the binding or not of a given self-peptide and thus the expression, or lack thereof, of the allo-determinants. (Clearly, certain amino acids of the α helix would be involved in peptide binding.)

These studies then provide evidence, which must be considered preliminary in terms of defining the parts of the class II molecule that are recognized by T cells, both for the face of the α helix that is involved in allo-determinant recognition by the T cell, as well as raising the possibility that polymorphism in the second HVR may be responsible for binding of self-peptide(s).

VI. CONCLUSION

We have discussed the approach that we have taken during the past more than two decades developing tests for the study of HLA class II products viewed from a T-cell perspective. At the very beginning of these studies we did not, of course, realize that we were studying a major histocompatibility complex in man; however, recognition of this fact and the evolving body of evidence that for both allo-recognition as well as restricted recognition, there is an all important, and somewhat magical, relationship between antigens of the MHC and T lymphocytes, persuaded us to follow our investigations with this emphasis. Our continued interest in the class II, as opposed to the class I, region no doubt arose out of the correlation between MLC and proliferation, the assay of that test, and the finding that it was the HLA-D region that stimulated that proliferation. We have herein reviewed not only the development of the tests that provided the technical basis for these studies (the MLC, CML, PLT, and cloned T lymphocyte approaches), but also the series of studies that, in part, led to our present-day understanding of this region.

It would seem to us that we are currently at the beginning of one of the most exciting eras in this work: namely, to understand the structure vis-a-vis function for the class II molecules in their relationship to T lymphocytes. Advances in our understanding of the three-dimensional structure of the molecules, the elucidation of the structure and genetics of the T cell receptor, and the availability of T-cell clones that define determinants on these molecules with great precision, as well as the known sequences of these molecules, promises a rapidly moving and exciting few years ahead. The possibility that different products, such as DR and DQ, can differentially activate what are functionally disparate subpopulations of T lymphocytes, invites the suggestion that we may gain understanding of the very important, yet poorly understood at a mechanistic level, relationship of certain HLA haplotypes with a variety of different diseases.

REFERENCES

1. **Bain, B., Vas, M. R., and Lowenstein, L.,** The development of large immature mononuclear cells in mixed leukocyte culture, *Blood,* 23, 108, 1964.
2. **Bach, F. H. and Hirschhorn, K.,** Lymphocyte interaction a potential *in vitro* histocompatibility test, *Science,* 143, 813, 1964.
3. **Bach, F. H. and Voynow, N. K.,** One-way stimulation in mixed leucocyte cultures, *Science,* 153, 545, 1966.

4. **Amos, D. B. and Bach, F. H.,** Phenotypic expressions of the major histocompatibility locus in man (HL-A): leukocyte antigens and mixed leukocyte culture reactivity, *J. Exp. Med.,* 128, 623, 1968.
5. **Bach, F. H. and Amos, D. B.,** Hu-1: major histocompatibility locus in man, *Science,* 156, 1506, 1967.
6. **Bach, F. H., Albertini, R. J., Amos, D. B., Ceppellini, R., Mattiuz, P. L., and Miggiano, V. C.,** *Transplant Proc.,* 1, 339, 1969.
7. **Yunis, E. J. and Amos, D. B.,** Three closely linked genetic systems relevant to transplantation, *Proc. Natl. Acad. Sci. U.S.A.,* 68, 3031, 1971.
8. **Bach, F. H., Widmer, M. B., Bach, M. L., and Klein, J.,** Serologically defined and lymphocyte-defined components of the major histocompatibility complex in the mouse, *J. Exp. Med.,* 136:1430, 1972.
9. **Widmer, M. B., Omodei-Zorini, C., Bach, M. L., Bach, F. H., and Klein, J.,** Importance of different regions of H-2 for MLC stimulation, *Tissue Antigens,* 3, 309, 1973.
10. **Meo, T., Vives, J., Miggiano, V., and Shreffler, D.,** A major role for the Ir-1 region of the mouse H-2 complex in the mixed leukocyte reaction, *Transplant. Proc.,* 5, 377, 1973.
11. **Solliday, S. and Bach, F. H.,** Cytotoxicity: specificity after *in vitro* sensitization, *Science,* 170, 1406, 1970.
12. **Hodes, R. J. and Svedmyr, E. A. J.,** Specific cytotoxicity of H-2 incompatible mouse lymphocytes following mixed culture *in vitro, Transplantation,* 9, 470, 1970.
13. **Hayry, P. and Defendi, V.,** Mixed lymphocyte cultures produce effector cells: model *in vitro* for allograft rejection, *Science,* 168, 133, 1970.
14. **Bach, F. H.,** The major histocompatibility complex in transplantation immunology, *Transplant Proc.,* 5(1), 23, 1973.
15. **Bach, F. H., Bach, M. L., and Sondel, P. M.,** Differential function of major histocompatibility complex antigens in T-lymphocyte activation, *Nature (London),* 259, 273, 1976.
16. **Eijsvoogel, V. P., duBois, R. S., Melief, C. J. M., et al.,** Lymphocyte activation and destruction *in vitro* in relation to MLC and HLA, *Transplant Proc.,* 5, 1301, 1973.
17. **Schendel, D. J. and Bach, F. H.,** Genetic control of cell-mediated lympholysis in mouse, *J. Exp. Med.,* 140, 1534, 1974.
18. **Bach, F. H., Bach, M. L., Alter, B. J., Lindahl, K. F., Schendel, D. J., and Sondel, P. M.,** Recognition in MLC and CML: the LD-SD dichotomy, in *Immune Recognition,* Rosenthal, A. S., Ed., Academic Press, New York, 1975, 173.
19. **Schendel, D. J. and Bach, F. H.,** H-2 and non H-2 determinants in the genetic control of cell-mediated lympholysis, *Eur. J. Immunol.,* 5, 800, 1975.
20. **Wagner, H., Gotze, D., Ptchelinzew, L., and Rollinghoff, M.,** Induction of cytotoxic T lymphocytes against I-region-coded determinants: *in vitro* evidence for a third histocompatibility locus in the mouse, *J. Exp. Med.,* 142, 1477, 1975.
21. **Zeevi, A. and Duquesnoy, R. J.,** PLT specificity of alloreactive lymphocyte clones for HLA-B locus determinants, *Proc. Natl. Acad. Sci. U.S.A.,* 80, 1440, 1983.
22. **Reinsmoen, N. L., Anichini, A., and Bach, F. H.,** Clonal analysis of T lymphocyte response to an isolated class I disparity, *Hum. Immunol.,* 8, 195, 1983.
23. **Flomenberg, N., Naito, K., Duffy, E., Knowles, R. W., Evans, R. L., and Dupont, B.,** Allocytotoxic T-cell clones: both leu 2+3− and 2−3+ T cells recognize class I histocompatibility antigens, *Eur. J. Immunol.,* 13(11), 905, 1983.
24. **Widmer, M. B., Alter, B. J., Bach, F. H., Bach, M. L., and Bailey, D. W.,** Lymphocyte reactivity to serologically undetected components of the major histocompatibility complex, *Nature (London),* 242, 239, 1973.
25. **Festenstein, H. and Oliver, R. T. D.,** Cellular testing, in *Histocompatibility Testing 1977,* Bodmer, W. F., Ed., Munksgaard, Denmark, 1977, 85.
26. **Reinsmoen, N. L. and Bach, F. H.,** Five HLA-D clusters associated with HLA-DR4, *Hum. Immunol.,* 4, 249, 1982.
27. **Dupont, B., Jersild, C., Hansen, G. S., Nielsen, S., Thomsen, M., and Svejgaard, A.,** Typing for MLC determinants by means of LD-homozygous and LD-heterozygous test cells, *Transplant. Proc.,* 5, 1543, 1973.
28. **Jorgensen, F., Lamm, L., and Kissmeyer-Nielsen, F.,** Mixed lymphocyte cultures with inbred individuals: an approach to MLC typing, *Tissue Antigens,* 4, 323, 1973.
29. **Van den Tweel, J. G., Blusse van Oud Alblas, A., Keuning, J. J., Goulmy, E., Termijtelen, A., Bach, M. L., and van Rood, J. J.,** Typing for MLC(LD). I. Lymphocytes from cousin marriage offspring as typing cells, *Transplant. Proc.,* 5, 1535, 1973.
30. **Dausset, J., Sasportes, M., and Lebrun, A.,** Mixed lymphocyte culture (MLC) between HL-A serologically identical parent-child and between HLA homo and heterozygous individuals, *Transplant. Proc.,* 5, 1511, 1973.

31. **Mempel, W., Grosse-Wilde, H., Baumann, P., Netzel, B., and Albert, E. D.,** Population genetics of the MLC response: typing for MLC determinants using homozygous and heterozygous reference cells, *Transplant Proc.,* 5, 1529, 1973.

32. **Sheehy, M. J., Sondel, P. M., Bach, M. L., Wank, R., and Bach, F. H.,** HLA-A LD (lymphocyte defined) typing: A rapid assay with primed lymphocytes, *Science,* 188, 1308, 1975.

33. **Sheehy, M. J. and Bach, F. H.,** Primed LD typing (PLT) — technical considerations, *Tissue Antigens,* 8, 157, 1976.

34. **Fathman, C. G. and Hengartner, H.,** Clones of alloreactive T cells, *Nature (London),* 272, 617, 1978.

35. **Bach, F. H., Inouye, H., Hank, J. A., and Alter, B. J.,** Human T lymphocyte clones reactive in primed lymphocyte typing and cytotoxicity, *Nature (London),* 281, 307, 1979.

36. **Bach, F. H., Reinsmoen, N. L., and Segall, M.,** Definition of HLA antigens with cellular reactants, *Transplant. Proc.,* 4, 102, 1983.

37. **Reinsmoen, N. and Bach, F. H.,** T cell clonal analysis of HLA-DR2 haplotypes, *Hum. Immunol.,* 20, 13, 1987.

38. **Shaw, S., Johnson, A. H., and Shearer, G. M.,** Evidence for a new segregant series of B cell antigens that are encoded in the HLA-D region and that stimulate secondary allogeneic proliferative and cytotoxic responses, *J. Exp. Med.,* 152, 565, 1980.

39. **Shaw, S., Pollack, M. S., Payne, S. M., and Johnson, A. H.,** HLA linked B-cell alloantigens of a new segregant series: population and family studies of the SB antigens, *Hum. Immunol.,* 1, 177, 1980.

40. **Bach, F. H. and Reinsmoen, N. L.,** Cloned cellular reagents to define antigens encoded between HLA-DR and glyoxylase, *Hum. Immunol.,* 5, 133, 1982.

41. **Reinsmoen, N. L. and Bach, F. H.,** Clonal analysis of HLA-DR and -DQ associated determinants: their contribution to Dw specificities, *Hum. Immunol.,* 16, 329, 1986.

42. **Quigstad, E., Thorsby, E., Reinsmoen, N. L., and Bach, F. H.,** Close association between the Dw14 (LD40) subtype of HLA-DR4 and a restriction element for antigen-specific T cell clones, *Immunogenetics,* 20, 583, 1984.

43. **Reinsmoen, N. L., Volkman, D. J., Bach, F. H., and Fauci, A. S.,** Soluble antigen specific recognition of the DR4/LD40 determinant by human T cell clones producing multiple lymphokines, *Fed. Proc.,* 43, 1918, 1984.

44. **Linner, K. M., Monroy, C., Bach, F. H., and Gehrz, R. C.,** Dw subtypes of serologically defined DR-DQ specificities restrict recognition of cytomegalovirus, *Hum. Immunol.,* 17, 79, 1986.

45. **Horibe, K., Flomenberg, N., Pollack, M. S., Adams, T. E., Dupont, B., and Knowles, R. W.,** Biochemical and functional evidence that an MT3 supertypic determinant defined by a monoclonal antibody is carried on the DR molecule on HLA-DR7 cell lines, *J. Immunol.,* 133, 3195, 1984.

46. **Groner, J., Watson, A., and Bach, F. H.,** Dw/LD related molecular polymorphism of DR4 β chains, *J. Exp. Med.,* 157, 1687, 1983.

47. **Nepom, B. S., Nepom, G. T., Mickelson, E., Antonelli, P., and Hansen, J. A.,** Electrophoretic analysis of human HLA-DR antigens from HLA-DR4 homozygous cell lines: correlation between β-chain diversity and HLA-D, *Proc. Natl. Acad. Sci. U.S.A.,* 80, 6962, 1983.

48. **Reinsmoen, N. L., Layrisse, Z., Betuel, H., and Bach, F. H.,** A study of HLA-DR2 associated HLA-Dw/LD specificities, *Hum. Immunol.,* 11, 105, 1984.

49. **Goyert, S. M., Gatti, R., and Silver, J.,** Peptide map comparisons of similar serologically defined HLA-DR antigens isolated from different lymphoblastoid cell lines, *Hum. Immunol.,* 5, 205, 1982.

50. **Bach, F. H. and Watson, A. J.,** Dw/LD associated molecular polymorphism of DR4 β-chains: intra-molecular localization of polymorphic sites, *J. Immunol.,* 131, 1622, 1983.

51. **Kaufman, J. F. and Strominger, J. L.,** The extracellular region of light chains from human and murine MHC class II antigens consists of two domains, *J. Immunol.,* 130, 808, 1983.

52. **Bach, F. H., Linner, K. M., Choong, S. A., and Groner, J. P.,** Molecular polymorphism of DR and DC products, in *Histocompatibility Testing 1984,* Albert, E. D., Baur, M. P., and Mayr, W. R., Eds., Springer-Verlag, New York, 1984, 516.

53. **Cairns, S., Curtsinger, J. M., Dahl, C. A., Freeman, S., Alter, B. J., and Bach, F. H.,** Sequence polymorphism of HLA-DR β1 alleles relating to T cell-recognized determinants, *Nature (London),* 317, 166, 1985.

54. **Nicklas, J. N., Noreen, H. N., Segall, M., and Bach, F. H.,** Southern analysis of DNA polymorphism between Dw subtypes of DR4, *Hum. Immunol.,* 13, 95, 1985.

55. **Bach, F. H.,** The HLA class II genes and products: the HLA-D region, *Immunol. Today,* 6, 89, 1985.

56. **Segall, M., Noreen, H., Schluender, L., and Bach, F. H.,** DNA restriction fragment length polymorphisms characteristic for Dw subtypes of DR2, *Hum. Immunol.,* 15, 336, 1986.

57. **Cohen, D., LeGall, I., Marcadet, A., Font, M. -P., Lalouel, J. -M., and Dausset, J.,** Clusters of HLA class II β restriction fragments describe allelic series, *Proc. Natl. Acad. Sci. U.S.A.,* 81, 7870, 1984.

58. **Biddison, W. E., Ward, F. E., Shearer, G. M., and Shaw, S.,** The self determinants recognized by human virus-immune T cells can be distinguished from the serologically defined HLA antigens, *J. Immunol.,* 124, 548, 1980.

59. **Goulmy, E., van Leeuwen, A., Blokland, F., van Rood, J. J., and Biddison, W. E.,** Major histocompatibility complex-restricted H-Y-specific antibodies and cytotoxic T lymphocytes may recognize different self determinants, *J. Exp. Med.,* 155, 1567, 1982.

60. **Nairn, R., Yamaga, K., and Nathenson, S. G.,** Biochemistry of the gene products from murine MHC mutants, *Annu. Rev. Genet.,* 14, 241, 1980.

61. **Gregersen, P. K., Shen, M., Song, Q., Merryman, P., Degar, S., Seki, T., Maccari, J., Goldberg, D., Murphy, H., Schwenzer, J., Wang, C. Y., Winchester, R. J., Nepom, G. T., and Silver, J.,** *Proc. Natl. Acad. Sci. U.S.A.,* 83, 2642, 1986.

62. **Wu, S., Saunders, T., and Bach, F. H.,** Polymorphism of human Ia antigens generated by reciprocal intergenic exchange between two DR β loci, *Nature (London),* 324, 676, 1986.

63. **Wu, S., Yabe, T., Madden, M., Saunders, T. L., and Bach, F. H.,** cDNA cloning and sequencing reveals that the electrophoretically constant DR β molecules from HLA-DR2 subtypes have different amino acid sequences including a hypervariable region for a functionally important epitope, *J. Immunol.,* 138, 2593, 1987.

64. **Liu, C., Bach, F. H., and Wu, S.,** Molecular studies of rare HLA class II haplotypes: multiple genetic mechanisms in the generation of polymorphic HLA class II genes, *J. Immunol.,* 140, 3631, 1988.

65. **Bjorkman, P. J., Saper, M. A., Samraoui, B., Bennett, W. S., Strominger, J. L., and Wiley, D. C.,** Structure of the human class I histocompatibility antigen, HLA-A2, *Nature (London),* 329, 506, 1987.

66. **Brown, J. H., Jardetzky, T., Saper, M. A., Samraoui, B., Bjorkman, P. J., and Wiley, D. C.,** A hypothetical model of the foreign antigen binding site of class II histocompatibility molecules, *Nature (London),* 332, 845, 1988.

67. **Claverie, J. -M. and Kourilsky, P.,** The peptidic self model: a reassessment of the role of the major histocompatibility complex molecules in the restriction of T cell response, *Ann. Inst. Pasteur/Immunol.,* 137D, 425, 1986.

68. **Wu, S., Saunders, T., and Bach, F. H.,** unpublished observations.

Chapter 3

HYBRID HLA CLASS II MOLECULES: EXPRESSION AND REGULATION

Dominique J. Charron, Vincent Lotteau, Pierre Hermans, David Burroughs, Pascale Turmel, Annick Faille, Li Ya Ju, and Luc Teyton

TABLE OF CONTENTS

INTRODUCTION*

Class II molecules of the major histocompatibility complex (MHC) are heterodimeric transmembrane glycoproteins composed of noncovalently associated α and β chains coded for by α and β genes. In the human there are three major expressed MHC class II dimers DR, DQ, and DP resulting from the corresponding α and β genes, while mice have only two expressed loci, I-E and I-A, the products of which, the I-E and I-A antigens, are respectively homologous to the DR and DQ molecules.[1,2] The murine equivalent of DP is lacking.[3] Additional class II genes have been identified, namely DZα, DOβ, and DXα and β.[4-7] However no products of these genes have yet been identified biochemically[6] and transcription is even questioned for DXα and β genes.[7] The dimers resulting from α-β genes present at loci within a subregion (DR, DQ, DP in man, I-E and I-A in mouse) can be referred to generically as isotypes. Three major isotypes are thus expressed in man, two in the mouse. The class II molecules must undergo intracellular assembly prior to their cell surface expression.[8] At an early stage of intracellular maturation a third molecular partner is found in association with the α-β chain dimers: the Ii chain[9] and its derivatives, a family of basic proteins which are not encoded by the MHC and whose function, despite their unique specificity of association with human leucocyte antigen (HLA) class II molecules, remains unexplained.

The most striking feature of the HLA class II molecules is their extensive polymorphism which is unique to this genetic system. Initially detected by cellular means (mixed lymphocyte reaction) and serology at the surface of B lymphocytes and monocytes, a large number of biochemical and molecular biology studies performed over the past 8 years have provided the structural basis for the allelic variation at individual loci within a subregion as well as for interallelic diversity.[10-12] The primary source of HLA class II diversity resides in the structure of the genes coding for the individual chains. Although the extent of nucleotide substitutions between different alleles can vary in number and position, the most meaningful variations appear concentrated in the NH_2 terminal domain of the individual chains.[13] This is in contrast with the interloci differences (DRvs DQvs DP) since the most divergent area between loci is found in the 3' UT region of the genes.[14] Multiplicity of alleles at most of the class II loci is regarded as beneficial to the species. Similarly the nonallelic diversity (isotypic and combinatorial) should provide some advantage to the individual. The number of DRβ genes varies according to the haplotypes.[15] A few haplotypes express only one DRβ chain while most haplotypes express two DRβ chains. Rare haplotypes may express three DRβ chains.[16] In addition to the functional genes one or two DRβ pseudogenes are present which, although they contribute to the genetic polymorphism, do not affect the antigenic and functional diversity of DR antigens. In addition to the number of loci combined two by two within a haplotype to form isotypes, a powerful way to increase the class II antigen repertoire in individuals would be to associate the α and β chains of the two haplotypes by *trans*-complementation within an isotype. These would be found only in heterozygous individuals. However, for the sake of simplicity most of the protein data on HLA class II molecules has been generated using cell lines homozygous at the HLA-D region, often of consanguinous origin. Thus, the opportunity for detecting class II molecules arising by *trans*-complementation has been operationally limited. This is in contrast with the murine situation in which I region recombinant animals and congenic strains were readily available. Moreover, studies on F1 animals could be easily realized in mice as opposed to rare opportunities for family studies in human. Finally, the reactivity of monoclonal anti-class II antibodies (mAbs) available is very different since in the mouse most of the mAbs are allele specific, while in

---

* Abbreviations: MHC = major histocompatibility complex; HLA = human leucocyte antigens; mAb = monoclonal antibody; 2D-PAGE = two-dimensional polyacrylamide gel electrophoresis; SDS = sodium dodecyl sulfate; HTC = homozygous typing cell.

man most are monomorphic (that is isotype specific and recognizing all haplotypes), some-times supertypic within one isotype, or even of broader reactivity including several different isotypes. An even more efficient way of generating additional diversity would be to combine α and β chains from different isotypes either in *cis* and/or *trans*. However it has long been considered a dogma that class II α-β pairing and expression occurs only within an isotype and that isotype mismatched α-β chain pairing is not permitted. Recent transfection exper-iments performed first, with mouse genes and later, with human genes have evaluated the possibility of such associations.[17,18]

In this paper we will review the data from our laboratory on α-β chain associations occurring in normal human cells which generate intra- and interisotypic hybrid molecules; we will describe some of the regulatory mechanisms underlying the formation of such associations, discuss the serological and functional implications, and explore their potential for the still elusive and puzzling problem of associations between HLA and disease.

II. INTRAISOTYPIC HYBRID HLA CLASS II MOLECULES

A. HLA-DQ HOMOLOGOUS HYBRID MOLECULES
1. Introduction

Hybrid HLA-DQ molecules consist of class II dimers created by α-β chain pairing resulting from gene *trans*-complementation within the HLA-DQ subregion. Such molecules include chains belonging to both the paternal and the maternal haplotype. Much was learned from studies in the mouse. Early genetic studies had indicated the presence of epitopes in F1 animals which were absent from both parental strains.[19] Such hybrid I region determinants were subsequently shown to contribute to the diversity of the MHC class II antigens and provide a molecular basis for Ir gene complementation. Extensive biochemical studies were performed to identify the molecular species resulting from gene *trans*-complementation within the IA subregion and bearing hybrid determinants.[20,21] Furthermore, such determinants can stimulate in the mixed lymphocyte reaction[19] and function as restriction elements for T-cell recognition in antigen-specific T-cell clones[22] and hybridomas.[23] In general, two re-quirements have to be fulfilled in order to identify MHC class II hybrid molecules within one isotype; first, is the ability to distinguish the allelic products of the α and β chain loci and second, an antibody which reacts solely with one chain (either α or β) of the dimer in an allelic (or haplotype restricted) manner is required. These requirements dictated the rationale of our approach.

The HLA-DQ products have the potential of forming hybrid molecules. Indeed both the DQα and the DQβ chains are polymorphic.[24-26] This has been documented at the protein level by a series of biochemical techniques, mainly two-dimensional gel electrophoresis, peptide mapping and NH_2 terminal amino acid sequencing and has been confirmed at the nucleotide level first by RFLP (restriction fragment length polymorphism) analysis of gen-omic DNA and more recently by gene sequencing.[7,27-29]

More problematic has been the search for allele (or supertypic) specific monoclonal antibodies which would be directed uniquely to the DQα or DQβ chain. The chain specificity of anti-HLA class II antibodies can be assessed by Western blotting after separation of the α and β chains.[30] Among several hundreds of anti-HLA class II mAbs which were produced since 1978, very few have fulfilled the criteria of being both chain specific and allele or haplotype (or supertypic) restricted.[31]

These technical constraints explain why in man the demonstration of hybrid molecules occurred several years later than in the mouse, where the use of allospecific sera and recombinant strains had led to the initial description of such molecules in F1 animals.[20,21] Our own approach included the use of two-dimensional gel electrophoresis (2D-PAGE) to detect and identify electrophoretic variants of DQ molecules. While nonequilibrium pH gradient gel electrophoresis (NEPHGE) provides good separation of class II β chains, noticeably the DQβ chains, the α chains are not well resolved by this procedure. We applied isoelectrofocusing in the first dimension (IEF) in order to maximize resolution of the acidic area of the gel. This gives clearly distinct electrophoretic patterns for DRα and several DQα alleles. The DQα alleles were initially identified using consanguinous homozygous cell lines. Mixing experiments were performed to confirm the ability of IEF 2D-PAGE to separate and distinguish DQα alleles. As expected the DQβ chains, which are more basic, were better resolved using NEPHGE 2D-PAGE. This procedure has previously been used to asess DRβ chain polymorphism.[32] These electrophoretic methods represent simple and reproducible ways to identify the haplotype of origin of each α or β subunit. The subunit reactivity of a panel of monoclonal anti-HLA class II antibodies was investigated by Western blotting. The α-β complex was first separated into α and β subunits by boiling in SDS prior to analysis by SDS-PAGE. However, when nondissociating conditions (0.6% SDS without boiling) were used, a significant although variable amount of nonassociated α and β chains could be detected in addition to the 55 kDa α-β complex. In some cases the antibody reactivity had to be confirmed by 2D PAGE analysis of the antigen.

2. Description

Two series of mAbs were used to demonstrate the presence of hybrid HLA-DQ molecules. CA 2.06[32] and SG 1 71[33] are two monomorphic monoclonal antibodies which, under strict experimental conditions, were shown independently by us and S. Goyert to react with all haplotypes; they immunoprecipitate the DR α-β complex from all haplotypes and also a large amount of DQw2 antigen when used in DR7 haplotypes. Western blot analysis shows that they both react with the β chain either alone or when complexed to the α chain (Figure 1a, 1b). Thus, in DR7 cells CA 2.06 and SG 1 71 can be considered operationally specific for the DQw2β chain when used in cells which have been previously depleted of all their DR α-β molecules. This was confirmed by the 2D PAGE Western blot analysis of such a cell lysate (Figure 1c). CA 2.06 and SG 1 71 were used to detect hybrid molecules containing a DQw2 (DR7) β chain.

A DQw2 (DR7) β DQw1 (DR1) α hybrid molecule was the first to be demonstrated in the human by precipitation of a ^{35}S-labeled DR1/7 heterozygous cell line whose DR α-β molecules had been previously removed by several cycles of sequential immunoprecipitations. MoAb CA 2.06 was able to immunoprecipitate a molecular complex consisting of a DQw2[7]* β chain and two DQα chains corresponding electrophoretically to DQw2[7] α and DQw1[1]** α. No DQw1[1] β chain was found in this immunoprecipitate. Since in such a cell lysate the antibody could only react with the DQw2[7] β chain we concluded that the DQw1[1] α chain was present because of its association with the DQw2[7] β chain.[25] Experiments with I[125] surface-labeled antigens demonstrated that these hybrid molecules are also expressed at the cell surface.[26] Moreover, the amount of hybrid molecules was substantial.[25,34] Although it was not possible to precisely determine the amount in such experiments the DQα-DQβ hybrid heterozygous product appeared to be present at approximately one half the level of the homozygous product (Figure 2a). Similar experiments were conducted in DR7-DR2 and DR7-DRw6 heterozygous cells (Figure 2b, c). In these combinations we were always able to distinguish the DQw1α chain from the DQw2[7] α chain by 2D PAGE. From the overall data we could conclude that the DQw2[7] β chain was able to pair with any of the DQw1α chains encoded by the non-DR7 haplotype (DQw1α of DR1, DQw1α of DR2 and DQw1α of DRw6).

A similar approach was undertaken using G2a5, a gamma 2a variant of Genox 3.53, which recognizes DQw1 molecules expressed in DR1, DR2, and most DRw6 haplotypes.[35]

---

*   DQw2[7] = DQw2 (DR7 haplotype)
** DQw1[1] = DQw1 (DR1 haplotype)

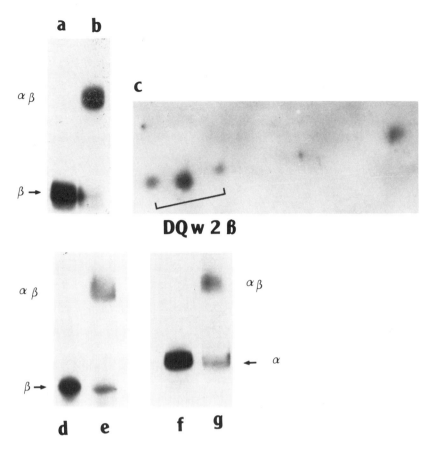

FIGURE 1.   Specificity of anti class II mAbs. 0.5% NP40 extracts of the cell lines were separated by SDS-PAGE or 2D-PAGE and transferred to nitrocellulose. a, d, f: extracts boiled with 2.3% SDS; b, e, g: extracts not boiled. a, β chain reactivity of CA 2.06; b, reactivity of CA 2.06 with a class II dimer; c, DQw2β chain specificity of CA 2.06; assessed on GM 3163 (DR7/ DQw2 HTC) cell lysate previously depleted of DR α-β complexes by several cycles of immunoprecipitation; d and e, DQw1β chain reactivity of G25a; assessed on SC-CA (DR1 HTC) cell lysate; f and g, DRα chain reactivity of DA6.147.

The primary use of this mAb was to detect interisotypic α-β associations in homozygous cells (as detailed in the next paragraph). However, when used in heterozygous cell lines it provides a tool to identify intraisotypic DQ hybrid molecules. The chain reactivity of G2a5 as assessed by Western blotting was found to be DQw1β (Figure 1d, e). G2a5 does not recognize DQα chain or any other α or β chain than DQw1β. Thus, any chain associated with the DQw1β chain in immunoprecipitation experiments with the anti-DQw1β chain mAb G2a5 will be subsequently precipitated on the basis of its noncovalent association with the DQw1β chain. Indeed, in addition to the expected DQw1α chain other DQα chains are seen when G2a5 is used in several DQw1/non-DQw1 heterozygous cells in which the cell lysate had previously been depleted extensively of DR molecules. Examples of DQα DQβ *trans*-associations were obtained in two different combinations of haplotypes. In a DR 1/7 cell a DQw2α chain was present in a G2a5 immunoprecipitate (DQw2α-DQw1β hybrid molecule) and, similarly, in a DR1/DR4 cell a DQw3α chain was found in addition to the DQw1α chain (DQw3α-DQw1β hybrid molecule). Overall, no reproducible or important differences were observed between these combinations regarding the amount of hybrid material present.

    Limited work has been published since our first demonstration of hybrid HLA-DQα-

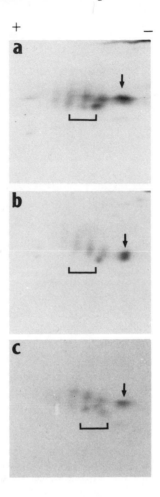

FIGURE 2.   Hybrid HLA-DQ molecules. ^{35}S methionine-la-
beled cell lysates previously depleted of DR molecules were
immunoprecipitated with CA 2.06. The panels a, b, c show the
α chain area analyzed by IEF-2D-PAGE in: a DR1/7 cell (a);
a DR2/7 cell (b); a DRw6/7 cell (c). The vertical arrows indicate
the DQw2α chain present in the three cell lines. The second
DQα chain represents the DQw1α chain from the DR1 (a) DR2
(b) and DRw6 (c) haplotype. The IEF 2D-PAGE did not allow
us to distinguish between DQw1α from DR1 and DR2 cells.
The DQw1α from DRw6 is slightly more basic.

DQβ molecules in heterozygous individuals.[25] This is easily explained by the difficulty in
finding an adequate anti-class II antibody for a given combination of haplotypes. While
studying HLA-DQα polymorphism, Giles et al.[26] found that amino-terminal sequences of
labeled tyrosine residues of DQα chains differ at several positions in DR5 and DR7 homozygous cells enabling them to distinguish these two alleles. An antibody specific for the β
chain of DQw3 (IVD 12) was used to isolate DQ molecules from a DR5/7 heterozygous
cell line. In such immunoprecipitates, tyrosine residues were detected in the DQα chain in
positions characteristic of both the DR5 and DR7 haplotypes suggesting the presence of a
DQw3β-DQw2α dimer. Interestingly, a small amount of "contaminating" DRα was noted
which in retrospect suggest the presence of a DQw3β-DRα dimer in the immunoprecipitate.
Using the same anti DQw3 MAbs, Nepom et al.[36] were able to identify by 2D-PAGE and
peptide mapping, a DQw3β-DQw2α chain complex in DR3/4 individuals. Moreover, the

same DQα (DQw2) DQβ (DQw3) dimer was present in DR3-4 individuals whether or not they had insulin-dependent diabetes mellitus (IDDM) with the amounts being comparable within the limits of the experiment. Similar results have been obtained in our laboratory.[68] However a DRα chain was also present in a small amount suggesting the presence of interisotypic hybrid products (see next section) in addition to the intraisotypic hybrid HLA-DQ molecules found in these DR 3/4 heterozygous individuals.

3. Natural Intraisotypic Hybrid HLA-DQ Molecules

The ability to form hybrid DQ molecules by gene trans-complementation should not be so surprising. Indeed, when DQ α and β alleles from distinct haplotypes are compared, a series of "naturally occurring" hybrid DQ molecules are found which are identical in their α and β chain composition to the hybrid molecules found in heterozygotes, but differ only in the fact that their α and β genes are coded on the same haplotype (*cis*-configuration). Recent data have been provided at the nucleotide level that DQ dimers in DR7-DQ2 and DR7-DQ3 haplotypes share a DQα chain but differ in their DQβ chains.[37] Although the sequences provided are restricted to the first domain of the α and β chains, the data are highly suggestive that the DR7-DQ3 haplotype encodes in *cis* a natural hybrid molecule formed by a DQw3 α chain originating from a DR4 haplotype and the typical DQw2β chain present in every DR7-DQw2 haplotype. Additional data were recently obtained on a larger panel of HTC cells in which DQ α and β polymorphism was studied at the protein level by isoelectrofocusing analysis.[38] Many, but not all, of the possible combinations of α and β chains were found. Out of 30 hypothetical dimers which could arise as a result of the various combinations between five DQα and six DQβ chains only 20 dimers were observed experimentally. Whether the absence of a given α-β combination is due to the inability to associate (forbidden α-β pairing) or whether the population studies are, as yet, insufficiently extensive to include all possible combinations is unresolved. Indeed, because of strong linkage disequilibrium between alleles of the DR and DQ loci certain combinations of alleles predominate in a given population. Different combinations of genes which would generate *cis*-derived hybrid molecules are most likely to be found in different populations. Natural hybrid molecules thus represent minor variant haplotypes. Whether these "natural" hybrid DQ molecules evolved by recombination between two distinct (heterozygous) haplotypes and were then fixed is very likely. These types of data argue in favor of a recombination hot spot between DQα and DQβ[37] and justifies a search for sequences facilitating recombination events as has been described in the mouse I region.[39,40]

4. Discussion

Finally, each possible combination of DQα and DQβ pairing should be searched for experimentally, particularly with regard to the evolutionary selection bias which may differ for different combinations. However, limitations to this have to be considered. The number of mAbs presently available which are suitable for this type of analysis is very limited and the size of the population which needs to be studied is not easily predictable. Complete analysis of permissive and forbidden DQα-DQβ chain pairing may greatly benefit from transfection experiments. Moreover when DQ α-β dimers are expressed, such transfectants can be used to generate monoclonal antibodies specific for a given combination. This in turn will permit the screening of populations on a larger scale.

The demonstration in heterozygous individuals of cell surface expression of hybrid HLA-DQ molecules raises several questions.

a. Are Hybrid HLA-DQ Determinants Serologically Detectable?

So far there is no convincing report of alloantisera behaving as if they could recognize hybrid molecules. It may well be that such sera have been overlooked by serologists. Indeed such sera would only recognize heterozygous individuals and would not segregate with a given HLA haplotype in family studies. Alternatively, it may be that these determinants are less immunogenic in terms of an alloantibody response. Very few attempts to raise mono-clonal antibodies against hybrid determinants have been made and none of those have been successful. Moreover, the mouse-human immunization procedure may not favor formation of mAb against such determinants. Transfectants containing hybrid molecules will provide an exquisite tool for production and screening of hybrid-specific antibodies which can be subsequently used in serological, functional and biochemical studies.

b. Are Hybrid HLA-DQ Determinants Functional?

In the mouse, the Ir gene *trans*-complementation phenomenon has been unequivoquely explained on the basis of hybrid molecules.[22,23] Furthermore, among the determinants that stimulate in an MLR there appear to be some that represent conformational epitopes created by *trans*-association of α and β chains.[19] In contrast, very little functional data is available in the human. Several years ago an antigen-specific T-cell clone was reported that responded specifically to PPD only when the antigen was presented by accessory cells carrying both HLA-D region antigens present in the autologous cell donor (DR2/DR3). The absence of polymorphism in the DRα chain makes it unlikely that the DRα chain was involved in the restriction element of this clone. In contrast, a hybrid DQ molecule created *de novo* could function as a restriction element and allow participation of both the paternal or the maternal haplotype thus fulfilling the requirements for heterozygosity. This clone is likely to be the functional counterpart of hybrid HLA-DQ molecule.[41] However, particularly in light of the possibility of isotype mismatched class II molecules, one cannot exclude the alternative of a restriction element formed by a DQα DRβ complex whose chains are encoded by genes in *trans*.

c. Are Hybrid Determinants Involved in Disease Susceptibility?

While formal proof of direct involvement of any class II molecule (conventional or hybrid) in disease susceptibility is still lacking, the case for a role for hybrid HLA-DQ determinants requires particular discussion. Indeed, in numerous epidemiologic studies first in IDDM[42] and later in juvenile rheumatoid arthritis (JRA)[43] and in coeliac disease (CD)[44] an unexpectedly high incidence of the disease has seen observed in particular combinations of haplotypes (heterozygous effect).[45] As a consequence, the relative risk (RR) was dramatically higher in some heterozygous situations (DR3/4 in IDDM) than for any unique allele even when present in a homozygous state. Such data need to be explained by a model different from the monoallelic association. If HLA class II molecules are involved, it becomes logical to propose that the particular determinant derives from a combination of products, one from the DR3 and the other from the DR4 haplotype. Because of the lack of poly-morphism in the DRα chain it is very unlikely that HLA-DR molecules themselves would be the structural basis of this heterozygous effect. Furthermore, the absence of strong linkage disequilibrium between the DR and DP loci does not favor DP as a candidate since epide-miologic studies show the highest incidence of diseases for fixed combinations of DR alleles. In contrast the DQ loci fulfill the two requirements of being in strong linkage disequilibrium with DR alleles and having both subunits structurally polymorphic. *Trans*-association of DQα-DQβ chains creates hybrid molecules which can therefore be consistent with the observed heterozygous effect, since only these HLA-DQ hybrid molecules will bear con-formational epitopes unique to the combination of paternal and maternal haplotypes. In addition to DR 4/3, a higher incidence of IDDM is also found in DR4/1 individuals. Although DQα and DQβ are structurally different in DR3 halotypes vs. DR1 haplotypes, either one may well create identical or very similar conformational structures when associated with a DQ product from a DR4 haplotype. A recent report indicates that position 57 of the DQβ gene is critical for susceptibility to IDDM since the absence of Asp at this position is highly positively associated with the disease.[46] Full HLA susceptibility for IDDM was therefore considered as dependent on the presence of two non-Asparagine residues at position 57 in DQβ. This can be attained either in a homozygous situation (DR4-DQw3.2 or DR3-DQw2 or DR1-DQw1) or in a heterozygous situation such as DR4-DQw3.2/DR3-DQw2 or DR4-DQw3.2/DR1-DQw1. However this explanation is not sufficient to explain the fact that the RR is lower in the homozygous situation than in the heterozygous one. Thus, in addition to the absence of ASP at position 57 the observed heterozygous effect still must be taken into account. A most plausible explanation is that maximum susceptibility requires the absence of ASP 57 in DQβ along with some other conformational epitope or subsequent molecular folding which exists only in heterozygous cells but not in homozygous cells. The fact that the presence of ASP at position 57 of DQβ provides strong protection against IDDM argues for either a direct role of this residue in preventing the binding of some autoantigen or an indirect effect through a conformational change in another part of the DQ molecule which could prevent hybrid molecule formation.

Concerning JRA the heterozygous effect was hidden. As reported by Nepom, the RR was dramatically higher in DR4/4 homozygous individuals (RR 36,3) than in DR4/- heterozygous subjects.[43] However, when biochemical analysis of the class II molecules present in the DR4/4 individuals was performed it was found that most of the cases that were serologically homozygous were in fact biochemically heterozygous. These results were supported by the D typing data which revealed a RR of 116 for Dw4/Dw14 individuals compared to 12.9 and 11.3 for Dw4 and Dw14 homozygosity, respectively. Hence, again, a heterozygous effect is unveiled in which hybrid molecules coded in *trans* are likely to be involved. Whether these are intraisotypic DQ hybrid molecules or interisotypic DR-DQ molecules is not known.

Finally, a similar epidemiologic situation is observed in coeliac disease in which a DR3/DR7 heterozygous effect is found.[41,45]

d. How Is the Expression of Hybrid HLA-DQ Molecules Controlled?

Recent work in mice using appropriate combinations of transfected genes has shown that while haplotype matched Aα-Aβ genes (haplotypes k, b, d) resulted in the optimum cell surface expression, (as one would expect from the biochemical studies previously conducted in normal cells) the level of expression of haplotype-mismatched Aα-Aβ dimers was extremely variable and depended on the combination of haplotypes which were used. As an example, Aα^k-Aβ^b transfectants had poor surface Ia expression while Aα^k-Aβ^d transfectants had no detectable Ia at their surface. Furthermore, it was demonstrated that appropriate combinations of polymorphic sequences in the NH2-terminal half of the Aα and Aβ chains could control the pairing. This allele-specific control of Ia molecule surface expression was rather unexpected. It is however consistent with most functional and evolutionary features of and class II genes.[47,48]

The inability of certain DQα and DQβ chains to form functional dimers has yet to be demonstrated in the human. Although some DQα and DQβ genes have never been found to occur in *cis* in the population, this may not preclude the possibility that they occur in rare cases or even in *trans*.[38] It is also still not known if the pairing efficiency is the same for DQ α-β gene products coded in *cis* versus in *trans*. Transfection experiments are presently being performed with different DQα-DQβ genes in order to determine which, if any, combinations are forbidden. Rules underlying the formation of DQα-DQβ pairs are still enigmatic and no data are available on which part (or parts) of the α-β chain are important in order for the dimer to be correctly assembled and transported to the cell surface. Mechanisms similar to the one described in the mouse system are expected to be found in man. However, several other possibilities have not been investigated which could interfere with hybrid molecule formation. These include numerous post translational processing events and quantitative regulatory mechanisms some of which are noticeably different in mouse and man.[49]

B. NON-HLA-DQ HOMOLOGOUS HYBRID MOLECULES

While the total lack of structural polymorphism in the DRα gene and protein means that there is no difference in whether the constitutive α chain of a DRα-DRβ dimer is coded for by either parental chromosome, other class II subsets are potential candidates for generating intraisotypic hybrid molecules. This is clearly a possibility for the DP subset both subunits of which are structurally polymorphic. The number of allelic DPβ proteins and genes is at least as extensive as PLT typing had predicted. Moreover the high percentage of DP "blanks" is likely to reflect the existence of several additional DP variants. In contrast only two DPα chain variants have been demonstrated biochemically.[50] This is in agreement with the nucleotide sequence data.[51] We have shown that DPw2, w3 and w4 β chains are associated with an acidic DPα chain whereas DPw1 and DPw5 antigens express a slightly more basic DPα chain.[50] Since several structurally distinct DPβ chains can pair with the same DPα chain this provides another example of naturally occurring hybrid molecules whose genes are coded for in *cis*. Whether the two DPα chains can pair indiscriminately in heterozygous cells with any DPβ chain is unknown at the present time.

However the actual structural data on DP can provide some clues to previously unexplained functional and serological data. The fact that the same DPα can pair with distinct DPwβ chains defining different DP subtypes suggests that the DPβ chain may be immunodominant in cellular typing. Furthermore, a mAb has been described which only reacts with cells typing as DPw2 and DPw4 while in Western blots it recognizes the separated DPα chain in all haplotypes.[52] Thus, it likely recognizes a conformational determinant (or conformationally induced determinant) which is masked when the DPα is combined with some DPβ chains (DPw3) but is accessible when other DPβ chains (DPw4) are bound to the same DPα chain. This illustrates that allelism at one locus can influence cell surface expression of monomorphic determinants coded for by a second locus.

III. INTERISOTYPIC HYBRID HLA CLASS II MOLECULES

A. DRα-DQβ ISOTYPE MISMATCHED CLASS II MOLECULES
1. Introduction

It has been suggested that due to a presumed higher affinity between the α and β chains when they are coded for by loci of the same isotype, expression of class II dimers will occur only within one isotype. Isotype mismatched molecules were therefore considered forbidden pairs and were not expected to be found at the surface of normal cells.[53] A series of recent experiments using murine cells transfected with class II genes have indicated that unorthodox α-β pairing can occur and attention has therefore been given to the possibility of such pairing occurring in normal cells.[17,54] Early studies had shown that when a Hamster cell line was transfected with mouse I-A^k genes, combinatorial association of hamster and murine α-β Ia chains could be detected biochemically.[69] Recently, Norcross reported the pairing of DRα and I-Aβ^d chains after transfection of the murine Aβ^d gene into a human EBV transformed cell line.[18] Furthermore, the transfection of both murine I-E and I-A genes in L cells was shown to lead to the assembly and surface expression of I-Aβ-I-Eα complexes.[17,54] Although highly suggestive, the transfection approach could not conclusively demonstrate that mixed isotype pairing would occur in unmanipulated cells. Indeed transfected cells differ from normal cells in critical ways: the transfected genes are not under normal regulatory control, the gene copy number may vary and, the transfected chains have no alternative partner with which to pair.

A series of experiments were conducted in our laboratory using normal EBV transformed human B cell lines. These provide direct evidence for the assembly and cell surface expression of mixed isotypes consisting of DRα and DQβ chains.

2. Description

Two types of mAbs were used in immunoprecipitation experiments. G2a5 which specifically recognizes epitopes of the β chain of DQw1 molecules and HC2.1 and DA6.147 which are both uniquely directed against the DRα chain as shown previously[55] and using Western blotting (Figure 1).[56] When a DR2/2 (PGF) HTC is used in conjunction with G2a5 in immunoprecipitation, the α chain area of the 2D gel reveals a DQw1α chain and a DRα chain. The identification of these additional α spots as DRα was confirmed by Western blotting using HC2.1. No DRβ chain was observed in DQw1 immunoprecipitates. Moreover data leading to a similar conclusion were obtained with DA6.147.[31] When used in a DR7/7 HTC DA6.147 immunoprecipitates a DRα-DQβ complex in addition to the expected DRα-DRβ complex. Furthermore, when cells were surface labeled with ^{125}I by the lactoperoxidase catalyzed iodination method, a DQα-DQβ molecule could still be precipitated from a cell lysate first depleted of any DRα associated β chains by successive cycles of immunoprecipitation with DA6.147. The presence of DQβ in DA6.147 immunoprecipitates is thus due to pairing with DRα and not to a contaminating DQ complex. This strongly argues for the cell surface expression of this molecule as a dimer consisting of a DRα-DQβ complex.[56] Additional haplotypes have been analyzed using DA6 1.47 as illustrated in Figure 3(c, d). Quantitation of the isotype mismatched dimers vs. the conventional class II dimers using immunoprecipitation of labeled cell lysates is not precise. It appears, nonetheless, that there is a lesser amount of DRα associated with DQβ than with DRβ and similarly less DQβ associated with DRα than with DQα.

In the mouse, transfection of I-E and I-A genes into L cells has recently demonstrated the unexpected formation of mixed isotype pairs consisting of IEα and IAβ chains.[17,54] This nonorthodox pairing appears to be predominantly influenced by the allelic polymorphism of the I-Aβ chains since the I-Aβ-IEα complex could be detected as an I-Eα-IAβ^d dimer but not as an IEα/IAβ^k or IAβ^b dimer. Furthermore, this work emphasized the unexpected influence of the polymorphic NH$_2$-terminal domain of the IAβ molecule in permitting pairing since the IEα chain is virtually nonpolymorphic. Transfection experiments with recombinant genes suggest that only five to seven allelically variable residues of the β1 domain 50 NH$_2$-terminal amino acids are of critical importance in dictating mixed isotype pairing. In contrast the exchange of the 40 residues located in the carboxy terminal domain of the α or β chain for those of another allele do not affect the selection process in α-β pairing. This contrasts with the observation that the COOH-terminal regions of each chain are among the most conserved and locus specific structures.[13] The data therefore argue against the previous claim that the transmembrane and cytoplasmic portions are the only factors determining correct chain pairing.[53] Although studies in the mouse indicate that mixed isotype pairing is highly dependent upon the allotype of the chains, it should be remembered that these results were obtained by transfection of α and β chains which had no alternative partner with which to pair. Experiments in which three class II genes are transfected into the same cell have established that locus-matched pairing is substantially more efficient, although both locus-matched and mismatched products are expressed at the cell surface.[70] In normal cells the situation is more complex since each α or β chain has the opportunity to pair with its homologous chain (in *cis* or in *trans*) and eventually with chains of the other isotypes (in *cis* or in *trans*).

3. Regulation of DRα-DQβ Pairing

Although experiments designed to assess the role of DQβ chain polymorphism on the efficiency of DRα-DQβ pairing are still in progress we have nevertheless recently investi-

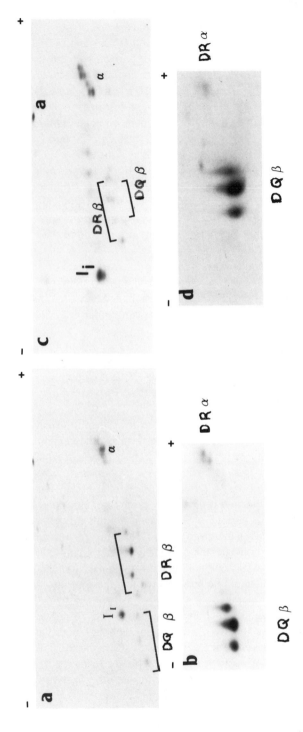

FIGURE 3. DRα DQβ hybrid class II molecules isolated from a DR7/7 HTC (panel a and b) and a DR4/4 HTC (panel c and d). Panel a and c: ³⁵S methionine-internally labeled cell lysates were immunoprecipitated with DA 6.147. Panels b and d: ¹²⁵I surface-labeled cell lysates were depleted of DR molecules prior to immunoprecipitation with DA 6.147.

TABLE 1
Schematic Representation of the Locus-Specific HLA Class II Probes Used in This Study

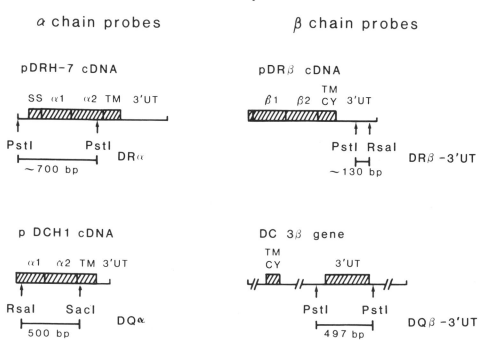

gated the role of another physiological parameter which has a direct role in allowing or preventing the formation of DRα-DQβ dimers. We have observed that a permissive haplotype does not automatically result in the appearance of an interisotypic heterodimer. We therefore investigated the amount of individual α-β chain mRNA present in EBV B cell lines in which biochemical analysis is sufficient to detect the presence or absence of isotype mismatched α-β pairs. Since class II α and β genes are highly homologous between loci, we used class II probes which were shown to be highly locus (isotype) and chain specific[57] (Table 1). No cross hybridization was found between α and β genes of either DR, DQ, or DP under the hybridization conditions used. Total RNA was extracted from the same number of cells by successive phenol-chloroform extractions and Li chloride precipitation. RNA was electrophoresed in agarose gels, transferred to nylon membranes, and the sheets were successively hybridized with the DRα, DQα, DRβ, and DQβ specific probes (Figure 4). The resulting signals were scanned to give a number proportional to the surface area and intensity of the band. Since the system is not calibrated with a reference curve for each probe, the ratios represent arbitrary numbers which do not correspond to a given amount of mRNA. However, the results can be interpreted by comparing these ratios between different cells (Table 2). When a biochemical analysis was performed on these same cells using the procedure described above (using the G 25a antibody) DRα-DQw1 β molecules could be detected only in two of the cells (ISa and Mic). They could not be detected in FJO and CA.

4. Regulation of DRα-DQβ Pair Formation in Transfectants

In order to address some of the mechanisms underlying the formation of mixed isotype dimers in a more direct manner we undertook experiments designed to specifically modulate the expression of the DR alpha gene in EBV transformed human B cell lines.[71] Two cell lines were chosen and analyzed biochemically for the presence or absence of DRα-DQβ dimers. SC-CA is a DR1 homozygous cell line which has been shown in previous experiments

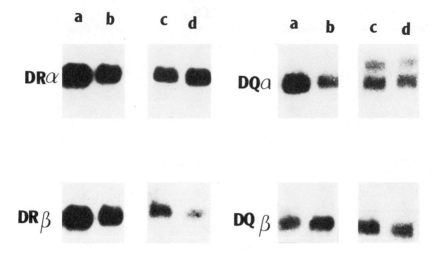

FIGURE 4. Northern blot analysis of HLA class II transcripts from four different cell lines FJO (a) CA (b) ISA (c) MIC (d) 20 μg of total RNA were electrophoresed and blotted onto filters; blots were probed successively with DRα, DQβ, DRα, and DQβ nick-translated and locus-specific probes.

TABLE 2
Presence of Mixed Isotypes as a Function of DRα/DRβ Ratio

Cell ratio	FJO	CA	Isa	Mic
DRα/DRβ	1.25	1.45	4.66	6.80
DQα/DQβ	2.53	0.63	0.87	0.84
2D-PAGE biochemistry DRα/DQβ	−	−	+	+

Note: Northern blots of mRNA from four different cell lines were successively hybridized with the DRα, DRβ, DQα, and DQβ probes. The films were scanned with an absorbance system which integrated surface area and intensity for each probe.

to be devoid of DRα-DQw1β molecules. Furthermore, we verified in other cell lines that a DQw1β chain structurally identical to the one found in SC-CA can physically associate with the DRα chain. The DRα-DQw1β association, therefore, is not prohibited and the absence of DRα-DQβ dimers in the SC-CA cell was very likely due to quantitative factors. In order to analyze this phenomenon, a series of transfections was performed using a vector derived from pHebo which contains a highly efficient CMV promotor.[58,72] Two selection systems were combined: those conferring ampicillin resistance to the plasmid and hygromycin resistance to the transfected cells. Full length DRα cDNA[59] was introduced at the BamHI sites. Cell surface expression was analyzed by flow cytometry and the molecules were analyzed by immunoprecipitation, 2D PAGE and/or Western blotting. The transfected cells were usually cultured for 2 to 3 weeks with 50 μg/ml of hygromycin prior to testing. In some cases a higher concentration (100 μg/ml to 300 μg/ml) of hygromycin was used. The DRα gene was alternatively integrated in the sense and anti-sense orientation. Results representative of several experiments are illustrated in Figure 5. SC-CA cells were transfected with DRα and assessed at 2 weeks after hygromycin selection at 50 μg/ml. Control transfectants contain the vector alone. The fluorescence pattern obtained with CA 1.41, a mon-

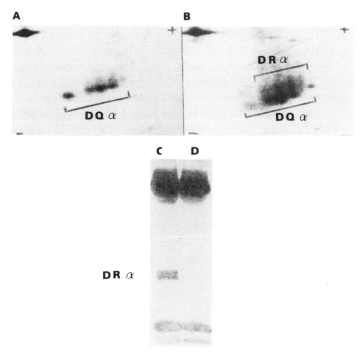

FIGURE 5. 2D gel and Western blot analysis of a B cell line transfected with a DRα cDNA.SC.CA, a DR/1 HTC was transfected with a DRα cDNA (vector: P Hebo/promotor: CMV/cDNA: p DRH-7). G2a5 immunoprecipitates a DQw1α-DQw1β molecule in SC.CA transfected with the vector alone (A/D) and an additional DRα-DQw1β hybrid molecule in the DRα transfected cell (panels B/C). Only the α chain areas of the 2D-PAGE are shown (A, B). Western blot analysis with HC2.1 detects a DRα chain in a G2a5 immunoprecipitate only in the transfected cell (C).

omorphic anti-DR mAb, does not vary significantly from the control cells (nontransfected SC-CA cell). However, immunoprecipitation with G2a5 (a mAb specific for the DQβ chain) of the DRα transfected CA cells vs. the control transfected CA cells reveals in the former an additional α chain not present in the latter. Both immunoprecipitates contain the expected DQw1 α chain. The identification of the additional spots as DRα is based on their electrophoretic pattern which is identical to the DRα pattern obtained with the DRα mAb HC2.1 (Figure 5).[55] From these experiments we can conclude that transfection of a DRα gene in to a cell line not expressing DRα-DQβ molecules is able to induce expression of DRα-DQβ dimers.

In a second series of experiments, GM 3163 cells (DR7/DQw2 homozygous) were transfected with a DRα anti-sense cDNA and assessed as above. In previous experiments we had shown that DRα-DQβ molecules were readily detectable biochemically in GM 3163 cells using the anti-DRα mAb DA6.147. After introduction of the DRα anti-sense cDNA, the transfected cells were cultured for 2 to 3 weeks with hygromycin (50 μg/ml). In most experiments, cell surface expression of the DRα-DRβ complex, as assessed with the anti-DR mAb 141, did not vary significantly. However at higher concentration of hygromycin (150 μg/ml) the fluorescence profile with mAb 141 decreased in intensity. The amount of DRα was reduced as assessed by Western blotting with DA6.147. Moreover, when DA6.147 was used to immunoprecipitate class II dimers from the transfected cells, 2D-PAGE analysis did not reveal any DRα-DQw2 β species although these are normally found in nontransfected cells. Such experiments show that transfection with anti-sense DRα cDNA has the ability to reduce the amount of DRα products synthesized. This in turn results in the extinction of the DRα-DQβ isotype mismatched pair. In addition, the preferential extinction of the DRα-

DQβ dimers with little or no decrease in the amount of DRα-DRβ molecules suggests a higher pairing affinity of the typical isotype paired α-β chains. In a series of other experiments DRα anti-sense transfectants were studied using SC-CA cells. We observed a reduction in the cell surface expression of DR α-β complexes as assessed by the binding of mAb 141. This was consistent with a noted decrease in the amount of total DRα chain found in these cells when tested by Western blotting with HC2.1. No variation in the total level of DRβ expression was found in those cells when they were analyzed by Western blotting using an anti-DRβ mAb (2.06) nor was the transcriptional level of DQ and DP decreased. From the above results we propose that there is likely to be an excess of DRβ chains in these transfected cells. Since we were unable to detect DRβ-DQα products when a DR mAb was used to immunoprecipitate class II dimers, two possibilities should be considered: (1) DR1β cannot pair with DQw1α, or (2) DQw1α may not be available for such pairing. Modulation of class II DQα and β expression in a manner similar to that described above should provide us with an exquisite tool to discriminate between these two possibilities. In conclusion, the absolute amount of each chain, the relative amount of each chain within an isotype, and the reciprocal affinity of each chain for one another are potential regulatory parameters.

5. Discussion

Whether isotype mismatched molecules are important in the physiology of the immune response is an open question. The fact that very little data have been published in this regard may reflect that such molecules have been overlooked since they were not thought to exist. In serological terms allele-"specific" human alloantisera are likely to have a broad reactivity within a haplotype rather than being strictly allele specific. These alloantisera may contain populations of antibodies recognizing interisotypic specificities (combination of DRα-DQβ, etc.) In this case the interisotypic hybrids would not be distinguished from the dominant allelic specificity since DR and DQ are in strong linkage disequilibrium and DRα is monomorphic. New tools can be designed to search for such sera and mutant cells (deleted of genes for one or several chains) should be very useful. Similarly, transfectants containing various combinations of mixed α-β isotypes pair should be suitable targets. This approach will be of great importance in the future especially if one or several of these combinatorial determinants appears to be directly implicated in the physiopathology of an HLA-associated disease.

The existence of functional isotypically mismatched determinants has been suggested on the basis of the unorthodox reactivity of some human T-cell clones. A hybrid HLA-DP-associated epitope has been proposed to explain the observation that TLC 56.94, a T-cell clone obtained from a DPw1-specific PLT cell line, when tested on a large panel reacts with most DPw1 individuals only when DPw1 is present in an heterozygous state and not with DPw1 HTC cells.[60] This suggests the presence of a determinant created by the association of a DPw1-β chain with another chain which is most probably (1) encoded in *trans* and (2) modifying the DPw1 chain in a constant manner. The partner of the DPw1 chain would have to be absent from the same haplotype at least in those described here. TA 1O is another T cell clone[61] specific for the influenza A virus which is restricted by the DRw8 haplotype but cross-reacts allospecifically with a *trans*-complementation product of the HLA DR2 and DR4 haplotype. This produces a dimer consisting of most likely a DQα chain from DR4 and a DRβ chain from DR2 since blocking studies suggest that a DRβ but not a DQβ mAb exerts a blocking effect. Unfortunately, the blocking study was not performed with an anti DQα reagent.

IV. GENERAL DISCUSSION AND PERSPECTIVES

Overall, the data from both mouse and human demonstrate that the HLA class II system acquires an additional level of diversity by mixing chains from different haplotypes and isotypes. This combinatorial diversity is, at this time, not directly predictable from the sequences of the genes of the α-β subunits involved in forming the dimeric molecules. However, the influence of polymorphic residues in the NH_2-terminal half of the β1 and α1 domains appears to be decisive in allowing conventional and hybrid pairing. The precise residues involved in the α1 β1 pairing and the type of secondary structures in which they are integrated (β pleated sheet, α helix, loops) are not yet known. However, the recent crystallographic resolution of an HLA class I antigen offers a schematic framework for the structure of an HLA class II molecule.[62] As suggested by Bjorkman et al. the spatial organization of the α1-β1 domains of class II may resemble that of the α1-α2 domains of class I molecules. The "intramolecular dimerization" observed between the α1 and α2 domains of class I molecules creates contacts between two anti-parallel β strands. This would be equivalent to the intermolecular contacts which may take place between the α1 β1 domains of class II molecules. In this case the interface between the α1, β1 domains would consist of hydrogen bonds between the N-terminal β strands of each domain. The intensity of the hydrogen bonds would be highly dependent on the amino acids present in the areas of the α1 and β1 domains resulting in a higher or lower affinity of association in different α1, β1 allelic, or isotypic combinations. Additional contacts between the C-terminal ends of the helical regions of one domain and β sheet residues in the other domain may also participate in stabilizing the α1, β1 pairing. When the crystallography data of a class II molecules become available it should be possible, by fitting the different residues at their appropriate positions, to deduce whether some associations will be feasible or unlikely to occur. Directed mutagenesis of the critical residues should provide further experimental verification.

Discussion of hybrid HLA class II molecule formation requires, at some point, considering the regulation of the individual α and β genes. While coordinate expression of the α and β chains appears, at a first evaluation, to be the rule for the known expressed class II genes,[63] it is conceivable that under special circumstances, in some tissues and/or during upregulation, a nonstoichiometric translation of α or β chains will occur and disturb the normal ratio of α/β chains. The α-β pairing would then be modified accordingly and noticeable amounts of unusual α-β pairs would emerge from and be brought to the cell surface. Such inbalanced transcription rates may represent a transient event following an increase of stimulating factors (IFN α, other lymphokines, etc.). It may ultimately become a more permanent state, either because of a sustained hyperstimulation of the cells or because of abnormal behavior of the regulatory genes themselves. In reality, coordinate expression of class II genes may not be as strict a rule as it is presently thought to be. A few years ago we investigated the transcription level of the individual α and β class II genes in a panel of leukemic cells and observed differences in the expression of each chain.[57] These results were consistent with what we had previously found at the protein level.[30] Dissociation of expression has also been documented for DR vs. DQ in some cell populations, principally those that are malignant[64,65] and during the early stage of normal hematopoietic differentiation.[65] It is conceivable, but purely hypothetical, at this time that some dimers are specifically expressed at discrete stages of a differentiation pathway. Loci which have never been found to be expressed in spite of their normal gene structure (DXα and DXβ) may then participate in forming some class II molecules of the conventional or hybrid type.[26] Among these loci DOβ and DZα have been found not to be coordinately expressed.[19] This feature has been used as an argument against the existence of DZα-DOβ dimers. The existence of hybrid molecules and a less strict coordination of expression of the other α and β genes in some cells makes this statement less definitive. If DOβ and DZα are demonstrated to be expressed proteins in some cell types they may indiscriminately pair with each other or other class II alpha beta partners.

On the basis of our observations that there exists quantitative variation in the level of DRα transcripts in different cell lines, we believe that there may be some important functional polymorphism in the regulation of class II expression. The controlling element(s) are very likely located in the flanking sequences of the genes. These elements would affect the level of transcription of a given gene and consequently could determine the availability of that particular gene product for the formation of conventional or hybrid molecules. Alternatively, regulatory proteins (such as DNA binding factors) could vary in their level of expression in different cells sharing the same class II haplotype. This could also explain differences in the transcription rate of a given class II gene in different individuals of the same DR type. In addition to the transcription rate the stability of the mRNA and post-translational events may affect the amount of and the capacity to form hybrid dimers. Although class II molecules are differentially glycosylated in different cell types no isotype or allele specific oligosaccharidic profile is apparent.[49] A role for Ii is still awaited.[9,66] It may stabilize the α-β pairing very early during biosynthesis; an α-β Ii complex can be found in *in vitro* translation experiments before any glycosylation has occurred. However, the affinity of Ii for the α-β complexes may vary with the α-β combinations as was suggested several years ago.

The concept of regulatory factors of either genetic (regulatory polymorphism) or somatic origin (lymphokines, etc.) has important implications, since such factors might directly modulate both quantitatively and/or qualitatively the repertoire of HLA class II determinants which are present at any particular site at a given time in the life of an individual. One has first to consider the level and type of class II dimers (conventional or hybrid, intra- or interisotypic) which are expressed during thymic education. This will depend on the regulation status (constitutive or induced) of class II gene expression in the thymic epithelium during this process. Later on in life hyperexpression of some hybrid molecules intra- and/or interisotypic, may occur locally in a disease tissue. This aberrantly high level of expression of a given set of hybrid determinants could initiate an auto-immune response by increasing the autoreactivity to MHC, MHC plus self-antigen or MHC plus foreign antigen. Such mechanisms will be favored by the upregulation of class II expression. The recent observation that class II expression is increased in aged mice is thus consistent with the finding of a higher incidence of auto-immunity.[67]

The relevance of hybrid determinants to the physiology and physiopathology of the immune response is still enigmatic. Clearly, they offer the potential of increasing the repertoire of Ia molecules. The role of these hybrid molecules in antigen presentation and disease susceptibility is a crucial question for the future. In addition, one should consider the possibility that hybrid molecules may have other specialized functions. For examples, hybrid molecules may have been selected to perform some other functions assigned to MHC class II molecules such as transmembrane signaling or control of cellular differentiation. In conclusion the finding of interisotypic hybrid HLA class II molecules requires a revision in our current concept of HLA and disease associations, since it appears that in some diseases the associations are stronger with haplotypes than with alleles.

ACKNOWLEDGMENTS

We thank N. Mooney for reading of the manuscript. We are also grateful to Ch. Auffray, and D. Piatier. We thank G. F. Bottazzo and J. F. Bach for informal discussion of some of the concepts developed in this manuscript and M. Brandel for excellent secretarial assistance. This work was supported by grants from INSERM, ARC, LNFCC, ARSEP, GEFLUC, and Fondation de France.

REFERENCES

1. **Kaufman, J. F., Auffray, C., Korman, A. J., Shackelford, D. A., and Strominger, J.,** The class II molecules of the human and murine major histocompatibility complex, *Cell* 36, 1, 1984.
2. **Hardy, D. A., Bell, J. I., Long, E. O., Lindsten, T., and McDevitt, H. O.,** Pulsed-field gel electrophoresis mapping of the class II region of the human major histocompatibility complex, *Nature (London)*, 323, 453, 1986.
3. **Long, E. O., Gorski, J., and Mach, B.,** Structural relationship of the SB beta-chain gene to HLA-D region genes and murine I-region genes, *Nature (London)*, 310, 233, 1984.
4. **Inoko, H., Ando, A., Kimura, M., and Tsuji, K.,** Isolation and characterization of the cDNA clone and genomic clones of a new HLA class II antigen heavy chain DO alpha, *J. Immunol.*, 135, 2156, 1985.
5. **Trowsdale, J. and Kelly, A.,** The human HLA class II alpha chain gene DZ alpha is distinct from genes in the DP, DQ and DR subregions, *EMBO J.*, 4, 2231, 1985.
6. **Tonnelle, C., DeMars, R., and Long, E. O.,** DQ beta: a new beta chain gene in HLA-D with a distinct regulation of expression, *EMBO J.*, 4, 2839, 1985.
7. **Auffray, C., Lillie, J. W., Korman, A. J., Boss, J. M., Frechin, N., Guillemot, F., Cooper, J., Mulligan, R. C., and Strominger, J. L.,** Structure and expression of HLA-DQ alpha and DX alpha genes: interallelic alternate splicing of the HLA-DQ alpha gene and functional splicing of the HLA-DX alpha gene using a retroviral vector, *Immunogenetics*, 26, 63, 1987.
8. **Charron, D. J. and McDevitt, H. O.,** Characterization of HLA-D region antigens by two-dimensional gel electrophoresis. Molecular genotyping, *J. Exp. Med.*, 152, 18s, 1980.
9. **Charron, D. J., Aellen-Schulz, M. P., St. Geme, J., Erlich, H. A., McDevitt, H. O.,** Biochemical characterization of an invariant polypeptide associated with Ia antigens in human and mouse, *Mol. Immunol.*, 20, 21, 1983.
10. **Gustafsson, K., Wiman, K., Emmoth, E., Larhammar, D., Bohme, J., Hyldig-Nielsen, J. J., Ronne, H., Peterson, P. A., and Rask, L.,** Mutations and selection in the generation of class II histocompatibility antigen polymorphism, *EMBO J.*, 3, 1655, 1984.
11. **Wu, S., Saunders, T. L., and Bach, F. H.,** Polymorphism of human Ia antigens generated by reciprocal intergenic exchange between two DR beta loci, *Nature (London)*, 324, 676, 1986.
12. **Gorski, J. and Mach, B.,** Polymorphism of human Ia antigens: gene conversion between two DR beta loci results in a new HLA-D/DR specificity, *Nature (London)*, 322, 67, 1986.
13. **Auffray, C., Lillie, J. W., Arnot, D., Grossberger, D., Kappes, D., and Strominger, J. L.,** Isotypic and allotypic variation of the class II human histocompatibility antigen alpha chain genes, *Nature (London)*, 308, 327, 1984.
14. **Auffray, C. and Strominger, J. L.,** Molecular genetics of the human histocompatibility complex, *Adv. Hum. Genet.*, 15, 197, 1986.
15. **Andersson, G., Larhammar, D., Widmark, E., Servenius, B., Peterson, P. A., and Rask, L.,** Class II genes of the human major histocompatibility complex. Organization and evolutionary relationship of the DR beta genes, *J. Biol. Chem.*, 262, 8748, 1987.
16. **Charron, D. J., Haziot, A., Lotteau, V., Neel, D., Merlu, B., and Teyton, L.,** Biochemical diversity of the human MHC class II antigens, in *Regulation of Immune Gene Expression*, Feldmann, M. and McMichael, A., Humana Press, Clifton, NJ, 1986, 61.
17. **Germain, R. N. and Quill, H.,** Unexpected expression of a unique mixed-isotype class II MHC molecule by transfected L cells, *Nature (London)*, 320, 72, 1986.
18. **Norcross, M. A., Raghupathy, R., Strominger, L., and Germain, R. N.,** Transfected human B lymphoblastoid cells express the mouse IA beta[d] chain in association with DR alpha, *J. Immunol.*, 137, 1714, 1986.
19. **Fathman, C. G.,** Hybrid I region antigens, *Transplantation*, 30, 1, 1980.
20. **Jones, P. P., Murphy, D. B., McDevitt, H. O.,** Two genes control of the expression of a murine Ia antigen, *J. Exp. Med.*, 148, 925, 1978.
21. **Silver, J., Swain, S. L., and Hubert, J. J.,** The small subunit of I-A subregion antigens determines the allospecificity recognized by a monoclonal antibody, *Nature (London)*, 286, 272, 1980.
22. **Kimoto, M. and Fathman, G.,** Antigen reactive T cell clones: transcomplementing hybrid I-A region gene products function effectively in antigen presentation, *J. Exp. Med.*, 152, 759, 1980.
23. **Reske-Kunz, A. B. and Rude, E.,** Insulin-specific T cell hybridomas derived from (H-2b x H-2k) mice preferably employ F1-unique restriction elements for antigen recognition, *Eur. J. Immunol.*, 15, 1048, 1985.
24. **Goyert, S. M. and Silver, J.,** Further characterization of HLA-DS molecules: implications for studies assessing the role of human Ia molecules in cell interactions and disease susceptibility, *Proc. Natl. Acad. Sci. U.S.A.*, 80, 5719, 1983.
25. **Charron, D., Lotteau, V., and Turmel, P.,** Hybrid HLA-DC antigens provide molecular evidence for gene *trans*-complementation, *Nature (London)*, 312, 157, 1984.

26. **Giles, R. C., DeMars, R., Chang, C. C., and Capra, J. D.,** Allelic polymorphism and transassociation of molecules encoded by the HLA-DQ subregion, *Proc. Natl. Acad. Sci. U.S.A.,* 82, 1776, 1985.

27. **Jonsson, A. K., Hyldig-Nielsen, J. J., Servenius, B., Larhammar, D., Andersson, G., Jörgensen, F., Peterson, P. A., and Rask, L.,** Class II genes of the human major histocompatibility complex. Comparisons of the DQ and DX alpha and beta genes, *J. Biol. Chem.,* 262, 8767, 1987.

28. **Gregersen, P. K., Shen, M., Song, Q. L., Merryman, P., Degar, S., Seki, T., Maccari, J., Goldberg, D., Murphy, H., Schwenzer, J., Wang, C. Y., Winchester, R. J., Nepom, G. T., and Silver, J.,** Molecular diversity of HLA-DR4 haplotypes, *Proc. Natl. Acad. Sci. U.S.A.,* 83, 2646, 1986.

29. **Spielman, R. S., Lee, J., Bodmer, W. F., Bodmer, J. G., and Trowsdale, J.,** Six HLA-D region alpha-chain genes on human chromosome 6: polymorphisms and associations of DC alpha-related sequences with DR types, *Proc. Natl. Acad. Sci. U.S.A.,* 81, 3461, 1984.

30. **Lotteau, V., Boyer, B., Debre, P., and Charron, D. J.,** Non-associated HLA-DR alpha and beta chains: molecular detection and differential cellular expression, in *Histocompatibility Testing 1984,* Albert, E. D., et al., Eds., Springer-Verlag, Berlin, 1984, 544.

31. **Steel, C. M.,** Table of workshop antibodies and their characteristics, *Dis. Markers,* 2, 263, 1984.

32. **Charron, D. J. and McDevitt, H. O.,** Analysis of HLA-D region-associated molecules with monoclonal antibody, *Proc. Natl. Acad. Sci. U.S.A.,* 76, 6567, 1979.

33. **Goyert, S., Shively, J. E., and Silver, J.,** Biochemical characterization of a second family of human Ia molecules, HLA-DS equivalent to murine-IA subregion molecules, *J. Exp. Med.,* 156, 550, 1988.

34. **Charron, D. J., Lotteau, V., and Turmel, P.,** Hybrid HLA-DQ antigens: molecular expression, in *Histocompatibility Testing 1984,* Albert, E. D. et al., Eds., Springer-Verlag, Berlin, 1984, 539.

35. **Brodsky, F. M., Parham, F. M., Barnstable, P., Crumpton, C. J., and Bodmer, W. F.,** Monoclonal antibodies for analysis of the HLA system, *Immunol. Rev.,* 47, 3, 1979.

36. **Nepom, B. S., Schwartz, D., Palmer, J. P., and Nepom, G. T.,** Transcomplementation of HLA genes in IDDM, *Diabetes,* 36, 114, 1987.

37. **Song, Q. L., Gregersen, P. K., Karr, R. W., and Silver, J.,** Recombination between DQ alpha and DQ beta genes generates human histocompatibility leukocytes antigen class II haplotype diversity, *J. Immunol.,* 139, 2993, 1987.

38. **Bontrop, R. E., Baas, E. J., Otting, N., Schreuder, G. M., and Giphart, M. J.,** Molecular diversity of HLA-DQ: DQ alpha and beta chain isoelectric point differences and their relation to serologically defined HLA-DQ allospecificities, *Immunogenetics,* 25, 305, 1987.

39. **Steinmetz, M., Stephan, D., and Lindahl, K. F.,** Gene organization and recombinational hotspots in the murine MHC, *Cell,* 44, 895, 1986.

40. **Smith, G. R., Kunes, S. M., Schultz, D. W., Taylor, A., and Trinan, K. L.,** Structure of chi hotspots of generalized recombination, *Cell,* 24, 429, 1981.

41. **Hansen, G. S., Svejgaard, A., and Claeson, M. H.,** T cell clones restricted to "hybrid HLA-D antigens", *J. Immunol.,* 128, 2497, 1982.

42. **Svejgaard, A., Platz, P., and Ryder, L.,** HLA and disease 1982 a survey, *Immunol. Rev.,* 70, 193, 1982.

43. **Nepom, B. S., Nepom, G. T., Mickelson, E. S., Challer, J. G., Antonelli, P., and Hansen, J. A.,** Specific HLA-DR4 associated histocompatibility molecules characterize patients with seropositive juvenile rheumatoid arthritis, *J. Clin. Invest.,* 74, 287, 1984.

44. **Betuel, H., Gebuhrer, L., Descos, L., Percebois, H., Winaire, Y., and Bertrand, J.,** Adult coeliac disease associated with HLA-DRw3 and DRw7, *Tissue Antigens,* 5, 231, 1980.

45. **Charron, D.,** Un modèle et trois exemples pour comprendre HLA et maladies: *cis* et *trans* associations moléculaires de classe II du CMH dans le diabète juvénile insulino-dépendant, l'arthrite rhumatoide de l'enfant et la maladie coeliaque, *Pathol. Biol.,* 34, 795, 1986.

46. **Todd, J. A., Bell, J. I., and McDevitt, H. O.,** HLA-DQ alpha gene contributes to susceptibility and resistance to insulin-dependent diabetes mellitus, *Nature (London),* 329, 599, 1987.

47. **Braunstein, N. and Germain, R. N.,** Allelic-specific control of Ia molecule surface expression and conformation: implications for a general model of Ia structure-function relationship, *Proc. Natl. Acad. Sci. U.S.A.,* 84, 2921, 1987.

48. **Sant, A. J., Braunstein, N. S., and Germain, R. N.,** Predominant role of amino-terminal sequences in dictating efficiency of class II major histocompatibility complex alpha beta dimer expression, *Proc. Natl. Acad. Sci. U.S.A.,* 84, 8065, 1987.

49. **Neel, D., Merlu, B., Turpin, E., Rabourdin-Comble, C., Mach, B., Goussault, Y., and Charron, D. J.,** Characterization of N-linked oligosaccharides of an HLA-DR molecule expressed in different cell lines, *Biochem. J.,* 244, 433, 1987.

50. **Lotteau, V., Teyton, L., Tongio, M. M., Soulier, A., Thomsen, M., Sasportes, M., and Charron, D.,** Biochemical polymorphism of the HLA-DP heavy chain, *Immunogenetics,* 25, 403, 1987.

51. **Ando, A., Hidetoshi, I., Kimura, M., Ogata, S., and Tsuji, K.,** Isolation and allelic polymorphism of cDNA clones and genomic clones of HLA-DP heavy and light chains, *Hum. Immunol.,* 17, 355, 1986.

52. **Heyes, J., Austin, P., Bodmer, J., Bodmer, W., Madrigal, A., Mazzili, M. C., and Trowsdale, J.,** Monoclonal antibodies to HLA-DP transfected mouse L cells, *Proc. Natl. Acad. Sci. U.S.A.,* 83, 3417, 1986.

53. **Travers, P., Blundell, T. L., Sternberg, M. J. E., and Bodmer, W. F.,** Structural and evolutionary analysis of HLA-D region products, *Nature (London),* 310, 235, 1984.

54. **Malissen, B., Shastri, N., Pierres, M., and Hood, L.,** Cotransfer of the E alpha^d and A beta^d genes into L cells results in the surface expression of a functional mixed isotype Ia molecule, *Proc. Natl. Acad. Sci. U.S.A.,* 83, 3958, 1986.

55. **Knudsen, P. J. and Strominger, J. L.,** A monoclonal antibody that recognizes the alpha chain of HLA-DR antigens, *Hum. Immunol.,* 15, 150, 1986.

56. **Lotteau, V., Teyton, L., Burroughs, D., and Charron, D. J.,** A novel HLA class II molecule (DR alpha-DQ beta) created by mismatched isotype pairing, *Nature (London),* 329, 339, 1987.

57. **Piatier-Tonneau, D., Trumel, P., Auffray, C., and Charron, D.** Construction of chain and locus specific HLA class II DNA probes. Study of HLA class II transcripts in leukemia, *J. of Immunol.,* 137, 2050, 1986.

58. **Sugden, B., Marsh, K., and Yates, J.,** *Mol. and Cell Biol.,* 5, 410, 1985.

59. **Korman, A. J., Knudsen, P. J., Kaufman, J. F., and Strominger, J. L.,** cDNA clones for the heavy chain of HLA-DR antigens obtained after immunopurification of polysomes by monoclonal antibody, *Proc. Natl. Acad. Sci. U.S.A.,* 79, 1844, 1982.

60. **Eckels, D. D., Hartzman, R. J., and Johnson, A. H.,** Recognition of a hybrid HLA-DP associated determinant by a human T lymphocyte clone, *J. Immunol.,* 136, 2515, 1986.

61. **Gomard, E., Henin, Y., Sterkers, G., Masset, M., Fauchet, R., and Levy, J. P.,** An influenza A virus-specific and HLA-DRw8-restricted T cell clone cross-reacting with a transcomplementation product of the HLA-DR2 and DR4 haplotypes, *J. of Immunol.,* 136, 3961, 1986.

62. **Bjorkman, P. J., Saper, M. A., Samraoui, B., Bennett, W. S., Strominger, J. L., and Wiley, D. C.,** Structure of the human class I histocompatibility antigen, HLA-A2, *Nature (London),* 329, 506, 1987.

63. **Berdoz, J., Gorski, J., Termijtelen, A. M., Dayer, J. M., Irle, C., Schendel, D., and Mach, B.,** Constitutive and induced expression of the individual HLA-DR beta and alpha chain loci in different cell types, *J. Immunol.,* 139, 1336, 1987.

64. **Faille, A., Turmel, P., and Charron, D. J.,** Differential expression of HLA-DR and HLA-DC/DS molecules in a patient with hairy cell leukemia: restoration of HLA-DC/DS expression by 12-*O*-tetradecanoyl phorbol-13-acetate, 5-azacytidine and sodium butyrate, *Blood,* 64, 33, 1984.

65. **Radka, S. F., Charron, D. J., and Bordsky, F. M.,** Review: class II molecules of the major histocompatibility complex considered as differentiation markers, *Hum. Immunol.,* 16, 390, 1986.

66. **Long, E. O.,** In search of a function for the invariant chain associated with Ia antigens, *Surv. Immunol.,* 4, 27, 1985.

67. **Sidman, C. L., Luther, E. A., Marshall, J. D., Nguyen, K. A., Roopenian, D. C., and Worthen, S. M.,** Increased expression of major histocompatibility complex antigens on lymphocytes from aged mice, *Proc. Natl. Acad. Sci. U.S.A.,* 84, 7624, 1987.

68. **Ju, L. Y.,** unpublished observations.

69. **Jones, P.,** personal communication.

70. **Sant, A. J. and Germain, R. N.,** personal communication.

71. **Lotteau, V., Sands, L., Teyton, P., Turmel, D. J., Charron, J. L., and Strominger, J. L.,** submitted.

72. **Sands et al.,** in preparation.

Chapter 4

ISOTYPIC DIVERSITY AND FUNCTION OF HLA CLASS II ANTIGENS

Eric O. Long, Sandra Rosen-Bronson, Steven Jacobson, and Rafick P. Sekaly

TABLE OF CONTENTS

I. INTRODUCTION

The main purpose of this review is to discuss how diverse isotypic class II major histocompatibility complex (MHC) antigens are in the human leucocyte antigen (HLA) system, and to describe the available evidence for such diversity. The immune system is able to generate a huge diversity in the specificities of the T-cell receptor (TCR), yet T-cell-mediated recognition is restricted by MHC antigens. Although T cells have the potential to recognize virtually any foreign element, they will do so only if the foreign element is presented by a self-MHC antigen. A T cell specific for a particular antigen will generally not recognize that same antigen in the context of a nonself-MHC antigen. Are the limitations imposed by MHC restriction overcome by a large isotypic diversity or by a very low peptide-binding specificity of MHC antigens? Before addressing this important question we will summarize the information gained by molecular studies on the allelic polymorphism of class II MHC antigens. A good understanding of the origin and the significance of allelic poly-morphism is a useful basis for a discussion on isotypic diversity. This review will end with a description of recent experiments on class II antigen function.

The extreme allelic polymorphism of antigens encoded in the MHC is a feature unique to this genetic system. The extent of the polymorphism, both in the number of different alleles for each locus and in the divergence between allelic sequences, strongly suggests that this diversity was under selection during evolution. Two fundamental questions that have not been fully answered are (1) what is the functional significance of allelic poly-morphism in MHC antigens and (2) how was the allelic polymorphism generated? The first question has been the source of much debate because the relationship between evolution and function cannot be tested directly. On the other hand, the molecular analysis of MHC genes has yielded important information on the structural basis for allelic polymorphism and has led to models on the mechanisms that generated the diversity.

The MHC was originally discovered because allelic polymorphism in MHC antigens led to allograft rejection. The same property of MHC antigens is also responsible for the phe-nomenon of mixed lymphocyte reaction. These important observations were only manifes-tations of MHC polymorphism but did not immediately reveal the true function of MHC antigens, because grafts and cultures of mixed lymphocytes do not occur in nature. Many years after these initial discoveries, a fairly good understanding of MHC antigen function has finally emerged. MHC antigens act as receptors for processed polypeptides that are presented to specific T cells. Both class I MHC antigens (HLA-A, -B, -C in man) and class II MHC antigens (HLA-DP, -DQ, -DR in man) fulfill this role. The distinction between class I- and class II-restricted presentation is rather at the level of the responding T cells: class I-restricted T cells express the CD8 molecule and are mostly cytotoxic, whereas class II-restricted T cells express the CD4 molecule. The original observations on Ir gene control of immune responses were also manifestations of allelic polymorphism in class II MHC antigens. It is now known that synthetic antigenic peptides bind to different MHC antigens with different affinities. Nonresponder phenotypes are due to class II MHC antigens that do not bind a peptide well enough to generate a T-cell response. Both class I and II antigens have an essential and similar function, namely, the binding of peptides and their presentation to T cells, despite the very different manifestations of their function which led to their discovery.

There are also important differences in the function of class I and class II antigens. Helper T cells, responsible for the initiation of immune responses, are stimulated by processed antigen presented mostly by class II antigens. In contrast, cytotoxic T cells can be restricted by either class I or class II antigens. This functional distinction may be controlled by the cellular pathways of antigen processing and presentation, as will be discussed. Furthermore, there is little doubt that antigen presentation is not the sole function of class I and class II antigens. Both classes of MHC antigens play additional and distinct roles in the immune system.

Knowledge of MHC antigen function is an obvious prerequisite to the understanding of MHC polymorphism. After discovering that cytotoxic T cells could recognize viral antigens only in the context of self class I MHC antigens, Zinkernagel and Doherty proposed a biological explanation for allelic polymorphism.[1] The idea is essentially that diversity generated in a population by allelic polymorphism is advantageous to the species. The risk that processed polypeptides derived from a given virus would not bind to host MHC molecules and would thereby escape the immune response, is reduced by the polymorphic diversity. The same interpretation is most often provided as the basis for polymorphism in class II MHC antigens. The range of immune responses that can be mounted against foreign elements by a species is vastly increased by the polymorphism. This interpretation is fully compatible with the observation that nonresponse to a particular antigen is more likely due to a lack of peptide binding to class II antigen rather than the failure to generate specific T cells.[2]

The selective advantage of polymorphic MHC antigens in a species, and of the resulting heterozygosity in individuals, is a quite satisfying and simple explanation for the functional significance of this diversity. Yet there are unresolved problems; the main one is that selection really operates on individuals and not on the species. Furthermore, several animal species have very low MHC polymorphism. Other theories on the significance of MHC polymorphism have been advanced. Klein proposes that the polymorphism currently present in MHC antigens is mostly a relic from earlier stages in evolution.[3] His argument is based primarily on the documented fact that existing allelic sequences predate speciation. The evolutionary selection for polymorphism may have operated in the past but no longer applies today. A radically different view of the significance of MHC polymorphism has been proposed by Andersson et al.[4] Much like the artificial phenomenon of graft rejection, the natural function of MHC polymorphism would be to ensure rejection of pathogens that passively acquire host antigens. The recognition of nonself-MHC antigens transmitted by such pathogens would lead to their rejection. According to this hypothesis the selection for polymorphism would operate at the population level. It is not clear, however, whether pathogens that acquire host antigens represent a sufficient selective force to be responsible for the enormous polymorphism of MHC antigens, and why class II MHC antigens would be polymorphic. The debate about the functional significance of MHC polymorphism continues.

If the diversity in MHC antigens represents a selective advantage, then it should have evolved within individuals and not only within species. It is, therefore, important to define the extent of the isotypic (or nonallelic) diversity in MHC antigens. Most of the attention has been focused in the past on allelic diversity because of practical reasons. There is a need for reliable techniques to determine individual HLA specificities in order to match allografts. The correlation between certain HLA specificities and the susceptibility to particular diseases has also led to intensive studies of allelic sequences. We consider the analysis of isotypic MHC antigens no less important because the immune responses that each individual can mount are entirely dependent on them.

II. DIVERSITY OF CLASS II MHC ANTIGENS

A. ALLELIC POLYMORPHISM

Amino acid and nucleotide sequences of class II antigens and restriction fragment length polymorphism (RFLP) patterns in the HLA-D region indicate that the polymorphism evolved primarily by point mutations and selection.[5-9] The origin of alleles is most likely very old because allelic sequences are conserved between different species of mice.[10] As a result, a fairly stable linkage exists between polymorphic sequences in the coding regions of class II genes and the sequences in introns and in flanking regions. This linkage results in a good correlation between RFLP patterns and class II antigenic specificities. Polymorphic sequences in class II β chains and in the DQα chain are clustered in several highly variable regions. Although the majority of polymorphic alleles probably evolved in the distant past and has been stable in the human population, more recent changes must have also occurred. Several HLA-DR serological specificities can be subdivided by cellular typing. These "splits" of DR are due to limited sequence divergence, some of which are found outside the three highly variable regions of DRβ chains.[11,12] In contrast to the major allelic specificities, these DR splits are not easily defined by RFLP analysis.[13] These sequence variations have probably occurred too recently to show strong linkage with flanking DNA polymorphism.

A mechanism which has been often invoked to explain rapid diversification in multigene families is gene conversion. The term is used loosely to mean genetic exchange between related genes, but without implying any specific molecular mechanism. The first evidence for gene conversion in class II MHC genes came from the murine mutant bm12.[14] In human class II MHC genes several gene sequences have been described which strongly suggest gene conversion events. The first case was the finding that the DR3β chain allelic sequence was totally identical to a DRw6β chain sequence except for a stretch of nucleotides that were in turn totally identical to another β chain gene in the DRw6 haplotype.[15] Thus, most likely, DR3 evolved by a single genetic event from the DRw6 haplotype. Similar examples have been described[12,16] with other DRβ sequences. Was gene conversion a driving force in the generation of class II MHC polymorphism? Probably not. A comparison with class I MHC genes is useful to discuss this point. Multiple examples of gene conversion have been documented in class I MHC genes.[17] These genetic exchanges have occurred so frequently that the relationship between class I MHC genes from different species is no longer apparent. Even though man and mouse share common ancestral class I genes, gene conversion events that have taken place after mammalian radiation led to independent intraspecies divergence. As a result, class I genes within a species are as related to each other as they are to class I genes in other species.

The situation with class II genes is in total contrast. Class II genes that share a common ancestor (i.e., DRβ in man and I-Eβ in mouse) are more related to each other than to other class II genes within the species.[18,19] This fact is incompatible with frequent gene conversion events within the class II gene family. Furthermore, all cases of potential gene conversion in human class II genes represent exchanges between DRβ genes and not between genes in different subregions (e.g. DRβ and DQβ). In summary, the molecular evolution of polymorphism in human class II genes occurred primarily by point mutations even though genetic exchanges between genes have also contributed to allelic diversity.

B. ISOTYPIC DIVERSITY
1. Introduction

If allelic polymorphism is necessary to provide a species with a wide range of potential immune responses, and if it was selected for on that basis, one might expect a similar selection for isotypic diversity in individuals. On the other hand, if allelic polymorphism does not, or does no longer, represent a selective advantage there is no reason to expect isotypic diversity. In the latter situation, isotypic diversity might still exist for functional reasons unrelated to the number of peptides that could potentially bind to class II MHC antigens. To address these questions we decided to determine the extent of isotypic diversity in human class II antigens.

A limitation on the extent of the isotypic diversity will be imposed by the need to distinguish self and nonself. A very large diversity of self peptides binding to multiple class II antigens would reduce the ability of the immune system to react to foreign determinants. One might also argue against the need for a large isotypic diversity on the basis that DR antigens clearly predominate over DQ and DP antigens in class II responses, and that certain

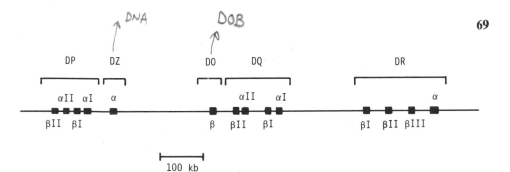

FIGURE 1. Molecular map of the HLA-D region derived from a combination of genomic DNA cloning and pulsed-field gel electrophoresis mapping. Boxes represent genes; they are not drawn to scale. The arrangement of genes in the DR subregion is that of the DR3 haplotype. αII and βII are either pseudogenes or inactive genes.

strains of mice express only a single class II antigen. However, there is no evidence to suggest that DR antigen alone would ensure the survival of man. It remains important to determine what the isotypic diversity of class II antigens is and what the functional significance of this diversity might be.

2. Multiple Functional HLA Class II Genes

DR and DQ antigens were first characterized serologically and biochemically. DP antigens were originally evidenced by cellular reactions and genetic studies. The complexity of class II antigens was suspected to be larger because of "supertypic" specificities that represented either additional antigens linked to certain DR specificities or determinants located on the DR molecules. The existing complexity of the HLA-D region at the genetic level was finally established by molecular cloning. The complete linkage of the different subregions in HLA-D required the combined use of gene cloning and pulsed-field gel electrophoresis.[20] A current map is shown in Figure 1.

There is one DRα gene but multiple DRβ genes. The number of DRβ genes even varies between haplotypes from 1 to 4. The two haplotypes characterized at the molecular level, DR3 and DR4,[21,22] contain respectively three and four DRβ genes. That two nonallelic DRβ genes can be expressed was first established by the isolation of four different cDNA clones from a heterozygous cell line.[5] Certain haplotypes, such as DR1 and DR8, probably express only one DRβ chain. Both expressed DRβ genes are coordinately regulated[23] and the two DRβ chains pair with the same DRα chain. Transfection experiments with the second expressed DRβ gene demonstrated that it encoded one of the supertypic specificities, now called DRw52.[24] The products of the second expressed DRβ genes, carrying the DRw52 or DRw53 specificities, can be involved in restricted antigen presentation but the functional significance of this gene duplication is not known. One or two of the DRβ genes, depending on the haplotype, are pseudogenes and additional incomplete DRβ gene fragments have been reported. These nonfunctional or incomplete genes contribute to the genetic but not the antigenic diversity of HLA-DR.

The DQ subregion consists of two pairs of DQα and DQβ genes. One pair encodes the DQ antigen. The other, often called DX, does not appear to be expressed even though no defect has been found in the genes by sequencing.[25,26] Therefore, only one DQ antigen per haplotype is expressed in B cells. The DP subregion consists also of two pairs of DPα and DPβ genes. One encodes the DP antigen while the other represents pseudogenes.[27]

Genes within a subregion are much more related to each other than they are to genes in other subregions. The duplications that generated DP, DQ, and DR genes preceded mammalian radiation, as evidenced by the strong structural conservation between these genes and their murine counterparts. On the other hand, duplications within subregions must be more recent.

DZα[28,29] and DOβ[19,30] represent additional genes that have been mapped in the HLA-D region. They will be described in a separate section. Because of the limitations inherent to nucleic acid hybridizations it remains possible that additional class II genes exist that have so far escaped detection by hybridization techniques. The large size of the HLA-D region as it is known now (~1 megabase or 1000 kb) can still accommodate other genes (Figure 1). Other class II genes may also lie outside of the current boundaries of HLA-D.

3. Biochemical Evidence

With the availability of monoclonal antibodies specific for monomorphic determinants on DR, DQ, and DP molecules, it became possible to obtain direct biochemical evidence for the existence of these distinct αβ pairs. However, much of the data reported on the biochemical characterization of class II antigens cannot be reconciled with the current molecular map of HLA-D. The main discrepancy is that the number of isotypic products defined biochemically often exceeds the number of functional genes. One explanation could be that cell lines considered homozygous by typing may in fact be heterozygous genetically. Even HLA-homozygous cells by consanguinity may carry recombinations undetected serologically. Another explanation could be that the number of functional genes varies between haplotypes and that some of the cell lines analyzed biochemically have additional class II genes. The finding of seven distinct class II β chain peptides in a DR2 homozygous cell line[7] can only be compatible with molecular data by assuming allelic heterogeneity or additional isotypic genes in this line. Several studies have reported the existence of two isotypic DQ or DQ-like antigens.[31-33] In these cases, it is unlikely that all three independent cell lines have undetectable heterozygosity at the DQ locus. Furthermore, the second pair of DQ genes is not detectably expressed in B-cell lines. It is not clear whether post-translational changes account for the biochemical differences or whether these products are derived from some as yet unknown genes.

Evidence for a fourth antigen distinct from DP, DQ, and DR has been presented based on immunodepletion and peptide maps.[34] The possibility that this antigen represents a mixed isotypic pair will be discussed below. Alternatively, it may represent an antigen encoded by other class II genes. The possibility that additional class II genes have remained undetected is supported by the observation that a large homozygous deletion of all known functional class II genes does not prevent stimulation of a mixed lymphocyte reaction.[35]

4. DZα and DOβ
a. Expression

The finding of a new pair of α and β genes in the HLA-D region, both of which are expressed at the RNA level, suggested that DZα and DOβ might encode the two polypeptide chains of a new class II antigen αβ heterodimer. If this were the case, one might expect these two genes to be coordinately expressed, as is the case for other class II α and β genes.[36,37] DZα and DOβ are not coordinately expressed.[19] DZα transcripts were induced in fibroblasts by treatment with interferon-γ, as were other class II genes, and were drastically reduced in a B-cell mutant that is defective in a *trans*-acting factor necessary for class II gene expression. In contrast, DOβ transcripts were not interferon-γ inducible and were still present in the class II-defective mutant. To test whether this distinct regulation might be reflected in the cellular pattern of expression we analyzed DZα and DOβ transcripts in various cells of the immune system.[89] The expression of DZα always paralleled that of other class II genes. In contrast, expression of DOβ was much more restricted. DOβ RNA was not detectable in pre-B cell lines that were positive for expression of DP, DQ, DR, and DZ genes. DOβ RNA was present in mature B-cell lines, either EBV-transformed or derived from Burkitt's lymphomas, and in total peripheral blood lymphocytes. B-cell lines representing late stages of B-cell development did not express any class II genes, including DZα and DOβ. Therefore, DOβ expression appears later in B-cell development than that of other class II genes. Resting T cells were negative for class II gene expression. An alloreactive T-cell clone expressed the DP, DQ, and DR genes, the DZα gene but not the DOβ gene. The very distinct expression of DOβ suggests, first, that DZα and DOβ are not likely to be partners in an αβ heterodimer and, second, that the function of DOβ may be different from that of other class II antigens. There is no evidence at present that DZα or DOβ encodes functional proteins but the fact that both genes produce translatable mRNA suggests strongly that there is a product *in vivo*. It will be necessary to develop antibodies specific for these chains to demonstrate their existence. A biochemical analysis of DZα and DOβ would also reveal what subunits may be associated with them. The DZα and DOβ genes are separated by at least 125 kb.[20] Additional genes may still exist in this region that could encode partner chains for DZα or DOβ.

The DZα and DOβ genes are probably not polymorphic. Both lie in DNA regions with low RFLP.[38,39] Furthermore, amino acid sequences derived from a DZα gene and of a DZα cDNA isolated from unrelated cells were identical, and so were DOβ sequences derived from a cDNA and an unrelated gene.

b. Alternative Splicing in DZα Generates a Short Non-Class II Polypeptide

Two mature transcripts are produced from DZα; one of ~1.2 kb and another of ~3.5 kb. The original DZα cDNA clone[29] had a 3′ poly(A) extension following an unusual signal ACTAAA.[28] The use of this signal would generate a spliced mature transcript of ~1.2 kb. To explore the relationship between these two transcripts we isolated several DZα cDNA clones one of which was 3.5 kb long.[89] Sequencing of all exon boundaries and of both ends revealed that the longer DZα transcript is fully spliced and that it terminates with a poly(A) tail following a typical signal AATAAA located further 3′ of the first signal. The long transcript is not a precursor of the short one but results from readthrough at the first poly(A) addition signal. Use of multiple poly-adenylation signals has been reported in several other genes and is of unknown functional significance.

The analysis of DZα cDNA clones revealed also an alternative splice at the 5′ end of the transcript. Two independent clones represented transcripts which had incompletely spliced the first intron by use of a splice acceptor located 5′ upstream of the α1 domain exon. This transcript encodes the signal sequence of DZα which is followed by an open reading frame of 42 amino acids. The reading frame is different from that of the class II α sequence and a termination codon occurs just 5′ upstream of the α1 domain exon. *In vitro* transcription and translation experiments demonstrated that a polypeptide of the expected size was produced from this alternatively spliced RNA. The conventionally spliced RNA produced *in vitro* a polypeptide of the size expected for the class II DZα chain.

A DNA probe specific for the alternatively spliced transcript was used to show that for all cells tested, including untransformed cells, this RNA was present in cells normally expressing the DZα gene. The alternative splice occurs in about 1/10th of the mature DZα RNA in all cells examined.

Alternative splicing in eukaryotic genes is not uncommon and is a useful mechanism to generate isoforms of a protein from a single gene.[40] The case of DZα is unique because the same gene encodes either a class II α chain or a totally unrelated short polypeptide. It is not known at present what the function of this potentially secretable peptide might be. It remains possible that inaccurate splicing occurs in some genes and that the product of such aberrant alternative splices has no function. If the product is not harmful it will not be under a strong negative selection pressure. Analysis of RNA from other species with a DZα gene counterpart will help elucidate this point.

5. Functional Diversity

It is undisputed that class II MHC antigens can bind peptides and present them to T cells. The DP, DQ, and both DR antigens are all involved in antigen presentation to T cells. But is antigen presentation their unique function? Furthermore, are these isotypic class II antigens functionally equivalent, in that they serve to expand the range of potential immune responses, or do they also fulfill distinct functions? These essential questions remain unanswered, but much evidence has accumulated in favor of a functional diversity in class II antigens. A divergence of class II antigen function occurred also during evolution because the antigens used predominantly in immune responses vary between species; it is DR in man, I-A in mouse (the DQ counterpart), and probably the DP counterpart in the mole rat.[41]

Resting T cells do not express class II antigens. After mitogenic or antigen-specific stimulation T cells actively synthesize DP, DQ, and DR antigens, as evidenced by RNA synthesis.[90] Although class II antigens expressed by T cells are able to present antigens to other T cells, it is probably not the reason why the class II antigens appear on T cells. One function of class II antigens on T cells may be to maintain these cells in an activated state. Antibodies to DR block the IL-2-mediated proliferation of T cells.[42] The understanding of class II antigen function on T cells is complicated by the fact that most murine activated T cells remain class II-negative. Three interpretations can be proposed. First, class II antigens on human T cells have no role, an argument difficult to defend. Second, class II antigen function on T cells is dispensable in mice. Finally, the function of class II antigens on human T cells is fulfilled by other molecules in murine T cells.

A specific role for DQ in the generation of effector T cells has been suggested by antibody-blocking experiments. Monoclonal antibodies to DQw1 inhibited the generation of allospecific cytotoxic T cells but, unlike monoclonal antibodies to DR, they did not inhibit the proliferation of T cells.[43,44] In a particular nonresponder haplotype to *Schistosoma japonicum* the response to the parasite antigen is controlled by the DR1 molecule but the DQw1 molecule stimulates the generation of specific suppressor T cells.[45] Functions uniquely associated with DP antigens have not been reported but remain a possibility.[46] The use of transfected cells expressing a single class II molecule provides a useful tool to test what type of T cells are stimulated by the different class II antigens.

Class II antigens may also serve functions that are unrelated to the stimulation of T cells through antigen presentation. The possibility most often considered for such a role is that of differentiation markers.[47] Class II antigen expression is coordinate in B cells, in activated T cells, and in a variety of cells treated with interferon-γ, but there is discordant expression of DP, DQ, and DR molecules in a number of hematopoietic leukemic cell lines. The distinct expression of DOβ during B-cell development suggests that this molecule, if expressed functionally at the surface of mature B cells, may serve as a differentiation marker specific to the B-cell linage.

Class II antigens can transmit signals to B cells. Monoclonal antibodies to DR antigen, in conjunction with pokeweed mitogen, provided helper signals to B cells.[48] In the murine system, a B-cell lymphoma was stimulated by helper T cells and by antibodies reactive with the I-E molecule only, even though both I-A and I-E molecules could stimulate helper T cells, suggesting a specific signaling role of I-E.[49] Direct biochemical evidence of class II-mediated transmembrane signaling was also obtained.[50] Interestingly, anti-Ia antibodies induced a translocation of protein kinase C to the nucleus rather than to the plasma membrane as observed with anti-Ig antibodies. The signal mediated by class II antigen may induce differentiation, as shown with murine Ly-1-positive B cells.[51]

Signaling in other class II-positive cells also occurs because anti-Ia antibodies induced IL-1 secretion by monocytes.[52] Differential expression of DP, DQ, and DR antigens in hematopoietic lineages may be due to distinct and regulated signaling via each class II antigen.

6. Mixed Isotypic Pairs

a. Introduction

If allelic polymorphism in class II MHC antigens represents a selective advantage to the species, isotypic diversity of class II antigens should be advantageous to the individual. Most HLA haplotypes encode at least four class II antigens: DPαDPβ, DQαDQβ, and two pairs of DRαDRβ each with a different β chain. A typical heterozygous individual would thus encode eight class II molecules. Additional functional genes could expand this diversity further. DZα and DOβ are candidates but so far these genes do not seem to be polymorphic, are expressing low levels of RNA, and have not been shown to be functional. Therefore, it seems unlikely that class II antigen diversity would increase significantly in functional terms by the contribution of DZα, DOβ, or any additional undetected gene.

A very efficient way to increase class II antigen diversity would be to mix α and β chains in different combinations. *Trans*-complementation between allelic forms of DQ antigens has been demonstrated.[53,54] Permissive *trans*-complementation between all DQ allelic forms would add two class II molecules per heterozygous individual (e.g., DQ1αDQ2β and DQ2αDQ1β). The same permissive associations may occur in DP antigens and contribute also two additional class II molecules. On the other hand, *trans*-complementation in DR antigens would not increase the diversity because DRα is not polymorphic. Therefore, a maximum of four additional class II molecules per individual may be contributed by *trans*-complementation.

A much more powerful source of diversity lies in mixed isotypic pairs, that is, in class II molecules made up of an α chain from one subregion (e.g., DRα) paired with a β chain from another subregion (e.g., DQβ). In a single haplotype, mixed isotypic pairs (MIPS) would contribute eight additional molecules. A heterozygous individual could therefore produce 16 MIPS by *cis*-combinations and another 12 MIPS by *trans*-combinations. This simple calculation reveals how significant MIPS could be in diversifying the range of restriction elements for antigen presentation because a heterozygous individual encodes 28 potential MIPS as compared to four *trans*-complementation products and the eight known isotypic αβ pairs. Only chains known to be expressed as DP, DQ and DR antigens were taken into account in the MIP calculations.

b. Transfections

The first question we asked was whether αβ pairing across isotypes of the DR1, DQ1, DP2 haplotype was possible. Even though isotypic α and β chains have retained approximately 65% and 70% identity in their amino acid sequences, respectively, it was possible that coevolution of αβ gene pairs and divergence between isotypes had led to a complete loss of ability to pair across isotypes. Transient cell-surface expression in transfected COS cells[55] was used as a rapid test for the formation of MIPS (Figure 2). cDNA clones encoding the α chains of DP and DR antigens, and the β chains of DP, DQ, and DR antigens had been constructed in an SV40-based expression vector.[19] Such vectors replicate to high copy numbers when introduced into simian COS cells. Expression of surface class II antigen was assayed by flow cytometry using the broadly reactive anti-class II antibody SG465 3 d after transfection. The level of the DRαDQβ, DRαDPβ, DPαDRβ, and DPαDQβ MIPS at the surface of transfected cells is similar to that of the matched DP and DR pairs. The percentage of transfected cells expressing surface class II antigen varied from one experiment to another and showed no correlation with expression of isotypic pairs or MIPS. The results shown in Figure 2 represent a single experiment. Other experiments have yielded a higher percentage of cells expressing the DPαDPβ antigen.[55] The transient expression system used in these experiments has the advantage of providing an answer in absence of selection. This is not only advantageous for its speed but also because the results cannot be attributed to an artifact of selection. Control transfections with a single expressible cDNA gave background levels

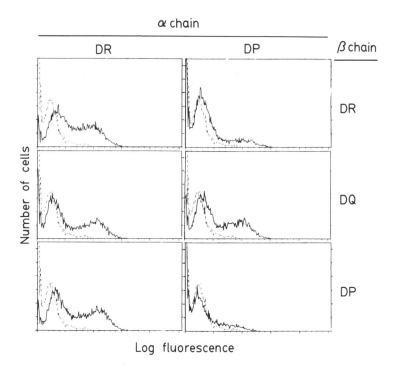

FIGURE 2.   Cell surface expression of class II antigens on transfected COS cells. COS cells were transfected with 10 μg of the indicated α cDNA, 10 μg of the indicated β cDNA and 10 μg of the invariant chain cDNA, as described.[55] Cells were harvested 72 h after transfection and stained by indirect immunofluorescence with the broadly class II-reactive monoclonal antibody SG465 (a kind gift of Dr. S. Goyert) and with FITC-conjugated goat anti-mouse antibodies. Background fluorescence was determined with cells transfected with the invariant chain cDNA only (dotted lines). The same background fluorescence was obtained with cells transfected only with an α cDNA or a β cDNA.

of fluorescence. We conclude that mixed isotypic pairing is physically possible and that the resulting MIPS can be detected serologically with broadly reactive monoclonal antibodies.

To analyze this phenomenon further we derived stably transfected murine L cells expressing MIPS.[91] Results obtained by transient expression were confirmed and the missing MIP combinations with DQα were also obtained. Serological and biochemical analysis of L cells expressing MIPS showed that MIPS represent unique entities that cannot be considered as only the sum of juxtaposed α and β chains. The apparent permissiveness of assembly between mixed isotypic chains was surprising. It is important to point out that all the genes used in these experiments were derived from an HLA-hemizygous mutant with a single haplotype carrying the DPw2, DQw1, and DR1 allelic specificities. An additional surprising finding was that the DOβ chain was expressed in association with class II α chains. The DOβ chain has an unusually hydrophobic first domain that was thought to prevent pairing with known α chains.[30]

The permissive assembly of MIPS in the DR1 haplotype contrasts with results obtained in the murine system.[56,57] The pairing of Eα and Aβ was detected at the surface of transfected L cells but not that of Aα with Eβ. As was the case for allelic pairing of Aα and Aβ chains in *trans*-complementation[58] the sequences that controlled the formation of MIPS were mapped in the first half of the amino terminal domain.[59] The more efficient pairing of Aα and Aβ chains from the same haplotype suggested that genes linked in *cis* coevolved to maintain a good fit between α and β chains despite their allelic diversification. The pairing of Eα with

Aβ may be possible because the Eα chain is not polymorphic. Analysis of mixed isotypic pairing with class II chains from different murine haplotypes should be informative. In any case our results with human class II MIPS have revealed a greater permissiveness in assembly than was found in the murine system. Further analysis with genes from other haplotypes is necessary to test how general permissive assembly might be and whether it has been selected for in humans to maintain a wider repertoire of restricting elements.

c. Evidence in vivo

Biochemical studies of class II MHC antigens have been numerous and, except for one recent report,[60] have not revealed the existence of MIPS. However, almost every study relied on the use of monoclonal antibodies to isolate class II antigens, and it is likely that in most cases MIPS would have been missed because their serological reactivities are distinct from those of isotype-matched class II antigens. Nevertheless, the lack of antibodies specific for a MIP suggests that if MIPS exist at all at the surface of class II-positive cells they may represent a small fraction of total class II molecules. The evidence reported[60] for a DRαDQβ dimer in human B cells relied on the specificity of monoclonal antibodies and is therefore indirect. Weak cross-reactions of monoclonal antibodies thought to be specific have often been noted when tested on transfected cells expressing only a single class II antigen. Whether MIPS exist in vivo and at what levels they may be expressed on class II-positive cells is still unknown. The formation of MIPS in transfected cells is an important observation because it shows that mixed isotypic pairing is possible, but it does not imply that they would form in normal B cells. A direct biochemical characterization of MIPS is also very difficult with existing reagents. To resolve this important issue it will probably be necessary to generate reagents specific for MIPS. Transfected cells expressing MIPS are a valuable material for that purpose.

Three observations can be interpreted as being compatible with the existence of MIPS. A fourth subset of human class II antigens has been found with a broadly reactive monoclonal antibody after immunodepletion with DP-, DQ-, and DR-specific monoclonal antibodies.[34] The peptide map of this molecule showed that it was distinct from the known class II antigens and also that the α chain was polymorphic. DZα is not a likely candidate because it does not appear to be polymorphic. This fourth subset could in fact be a MIP despite the unique peptide map because cell surface labeling of antigens is dependent on protein conformation. The DQα or DPα chains, both polymorphic, may well be folded differently in a MIP as they are normally in the DQ or DP antigen, respectively. Second, a potential new class II antigen, named DY, defined on the basis of cellular tests, can also be interpreted as a MIP.[61] DY stimulated autoreactive T-cell clones. Such DY-reactive clones suppressed allogeneic MLC. DY is expressed on B cells, suppressor T cells but not helper T cells. DY responses, as well as DY-induced suppressive activity, can be blocked by the broadly class II-reactive antibody Tu39 but not by DP-, DQ-, or DR-specific antibodies. Finally, the restriction pattern of a T-cell clone suggests that it may be cross-reacting with a MIP.[62] The TA10 clone is specific for the influenza virus hemagglutinin in the context of DR8 but also cross-reacts with DR2,4 cells in absence of viral antigen. This cross-reactivity requires the presence of both DR2 and DR4 haplotypes and can be blocked by anti-DR antibodies. Trans-complementation between DRα and DRβ chains is not a likely interpretation because DRα is not polymorphic. The determinant may thus be a MIP between DRβ and another α chain.

d. Discussion

MIPS either do not exist in vivo or, if they do, most likely represent a small fraction of class II molecules at the cell surface. How then does the sorting operate that brings isotypic αβ heterodimers together? The simplest mechanism to achieve this would be that differences in affinities between α and β chains are sufficient to sort out the proper pairs.

Transfection experiments only showed that MIPS can form but did not provide measurements of the relative affinities between α and β chains. The comparable level of MIPS and isotypic αβ pairs at the surface of transfectants (Figure 2) does not imply equivalent efficiency of pairing and of transport because the system may have been saturated. We have shown that transfected COS cells produce very high levels of RNA from the introduced class II cDNAs.[55] The only way to directly address the question of affinities would be by competition experiments. Sorting of αβ pairs based on their relative affinities seems like the most obvious mechanism even though alternative explanations such as specialized cellular compartments and "gene gating"[63,64] can be proposed.

The evidence to date does not support a major role for MIPS in antigen presentation. Apparently the ability of a single class II molecule to bind a large array of peptides compensates for the small isotypic diversity of class II antigens. The paradox of a high allelic diversity favoring the species, while selection operates on individuals with low isotypic diversity, therefore remains.

Instead of being useful by increasing the range of class II restriction elements, MIPS may be detrimental by causing autoimmune reactions. If normal class II-positive cells, in particular those in the thymus, do not express detectable amounts of MIPS, aberrant expression of MIPS on certain cells could trigger an autoreaction. MIP expression could result from an imbalance in class II gene regulation. As an example, defective expression of the DRα gene might result in free DRβ chains available for mixed isotypic pairing. This possibility of aberrant MIP expression becomes an important consideration in the interpretation of HLA and disease associations. Most diseases that show a correlation with class II specificities have an autoimmune component. Because MIPS might be involved in the presentation of self-peptides it becomes necessary to correlate disease associations with haplotype and even *trans*-combinations of specificities rather than with DP, DQ, or DR specificities only.

III. MOLECULAR ANALYSIS OF HLA CLASS II ANTIGEN FUNCTION

A. ANTIGEN PRESENTATION TO T LYMPHOCYTES

The first step of an immune response is the recognition of a foreign element by a helper T lymphocyte. Helper T cells express the CD4 surface molecule and an antigen-specific receptor that recognizes antigens bound to class II MHC antigens. The function of class II antigens is thus to bind peptides in such a way that T cells will recognize them and initiate an immune response. Cytotoxic T cells that appear during the effector phase of the cellular immune response can recognize foreign antigens bound either to class I or to class II MHC antigens. Class I-restricted cytotoxic T cells generally express the CD8 surface molecule, whereas class II-restricted cytotoxic T cells are primarily CD4-positive.

Direct binding of peptides derived from antigens and of synthetic peptides to class II MHC molecules has been demonstrated.[65] Restriction specificities can be explained essentially by the affinities of peptides for different allelic or isotypic forms of class II MHC antigens.[66] While impressive progress has been made in this area, it is still largely unknown how antigens are processed and how processed peptides reach MHC antigens in the presenting cells.

Antigen-presenting cells take up soluble proteins by endocytosis and, after processing events that most likely take place in endocytic vesicles, present processed peptides to class II-restricted T cells.[67] This exogenous and endocytic pathway of presentation can be blocked by lysosomotropic agents such as chloroquine. Class II-restricted cytotoxic T cells specific for influenza virus recognize target cells that have processed viral antigens through the exogeneous-endocytic pathway but not viral antigens synthesized by the target cells.[68] Interestingly, the processing requirements for class I-restricted cytotoxic T cells are totally different. Exogenous antigens added either purified or an inactivated virus could not be presented to class I-restricted T cells. In contrast, newly synthesized viral antigens were efficiently presented even though class II-restricted cytotoxic T cells are unable to recognize them. This dichotomy in processing requirements for class I- and class II-restricted presentation of influenza virus antigens led to the proposal that specific processing pathways are linked to presentation by either class I or class II antigens.[69,70] Antigens processed in endosomes would intersect the biosynthetic pathway of class II antigens only, whereas antigens synthesized in the target cell would be processed by a mechanism that would bring the resulting peptides in contact with only class I antigens. Evidence for endogenous cytoplasmic processing exists for class I-restricted presentation of influenza virus antigens.[71,72] The processing pathway is still unknown but it was suggested that it might be related to the ubiquitin-mediated protein degradation mechanism.[73]

We have studied the cytotoxic T-cell response to measles virus as a model for class II-restricted antigen presentation, because the cellular response to this virus is mediated essentially by CD4-positive class II-restricted T cells.[74] Unlike influenza viruses that fuse with the host cell membrane in the low pH environment of endosomes, paramyxoviruses such as measles can fuse with the host cell membrane at neutral pH. We therefore tested whether measles virus antigen processing and presentation by class II antigens would be sensitive to chloroquine treatment. As previously shown,[68] control cells infected with influenza virus in the presence of chloroquine were not lysed by class II-restricted influenza virus-specific CD4-positive cytotoxic T cells. In contrast, the chloroquine treatment during infection with measles virus had no effect on the presentation of viral antigens to class II-restricted cytotoxic T cells.[75] It is therefore possible that processing of viral antigens in endosomes is not an absolute requirement for class II-restricted presentation.

Target cells infected with measles virus were either an EBV-transformed B-cell line or a fibroblast line transfected with expressible DRα and DRβ cDNA clones.[75] Both targets were still presenting viral antigens to cytotoxic T cells after infection with measles virus in the presence of chloroquine. Most interestingly, the transfected fibroblast expresses DR antigen in the absence of the invariant chain. Because class II antigens are always associated with the invariant chain inside class II-positive cells, but not at their surface, it was thought that the invariant chain played a role in their transport to the cell surface. When it was shown that efficient cell-surface expression of class II antigens occurred in the absence of the invariant chain,[55,76] it was proposed that its role might be in class II antigen function. Our results now show that the invariant chain is not required for class II-restricted measles virus antigen presentation. The invariant chain may still be involved in other aspects of class II antigen function. For instance, it may be required for the interaction with processed peptides generated through the exogenous endocytic pathway. Presentation of measles virus antigens may be independent of that pathway, unlike presentation of influenza virus antigens. It is of interest to test the role of the invariant chain in the class II-restricted presentation of influenza virus antigens or other exogenous antigens.

To test whether endogenously synthesized measles virus antigens could be presented by class II antigens to cytotoxic T cells we supertransfected murine L cells expressing DR antigen with measles virus genes. Expressible cDNA clones for measles virus matrix and nucleocapsid proteins were used. These doubly transfected L cells were lysed very efficiently by measles virus-specific DR-restricted T-cell lines.[92] How are these cytoplasmic proteins processed and presented by class II antigens? Several pathways can be considered (Figure 3). Our data rule out the possibility that endogenously synthesized proteins are presented exclusively by class I antigens (pathway B). Cytoplasmic proteins, either in intact form or fragmented, could be brought to the cell surface (pathway A). Membrane recycling and endocytosis could then reintroduce these proteins through a typical exogenous-endocytic

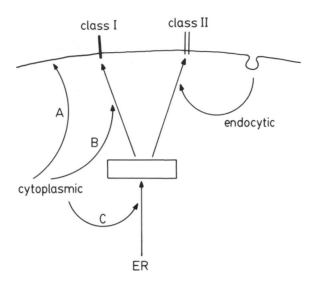

FIGURE 3.   Pathways of antigen presentation. The cell surface with a class I molecule, a class II molecule, and an endocytic vesicle is diagrammed. Straight arrows indicate the biosynthetic pathway of MHC antigens from the endoplasmic reticulum (ER) to a putative compartment (open rectangle) where class I and class II molecules might be directed separately to the cell surface. Curved arrows indicate antigen processing pathways as discussed in the text.

pathway and allow for their interaction with class II antigens. The lack of chloroquine inhibition of measles virus antigen presentation in infected cells argues against this possibility. Alternatively, pathway A could bring processed peptides to the cell surface where they may bind to class II antigens without internalization. Data in favor of cell surface interactions between peptides and class II antigens have been reported.[77,78] A totally distinct possibility is represented by pathway C. Measles virus antigens may be processed in the cytoplasm and the resulting peptides may be targeted to a cellular compartment containing newly synthesized class I and class II antigens. The mechanism of presentation would be very similar to the one involved in the presentation of influenza virus proteins by class I antigens. The different requirements for presentation of influenza virus antigens and measles virus antigens may be related to the biology of these two viruses rather than to distinct processing pathways for presentation by class I and class II antigens.[93] The big challenge for the future is to define at the cellular and biochemical levels where and how antigen processing takes place.

B. INTERACTION WITH THE CD4 MOLECULE

The specificity of an immune response is dictated by the interaction of the T-cell receptor with the foreign antigen bound to an MHC molecule. However, the interactions between antigen-presenting cells and responding T cells are more complex. Several accessory molecules are essential for conjugate formation between T cells and antigen-presenting cells. These molecules are not involved in the specificity of recognition but are nevertheless necessary. Two pathways of conjugate formation have been identified.[79] The CD2 T-cell surface molecule interacts with LFA-3 on the target cell and the LFA-1 molecule on T cells can interact with several molecules on target cells, one of which is ICAM-1.[80] These pathways provide the necessary adhesion forces to allow T-cell and target-cell interaction.

The CD4 and CD8 molecules, expressed on different subsets of T cells, are also accessory molecules for T-cell activation. Their role may be in specifying the class I or class II restriction

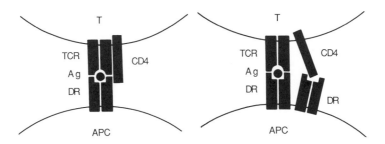

FIGURE 4. Models for the CD4-DR interaction. An antigen-presenting cell (APC) is depicted with surface DR antigen carrying a peptide antigen (Ag, shown as a dot). An antigen-specific CD4-positive responding T-cell is depicted in the two modes of CD4-DR interaction.

of the T cell. Most CD4-positive T cells are class II-restricted, whereas most CD8-positive T cells are class I-restricted. This correlation suggested a direct interaction between CD4 and class II molecules and between CD8 and class I molecules. An important role for CD4 and CD8 molecules in T-cell responses was suggested by antibody inhibition experiments. The interpretation of such experiments is complicated by the fact that antibodies to CD4 can inhibit the T-cell response even in absence of class II molecules.[81,82] Therefore, it was not clear whether anti-CD4 antibodies blocked normal T-cell responses by preventing their interaction with class II molecules on the antigen-presenting cell or by interfering with a class II-independent stimulation pathway.

We took advantage of an unusual murine T-cell hybridoma to obtain evidence for a direct functional interaction between CD4 molecules on the responding T-cell and DR molecules on the stimulating cell.[83] The CD4-DR interaction resulted in an enhanced T-cell response, as measured by IL-2 production. The murine hybridoma 3DT52.5 is specific for the murine class I molecule D^d and also expresses L3T4, the murine CD4 counterpart. This unusual phenotype may be due to the fact that the D^d-specific receptor on 3DT52.5 results from the pairing of the α chain from the thymoma with the β chain from the T-cell fusion partner. An important advantage of this cell is that class II antigen on stimulator cells can be manipulated without affecting the structure recognized by the T-cell receptor. Furthermore, a variant was obtained (3DT52.5.8) that had lost expression of L3T4. 3DT52.5.8 cells with a low level of surface receptor respond weakly to D^d-expressing cells, such as P815. The human CD4 molecule was expressed in 3DT52.5.8 by use of a recombinant retroviral vector. A tenfold increase in IL-2 production by CD4-positive 3DT52.5.8 cells was obtained only when P815 cells expressed DR antigen, suggesting strongly that a direct interaction between CD4 and DR took place. An important conclusion from this study, that cannot be made from similar reports on CD4-DR interactions,[84,85] is that TCR and CD4 need not interact with the same MHC molecule (Figure 4). This conclusion by no means excludes an interaction between TCR and CD4. Recent data have provided evidence in favor of a TCR-CD4 interaction upon specific stimulation by an antigen presented in the context of a class II MHC molecule.[86-88] Such an interaction could explain how a very weak CD4-class II interaction could still provide a strong stimulation to the responding T cell.

An obvious advantage of the cellular assay used here is that mapping of DR determinants involved specifically in the interaction with CD4 is possible, independently from determinants involved in antigen binding and presentation to the TCR. DRα and DRβ chains can be mutated, expressed in a D^d-positive cell and assayed for interaction with CD4 by measuring the enhancement in IL-2 production by the CD4-positive murine hybridoma. Because all the amino terminal domains of DR, CD4, and TCR molecules are members of the Ig superfamily, it will be very interesting to compare how they all interact with each other. Is there essentially one way in which all these molecules can interact or will completely different interactions be revealed that evolved independently after the divergence of Ig family members? Mutated DR molecules that can be assayed independently for their interaction with CD4 or with antigen and TCR will provide answers to that question.

C. PERSPECTIVES

The possibility of expressing class II antigens at the surface of cells by transfection provides a very powerful tool to analyze class II antigen function. Class II-positive cells normally express several class II antigens and are rarely completely homozygous. A single class II $\alpha\beta$ heterodimer can be expressed on a transfected class II-negative cell and the role of individual class II antigens in stimulating T cells can be assessed. Such experiments will finally establish whether the DP, DQ, and DR antigens specifically induce different subsets of T cells. The expression of a single DR antigen on a transfected cell is also very useful in studies of class II-restricted T-cell recognition, because it guarantees a DR-restricted response regardless of what effector cells are used. Furthermore, doubly transfected cells expressing DR antigen and individual viral genes can be used to define which viral antigens elicit class II-restricted responses, and to analyze the pathways of antigen processing and presentation.

The main advantage of expressible class II cDNAs, is the possibility of introducing mutations into the α and β chains. Mutated molecules can be expressed and tested for their ability to function. It will be important to test mutants in several assays in order to distinguish between specific effects and gross structural alterations in the class II molecules. A combination of mutagenesis and of assays for (1) assembly and cell surface expression, (2) interaction with the CD4 molecule, and (3) presentation of specific peptides to class II-restricted T cells will provide a detailed map of functional determinants in class II molecules.

NOTE ADDED IN PROOF

According to the new nomenclature for the HLA system the DZα gene is now named DNA, and the DOβ gene is now named DOB.

ACKNOWLEDGMENTS

We are grateful to Dr. Sanna Goyert for a generous gift of the monoclonal antibody SG465. We thank Dr. Graham Pawelec for communicating results prior to publication and Ginny Shaw for her kind editorial assistance.

REFERENCES

1. **Zinkernagel, R. M. and Doherty, P. C.,** MHC restricted cytotoxic T-cells: studies on the biological role of polymorphic major transplantation antigens determining T-cell restriction specificity, *Adv. Immunol.,* 27, 51, 1979.
2. **Ogasawara, K., Maloy, W. L., and Schwartz, R. H.,** Failure to find holes in the T-cell repertoire, *Nature (London),* 325, 450, 1987.
3. **Klein, J.,** Origin of major histocompatibility complex polymorphism: the *trans-* species hypothesis, *Hum. Immunol.,* 19, 155, 1987.
4. **Andersson, L. Paabo, S., and Rask, L.,** Is allograft rejection a clue to the mechanism promoting MHC polymorphism?, *Immunol. Today,* 8, 206, 1987.
5. **Long, E. O., Wake, C. T., Gorski, J., and Mach, B.,** Complete sequence of an HLA-DR β chain deduced from a cDNA clone and identification of multiple non-allelic DR β chain genes, *EMBO J.,* 2, 389, 1983.

6. **Gustafsson, K., Wiman, K., Emmoth, E., Larhammar, D., Bohme, J., Hyldig-Nielsen, J. J., Ronne, H., Peterson, P. A., and Rask, L.,** Mutations and selection in the generation of class II histocompatibility antigen polymorphism, *EMBO J.,* 3, 1655, 1984.

7. **Kratzin, H., Götz, H., Thinnes, F. P., Kruse, T., Barnikol, H. U., Wernet, P., and Hilschmann, N.,** Structure of human class II antigens expressed by a homozygous lymphoblastoid B cell line, in *HLA Class II Antigens,* Solheim, B. G., Moller, E., and Ferrone, S., Eds., Springer-Verlag, New York, 1986, 49.

8. **Cohen, D., LeGall, I., Marcadet, A., Font, M. P., Lalouel, J. M., and Dausset, J.,** Clusters of HLA class II β restriction fragments describe allelic series, *Proc. Natl. Acad. Sci. U.S.A.,* 81, 7870, 1984.

9. **Bell, J. I., Denney, D., MacMurray, A., Foster, L., Watling, D., and McDevitt, H. O.,** Molecular mapping of class II polymorphisms in the human major histocompatibility complex. I. DRβ, *J. Immunol.,* 139, 562, 1987.

10. **Arden, B. and Klein, J.,** Biochemical comparison of major histocompatibility complex molecules from different subspecies of *Mus musculus:* evidence for *trans*specific evolution of alleles, *Proc. Natl. Acad. Sci. U.S.A.,* 79, 2342, 1982.

11. **Cairns, J. S., Curtsinger, J. M., Dahl, C. A., Freeman, S., Alter, B. J., and Bach, F. H.,** Sequence polymorphism of HLA DRβ1 alleles relating to T-cell-recognized determinants, *Nature (London),* 317, 166, 1985.

12. **Gregersen, P. K., Shen, M., Song, Q. L., Merryman, P., Degar, S., Seki, T., Maccari, J., Goldberg, D., Murphy, H., Schwenzer, J., Wang, C. Y., Winchester, R. J., Nepom, G. T., and Silver, J.,** Molecular diversity of HLA-DR4 haplotypes, *Proc. Natl. Acad. Sci. U.S.A.,* 83, 2642, 1986.

13. **Segall, M., Cairns, J. S., Dahl, C. A., Curtsinger, J. M., Freeman, S., Nelson, P. J., Cohen, O., Wu, S., Nicklas, J. N., Noreen, H. J., Linner, K. M., Saunders, T. L., Choong, S. A., Ohta, N., Reinsmoen, N. L., Alter, B. J., and Bach, F. H.,** DNA and protein studies of HLA class II molecules: their relationship to T cell recognition, *Immunol. Rev.,* 85, 129, 1985.

14. **Mengle-Gaw, L. and McDevitt, H. O.,** Genetics and expression of mouse Ia antigens, *Annu. Rev. Immunol.,* 3, 367, 1985.

15. **Gorski, J. and Mach, B.,** Polymorphism of human Ia antigens: gene conversion between two DRβ loci results in a new HLA-D/DR specificity, *Nature (London),* 322, 67, 1986.

16. **Wu, S., Saunders, T. L., and Bach, F. H.,** Polymorphism of human Ia antigens generated by reciprocal intergenic exchange between two DRβ loci, *Nature (London),* 324, 676, 1986.

17. **Geliebter, J., Zeff, R. A., Melvold, R. W., and Nathenson, S. G.,** Mitotic recombination in germ cells generated two major histocompatibility complex mutant genes shown to be identical by RNA sequence analysis: Kbm9 and Kbm6, *Proc. Natl. Acad. Sci. U.S.A.,* 83, 3371, 1986.

18. **Long, E. O., Gorski, J., and Mach, B.,** Structural relationship of the SB β-chain gene to HLA-D-region genes and murine I-region genes, *Nature (London),* 310, 233, 1984.

19. **Tonnelle, C., DeMars, R., and Long, E. O.,** DOβ: a new β chain gene in HLA-D with a distinct regulation of expression, *EMBO J.,* 4, 2839, 1985.

20. **Hardy, D. A., Bell, J. I., Long, E. O., Lindsten, T., and McDevitt, H. O.,** Pulsed-field gel electrophoresis mapping of the class II region of the human major histocompatibility complex, *Nature (London),* 323, 453, 1986.

21. **Rollini, P., Mach, B., and Gorski, J.,** Linkage map of three HLA-DR β-chain genes: evidence for a recent duplication event, *Proc. Natl. Acad. Sci. U.S.A.,* 82, 7197, 1985.

22. **Andersson, G., Larhammar, D., Widmark, E., Servenius, B., Peterson, P. A., and Rask, L.,** Class II genes of the human major histocompatibility complex. Organization and evolutionary relationship of the DRβ genes, *J. Biol. Chem.,* 262, 8748, 1987.

23. **Berdoz, J., Gorski, J., Termijtelen, A. M., Dayer, J. M., Irle, C., Schendel, D., and Mach, B.,** Constitutive and induced expression of the individual HLA-DR β and α chain loci in different cell types, *J. Immunol.,* 139, 1336, 1987.

24. **Gorski, J., Tosi, R., Strubin, M., Rabourdin-Combe, C., and Mach, B.,** Serological and immunochemical analysis of the products of a single HLA-DRα and DRβ chain gene expressed in a mouse cell line after DNA-mediated cotransformation reveals that the β chain carries a known supertypic specificity, *J. Exp. Med.,* 162, 105, 1985.

25. **Jonsson, A. K., Hyldig-Nielsen, J. J., Servenius, B., Larhammar, D., Andersson, G., Jörgensen, F., Peterson, P. A., and Rask, L.,** Class II genes of the human major histocompatibility complex. Comparisons of the DQ and DX α and β genes, *J. Biol. Chem.,* 262, 8767, 1987.

26. **Auffray, C., Lillie, J. W., Korman, A. J., Boss, J. M., Fréchin, N., Guillemot, F., Cooper, J., Mulligan, R. C., and Strominger, J. L.,** Structure and expression of HLA-DQα and DXα genes: interallelic alternate splicing of the HLA-DQα gene and functional splicing of the HLA-DXα gene using a retroviral vector, *Immunogenetics,* 26, 63, 1987.

27. **Gustafsson, K., Widmark, E., Jonsson, A. K., Servenius, B., Sachs, D. H., Larhammar, D., Rask, L., and Peterson, P. A.,** Class II genes of the human major histocompatibility complex. Evolution of the DP region as deduced from nucleotide sequences of the four genes, *J. Biol. Chem.*, 262, 8778, 1987.

28. **Trowsdale, J. and Kelly, A.,** The human HLA class II α chain gene DZα is distinct from genes in the DP, DQ and DR subregions, *EMBO J.*, 4, 2231, 1985.

29. **Inoko, H., Ando, A., Kimura, M., and Tsuji, K.,** Isolation and characterization of the cDNA clone and genomic clones of a new HLA class II antigen heavy chain DOα, *J. Immunol.*, 135, 2156, 1985.

30. **Servenius, B., Rask, L., and Peterson, P. A.,** Class II genes of the human major histocompatibility complex. The DOβ gene is a divergent member of the class II gene family, *J. Biol. Chem.*, 262, 8759, 1987.

31. **Accolla, R. S.,** Analysis of the structural heterogeneity and polymorphism of human Ia antigens. Four distinct subsets of molecules are coexpressed in the Ia pool of both DR1,1 homozygous and DR3,w6 heterozygous B cell lines, *J. Exp. Med.*, 159, 378, 1983.

32. **Karr, R. W., Alber, C., Goyert, S. M., Silver, J., Duquesnoy, R. J.,** The complexity of HLA-DS molecules. A homozygous cell line expresses multiple HLA-DS alpha chains, *J. Exp. Med.*, 159, 1512, 1984.

33. **Johnson, J. P. and Wank, R.,** Identification of two *cis*-encoded HLA-DQ molecules that carry distinct alloantigenic specificities, *J. Exp. Med.*, 160, 1350, 1984.

34. **Carra, G. and Accolla, R. S.,** Structural analysis of human Ia antigens reveals the existence of a fourth molecular subset distinct from DP, DQ and DR molecules, *J. Exp. Med.*, 165, 47, 1987.

35. **DeMars, R., Chang, C. C., Shaw, S., Reitnauer, P. J., and Sondel, P. M.,** Homozygous deletions that simultaneously eliminate expression of class I and class II antigens of EBV-transformed B-lymphoblastoid cells, *Hum. Immunol.*, 11, 77, 1984.

36. **Long, E. O., Mach, B., and Accolla, R. S.,** Ia-negative B-cell variants reveal a coordinate regulation in the transcription of the HLA class II gene family, *Immunogenetics*, 19, 349, 1984.

37. **Collins, T., Korman, A. J., Wake, C. T., Boss, J. M., Kappes, D. J., Fiers, W., Ault, K. A., Gimbrone, M. A., Strominger, J. L., and Pober, J. S.,** Immune interferon activates multiple class II major histocompatibility complex genes and the associated invariant chain gene in human endothelial cells and dermal fibroblasts, *Proc. Natl. Acad. Sci. U.S.A.*, 81, 4917, 1984.

38. **Spielman, R. S., Lee, J., Bodmer, W. F., Bodmer, J. G., and Trowsdale, J.,** Six HLA-D region α-chain genes on human chromosome 6: polymorphisms and associations of DCα-related sequences with DR types, *Proc. Natl. Acad. Sci. U.S.A.*, 81, 3461, 1984.

39. **Amar, A., Nepom, G. T., Mickelson, E., Erlich, H., and Hansen, J. A.,** HLA-DP and HLA-DO genes in presumptive HLA-identical siblings: structural and functional identification of allelic variation, *J. Immunol.*, 138, 1947, 1987.

40. **Andreadis, A., Gallego, M. E., and Nadal-Ginard, B.,** Generation of protein isoform diversity by alternative splicing: mechanistic and biological implications, *Annu. Rev. Cell. Biol.*, 3, 207, 1987.

41. **Nizetic, d., Figueroa, F., Dembic, Z., Nevo, E., and Klein, J.,** Major histocompatibility complex gene organization in the mole rat Spalax ehrenbergi: evidence for transfer of function between class II genes, *Proc. Natl. Acad. Sci. U.S.A.*, 84, 5828, 1987.

42. **Moretta, A., Accolla, R. S., and Cerottini, J. C.,** IL-2-mediated T cell proliferation in humans is blocked by a monoclonal antibody directed against monomorphic determinants of HLA-DR antigens, *J. Exp. Med.*, 155, 599, 1982.

43. **Corte, G., Moretta, A., Cosulich, M. E., Ramarli, D., and Bargellesi, A.,** A monoclonal anti-DC1 antibody selectively inhibits the generation of effector T cells mediating specific cytolytic activity, *J. Exp. Med.*, 156, 1539, 1982.

44. **Navarrete, C., Jaraquemada, D., Fainboim, L., Karr, R., Hui, K., Awad, J., Bagnara, M., and Festenstein, H.,** Genetic and functional relationships of the HLA-DR and HLA-DQ antigens, *Immunogenetics*, 21, 97, 1985.

45. **Hirayama, K., Matsushita, S., Kikuchi, I., Iuchi, M., Ohta, N., and Sasazuki, T.,** HLA-DQ is epistatic to HLA-DR in controlling the immune response to schistosomal antigen in humans, *Nature (London)*, 327, 426, 1987.

46. **Sanchez-Perez, M. and Shaw, S.,** HLA-DP: current status, in, *HLA Class II Antigens*, Solheim, B. G., Moller, E., and Ferrone, S., Eds., Springer Verlag, New York, 1986, 83.

47. **Radka, S. F., Charron, D. J., and Brodsky, F. M.,** Review: class II molecules of the major histocompatibility complex considered as differentiation markers, *Hum. Immunol.*, 16, 390, 1986.

48. **Palacios, R., Martinez-Maza, O., and Guy, K.,** Monoclonal antibodies against HLA-DR antigens replace T helper cells in activation of B lymphocytes, *Proc. Natl. Acad. Sci. U.S.A.*, 80, 3456, 1983.

49. **Corley, R. B., LoCascio, N. J., Ovnic, M., and Haughton, G.,** Two separate functions of class II (Ia) molecules: T-cell stimulation and B-cell excitation, *Proc. Natl. Acad. Sci. U.S.A.*, 82, 516, 1985.

50. **Chen, Z. Z., McGuire, J. C., Leach, K. L., and Cambier, J. C.,** Transmembrane signaling through B cell MHC class II molecules: anti-Ia antibodies induce protein kinase C translocation to the nuclear fraction, *J. Immunol.,* 138, 2345, 1987.

51. **Bishop, G. A. and Haughton, G.,** Induced differentiation of a transformed clone of Ly1$^+$ B cells by clonal T cells and antigen, *Proc. Natl. Acad. Sci. U.S.A.,* 83, 7410, 1986.

52. **Palacios, R.,** Monoclonal antibodies against human I antigens stimulate monocytes to secrete interleukin 1, *Proc. Natl. Acad. Sci. U.S.A.,* 82, 6652, 1985.

53. **Charron, D., Lotteau, V., and Turmel, P.,** Hybrid HLA-DC antigens provide molecular evidence for gene *trans*-complementation, *Nature (London),* 312, 157, 1984.

54. **Giles, R. C., DeMars, R., Chang, C. D., and Capra, J. D.,** Allelic polymorphism and transassociation of molecules encoded by the HLA-DQ subregion, *Proc. Natl. Acad. Sci. U.S.A.,* 82, 1776, 1985.

55. **Sekaly, R. P., Tonnelle, C., Strubin, M., Mach, B., and Long, E. O.,** Cell surface expression of class II histocompatibility antigens occurs in the absence of the invariant chain, *J. Exp. Med.,* 164, 1490, 1986.

56. **Malissen, B., Shastri, N., Pierres, M., and Hood, L.,** Cotransfer of the E_α^d and A_β^d genes into L cells results in the surface expression of a functional mixed isotype Ia molecule, *Proc. Natl. Acad. Sci. U.S.A.,* 83, 3958, 1986.

57. **Germain, R. N. and Quill, H.,** Unexpected expression of a unique mixed-isotype class II MHC molecule by transfected L cells *Nature (London),* 320, 72, 1986.

58. **Braunstein, N. and Germain, R. N.,** Allele-specific control of Ia molecule surface expression and conformation: implications for a general model of Ia structure-function relationship, *Proc. Natl. Acad. Sci. U.S.A.,* 84, 2921, 1987.

59. **Sant, A. J., Braunstein, N. S., and Germain, R. N.,** Predominant role of amino-terminal sequences in dictating efficiency of class II major histocompatibility complex $\alpha\beta$ dimer expression, *Proc. Natl. Acad. Sci. U.S.A.,* 84, 8065, 1987.

60. **Lotteau, V., Teyton, L., Burroughs, D., and Charron, D.,** A novel HLA class II molecule (DRα-DQβ) created by mismatched isotype pairing, *Nature (London),* 329, 339, 1987.

61. **Pawelec, G., Fernandez, N., Schneider, E. M., Festenstein, H., and Wernet, P.,** DY determinants possibly associated with novel class II molecules stimulate autoreactive CD4$^+$ T cells with suppressive activity, *J. Exp. Med.,* 167, 243, 1989.

62. **Gomard, E., Henin, Y., Sterkers, G., Masset, M., Fauchet, R., and Levy, J. P.,** An influenza A virus-specific and HLA-DRw8-restricted T cell clone cross-reacting with a transcomplementation product of the HLA-DR2 and DR4 haplotypes, *J. Immunol.,* 136, 3961, 1986.

63. **Blobel, G.,** Gene gating: a hypothesis, *Proc. Natl. Acad. Sci. U.S.A.,* 82, 8527, 1985.

64. **de la Pena, P. and Zasloff, M.,** Enhancement of mRNA nuclear transport by promoter elements, *Cell* 50, 613, 1987.

65. **Babbitt, B. P., Allen, P. M., Matsueda, G., Haber, E., and Unanue, E. R.,** Binding of immunogenic peptides to Ia histocompatibility molecules, *Nature (London),* 317, 359, 1985.

66. **Buus, S., Sette, A., Colon, S. M., Miles, C., and Grey, H. M.,** The relation between major histocompatibility complex (MHC) restriction and the capacity of Ia to bind immunogenic peptides, *Science,* 235, 1353, 1987.

67. **Unanue, E. R.,** Antigen-presenting function of the macrophage, *Annu. Rev. Immunol.,* 2, 395, 1984.

68. **Morrison, L. A., Lukacher, A. E., Braciale, V. L., Fan, D. P., and Braciale, T. J.,** Differences in antigen presentation to MHC class I- and class II-restricted influenza virus-specific cytolytic T lymphocyte clones, *J. Exp. Med.,* 163, 903, 1986.

69. **Germain, R. N.,** The ins and outs of antigen processing and presentation, *Nature (London),* 322, 687, 1986.

70. **Braciale, T. J., Morrison, L. A., Sweetser, M. T., Sambrook, J., Gething, M. J., and Braciale, V. L.,** Antigen presentation pathways to class I and class II MHC-restricted T lymphocytes, *Immunol. Rev.,* 98, 95, 1987.

71. **Townsend, A. R. M., Rothbard, J., Gotch, F. M., Bahadur, G., Wraith, D., and McMichael, A. J.,** The epitopes of influenza nucleoprotein recognized by cytotoxic T lymphocytes can be defined with short synthetic peptides, *Cell,* 44, 959, 1986.

72. **Townsend, A. R. M., Bastin, J., Gould, K., and Brownlee, G. G.,** Cytotoxic T lymphocytes recognize influenza hemagglutinin that lacks a signal sequence, *Nature (London),* 324, 575, 1986.

73. **Townsend, A. R. M.,** Recognition of influenza virus proteins by cytotoxic T lymphocytes, *Immunol. Res.,* 6, 80, 1987.

74. **Jacobson, S., Richert, J. R., Biddison, W. E., Satinsky, A., Hartzman, R. J., and McFarland, H. F.,** Measles virus specific T4$^+$ human cytotoxic T-cell clones are restricted by class II HLA antigens, *J. Immunol.,* 133, 754, 1984.

75. **Sekaly, R. P., Jacobson, S., Richert, J. R., Tonnelle, C., McFarland, H. F., and Long, E. O.,** Antigen presentation to HLA class II-restricted measles virus-specific T-cell clones can occur in the absence of the invariant chain, *Proc. Natl. Acad. Sci. U.S.A.,* 85, 1209, 1988.

76. **Miller, J. and Germain, R. N.,** Efficient cell surface expression of class II MHC molecules in the absence of associated invariant chain, *J. Exp. Med.,* 164, 1478, 1986.

77. **Falo, L. D., Benacerraf, B., and Rock, K. L.,** Phospholipase treatment of accessory cells that have been exposed to antigen selectively inhibits antigen-specific Ia-restricted but not allospecific, stimulation of T lymphocytes, *Proc. Natl. Acad. Sci. U.S.A.,* 83, 6994, 1986.

78. **Falo, L. D., Haber, S. I., Herrmann, S., Benacerraf, B., and Rock, K. L.,** Characterization of antigen association with accessory cells: specific removal of processed antigens from the cell surface by phospholipases, *Proc. Natl. Acad. Sci. U.S.A.,* 84, 522, 1987.

79. **Shaw, S., Luce, G. E., Quinones, R., Gress, R. E., Springer, T. A., and Sanders, M. E.,** Two antigen-independent adhesion pathways used by human cytotoxic T-cell clones, *Nature (London),* 323, 262, 1986.

80. **Springer. T. A., Dustin, M. L., Kishimoto, T. K., and Marlin, S. D.,** The lymphocyte function-associated LFA-1, CD2, and LFA-3 molecules: cell adhesion receptors of the immune system, *Annu. Rev. Immunol.,* 5, 223, 1987.

81. **Wassmer, P., Chan, C., Lögdberg, L., and Shevach, E. M.,** Role of the L3T4-antigen in T cell activation. II. Inhibition of T cell activation by monoclonal anti-L3T4 antibodies in the absence of accessory cells, *J. Immunol.,* 135, 2237, 1985.

82. **Moldwin, R. L., Havran, W. L., Nau, G. J., Lancki, D. W., Kim, D. K., and Fitch, F. W.,** Antibodies to the L3T4 and Lyt-2 molecules interfere with antigen receptor-driven activation of cloned murine T cells, *J. Immunol.,* 139, 657, 1987.

83. **Gay, D., Maddon, P., Sekaly, R., Talle, M. A., Godfrey, M., Long, E., Goldstein, G., Chess, L., Axel, R., Kappler, J., and Marrack, P.,** Functional interaction between human T-cell protein CD4 and the major histocompatibility complex HLA-DR antigen, *Nature (London),* 328, 626, 1987.

84. **Sleckman, B. P., Peterson, A., Jones, W. K., Foran, J. A., Greenstein, J. L., Seed, B., and Burakoff, S. J.,** Expression and function of CD4 in a murine T-cell hybridoma, *Nature (London),* 328, 351, 1987.

85. **Doyle, C. and Strominger, J. L.,** Interaction between CD4 and class II MHC molecules mediates cell adhesion, *Nature (London),* 330, 256, 1987.

86. **Fazekas de St. Groth, B., Gallagher, P. F., and Miller, J. F. A. P.,** Involvement of Lyt-2 and L3T4 in activation of hapten-specific Lyt-2$^+$ L3T4$^+$ T-cell clones, *Proc. Natl. Acad. Sci. U.S.A.,* 83, 2594, 1986.

87. **Saizawa, K., Rojo, J., and Janeway, C. A.,** Evidence for a physical association of CD4 and the CD3:α:β T-cell receptor, *Nature (London),* 328, 260, 1987.

88. **Kupfer, A., Singer, S. J., Janeway, C. A., and Swain, S. L.,** Coclustering of CD4 (L3T4) molecule with the T-cell receptor is induced by specific direct interaction of helper T cells and antigen-presenting cells, *Proc. Natl. Acad. Sci. U.S.A.,* 84, 5888, 1987.

89. **Rosen-Bronson, S. and Long, E. O.,** in preparation.

90. **Long, E. O.,** unpublished observation.

91. **Sekaly, R. P. and Long, E. O.,** unpublished observation.

92. **Jacobson, S., Sekaly, R. P., Jacobson, C. L., McFarland, H. F., and Long, E. O.,** HLA class II-restricted presentation of cytoplasmic measles virus antigens to cytotoxic T cells, *J. Virol.,* 63, 1756, 1989.

93. **Long, E. O., and Jacobson, S.,** Pathways of viral antigen processing and presentation to CTL: defined by the mode of virus entry?, *Immunol. Today,* 10, 45, 1989.

Chapter 5

ANALYSIS OF HLA-CLASS II PRODUCTS BY DNA-MEDIATED GENE TRANSFER

David Wilkinson, Daniel M. Altmann, Hitoshi Ikeda, and John Trowsdale

TABLE OF CONTENTS

I. INTRODUCTION

In this review, we describe some of the work being undertaken in our laboratory to study HLA class II glycoproteins by means of DNA-mediated gene transfer. The chapter is broadly divided into three sections. We first highlight some of the questions that can be approached by the transfection of cloned HLA-class II genes into appropriate cells. We then describe the transfection strategies we have adopted in order to generate HLA-class II transfectants suitable for functional studies using human T cells. The final section summarizes the information we have gained on the HLA-DR products of the HLA-DR2Dw2 haplotype by the application of the techniques described.

A. GENETIC ORGANIZATION AND FUNCTION OF THE HLA-D REGION GENES

The HLA-D region encodes major histocompatibility complex (MHC) class II molecules, consisting of heterodimers of an α chain (MW about 34 kDa) and a β chain (MW about 28 kDa).[1-3] The region encompasses a complex array of α and β chain genes including three major expressed loci: DP, DQ, and DR (Figure 1). Each of these loci contains one functional α chain gene and one functional β chain gene, except in the case of DR where the number of β chain genes differs in different haplotypes. DR1 haplotypes contain only one expressed DRβ whereas most other haplotypes contain two such DRβ related sequences.[4-7] In addition to these expressed genes, there are a number of clearly defined pseudogenes which contain defects rendering them nonfunctional. The DPα2 and β2 genes, DVβ and two sequences in the DR locus are in this category.[5,8-10] Finally, there are a number of genes of undefined status, including DZα, DOβ and DXα, and β.[7,11,12] The former pair are expressed as mRNA transcripts, but, so far, no protein products have been identified for any of these genes. It has been documented that, generally, α and β chain products pair within loci (DRα with DRβ, DQα with DQβ, etc.), although there is recent evidence in both human and mouse that interlocus pairing may take place.[13-17] The generality and functional significance of this type of pairing *in vivo* remains to be explored.

The central function of the MHC class II genes is in the presentation of antigens to T lymphocytes. T helper cells recognize antigen in the context of an MHC class II product, this recognition being allele-specific for the class II product, leading to the term MHC restriction. A number of studies in the murine system, examining the binding of antigens to purified class II products, have clarified our understanding of the interaction between T-cell receptor (TCR), antigen and MHC class II product. In many cases at least, the antigen recognized is in the form of a short peptidic fragment, and such peptides can bind directly to a single site on solubilized class II molecules with dissociation constants in the micromolar range.[18,19] The observation that in some instances there is a correlation between the ability of particular MHC class II molecules to bind an antigen and the ability of the mouse strain, from which the class II molecule was purified, to mount an immune response to that antigen may explain some of the immune response (Ir) gene effects in mice in different H-2 types.[20]

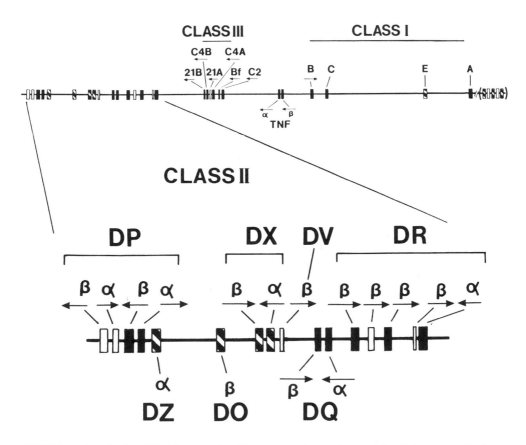

FIGURE 1. Organization of the HLA complex. The upper map is a scale map of the HLA complex. The lower map is an enlarged map of the class II region. Black boxes represent expressed genes; open boxes, pseudogenes; hatched boxes, genes of undefined status. The direction of transcription (5′ → 3′) is designated by an arrow. The number of DRβ genes is dependent upon the haplotype.

Recent studies on the determination of the three-dimensional structure of class I molecules have led to a much clearer understanding of how peptide antigens can interact with them. Although the three-dimensional structure of class II molecules is not yet known in such detail, it can be modeled on the class I structure. In such a model the α1 and β1 domains fold together to form a binding site for antigen. This binding site is probably a trough, the bottom of which is comprised of anti-parallel β-pleated sheets and the walls are α-helices, one from the α and one from the β chain. Again by analogy with the class I structure, most of the polymorphic residues would be expected to point into the trough, explaining the antigen binding difference observed between class II products from different alleles. Knowledge of the structure and sequences of different class II allelic products provides a foundation on which to base experiments analyzing structure-function relationships.

B. DNA-MEDIATED GENE TRANSFER AS A MEANS OF INVESTIGATING THE FUNCTION, BIOCHEMISTRY, AND SEROLOGY OF HLA-CLASS II PRODUCTS

The very complexity of the HLA-D region, in terms of the number of expressed genes and the linkage disequilibrium of some of them, makes functional, biochemical, and serological analysis of the individual gene products an onerous task. The transfection of defined HLA class II genes into cells not normally expressing these products is one means of dissecting this complexity.

For the murine system, a wealth of information exists concerning gene transfection as a tool for analyzing H-2 function.[22] In contrast, relatively little work has been reported on the use of this technique to investigate the function of human HLA class II gene products.[23-25] Although much of the information gained from the mouse studies is relevant to the study of human MHC class II structure and function, certain aspects of the biology of HLA made an independent study of HLA class II products an important endeavor. Specifically, the immunological roles of these products in resistance to human diseases and the basis of HLA/disease associations are questions of both fundamental and clinical relevance. Some of the questions that can be addressed by the transfection of cloned HLA class II genes into HLA class II negative cells are as follows.

1. Molecular Analysis of the Interaction Between The T-Cell Receptor, Antigen, and HLA Class II Product

The use of transfectants expressing a single, defined HLA class II product as antigen presenting cells for human T cells affords the opportunity of studying the molecular interaction between the TCR, antigen and HLA class II product in an unambiguous manner. More specifically, mutagenesis of the class II genes prior to transfection together with the use of substituted peptides as antigens, will allow the contribution of particular class II residues to antigen binding and T cell recognition to be investigated.

2. Functional Analysis of the Products Encoded by the Different HLA-D Subregions

It is clear from a number of studies that HLA-DP, -DQ, and -DR products can all act as restriction elements for human T cells, although for most antigen responses, DR products appear to be immunodominant.[26] The notion that these different products may serve functionally different roles has often been proposed as an explanation for the multiplicity of class II genes. One approach to answering this question is to use transfectants expressing HLA-DP, -DQ, or -DR products alone to selectively expand T-cell populations from peripheral blood and then to analyze their phenotype and antigen specificity. Similarly, the functional significance of haplotype mismatched heterodimers (e.g., DQw1α with DQw3β) and isotype mismatched pairs (e.g., DRα with DQβ) can be examined using transfectants expressing these products.

3. Requirements for Antigen Presentation

Gene transfection into appropriate cell types offers the opportunity of dissecting the requirements for antigen presentation. By sequential gene transfer, the relative roles of accessory molecules and lymphokines in antigen presentation to peripheral blood T cells and T-cell lines and clones can be examined.

4. Characterization and Generation of Monoclonal Antibodies

Transfectants expressing HLA class II products represent ideal reagents for characterizing unequivocally the specificities of existing monoclonal antibodies (mAbs). Additionally, if mouse cells are used as the recipient cells for transfection, the transfectants can be used to immunize mice of the same strain, thereby focusing the immune response of the mouse to the transfected gene products. This technique has already been used successfully to generate an anti-HLA-DP mAb (11.1) specific for DPw2 and DPw4 products (see below).[34]

Having outlined the types of problems that can be approached by transfection of cloned HLA class II genes, the strategies that we have adopted in the application of this technique will be briefly described, followed by a more detailed description of the information we have gained concerning the two expressed DRβ genes of the DR2Dw2 haplotype.

II. TRANSFECTION STRATEGIES

A. CHOICE OF RECIPIENT CELL TYPE FOR TRANSFECTION

The cell type to be used as the recipient for the transfection with HLA class II genes depends upon the experiments envisaged for the resulting transfectants. In studies of mouse class II gene products, the mouse L cell has been the cell type used almost exclusively, because of the ease with which it may be transfected as well as its stable class II negative phenotype. Furthermore, experiments by Austin et al. in this laboratory demonstrated that L cells transfected with HLA-DP genes could function as antigen presenting cells to an HLA-DP restricted, influenza-specific human T-cell clone.[23] For experiments investigating the specificity of human class II restricted T-cell clones, the L cell therefore provides a useful cell type for transfection.

T-cell clones represent a highly selected population of cells resulting from numerous rounds of *in vitro* stimulation and it is conceivable that their requirements for activation differ from those necessary for the generation of primary T-cells responses. Indeed, L cells transfected with mouse class II genes are comparatively poor stimulators in the mixed lymphocyte response (MLR). This is probably a reflection of the critical role of lymphokines and accessory molecule interactions in primary responses. For experiments demanding the use of transfectants as presenting cells to primary T cells, the L cell is consequently not an ideal choice. A variety of mutant human B-cell lines exist which do not express class II products due to deletion or aberrant regulation of their own class II genes.[27] These cell lines can be successfully transfected using electroporation or protoplast fusion and so provide useful recipient cell lines for transfection in those cases where the L cell is unsuitable.

B. CHOICE OF HLA CLASS II DNA CLONES FOR TRANSFECTION

The magnitude of T-helper cell proliferative responses is a function of both the antigen concentrations and the density of MHC class II products on the antigen presenting cell.[28] An important issue concerning the generation of transfectants, therefore, is the level of cell surface expression of the products of the transfected genes. We have adopted a number of approaches to producing transfectants useful for functional studies. These will be illustrated using the production of L cells expressing DQw3 and DR7 as examples.

1. Introduction of Cosmid Clones Containing DQα and DQβ Genes into Mouse L Cells

Two cosmid clones were isolated from a lung carcinoma cosmid library which together contain the genomic sequences for DQα (LC10) and DQβ (LC14) (Figure 2a). Both of these cosmid clones are in the vector pSAE which contains the herpes simplex virus (HSV) thymidine kinase (tk) gene. The two cosmids were digested with the restriction enzyme *NruI* to cleave the thymidine kinase gene and were cotransfected together with *EcoRI*-digested plasmid pOPF into mouse Ltk⁻ cells using the calcium phosphate technique.[29,30] (*EcoRI* digestion of pOPF removes some 5′ promoter sequences from its HSV-thymidine kinase gene, thereby reducing transcription from the gene, necessitating the uptake of several copies of this truncated fragment for expression of the tk⁺ phenotype.) This procedure introduces a primary selection for cells that have stably incorporated relatively large quantities of exogenous DNA. Transfectants were then selected using HMT- (hypoxanthine, methotrexate, and thymidine) containing medium. HMT-resistant colonies were pooled and subjected to several rounds of flow microfluorometric sorting using an anti-HLA-DQ mAb (Tu22) and fluorescein-isothiocyanate conjugated rabbit anti-mouse Ig, to obtain a homogenous cell population expressing cell surface HLA-DQ. Southern-blot analysis of DNA isolated from the transfectants at different stages of cell sorting indicated that this procedure was selecting cells with high numbers of integrated HLA-DQ α and β genes. Northern analysis demon-

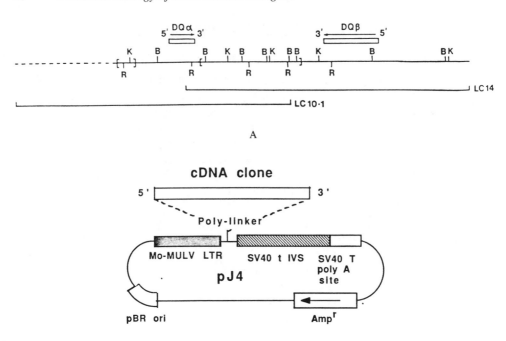

A

B

FIGURE 2.   (A) Schematic maps of cosmids LC10.1 and LC14 used for transfection. Both cosmids were isolated from a lung carcinoma library and together contain an intact DQα (LC10.1) and DQβ (LC14) gene. BamHI (B), EcoRI (E) and KpnI (K) restriction enzyme sites are shown. (B) Schematic map of the cDNA expression vector pJ4. This vector contains the Moloney murine leukemia virus long terminal repeat (MoMLV LTR) as promoter, the SV40 small t intervening sequence (SV40 t IVS) and SV40 large T polyadenylation site (SV40 T poly A site).

strated the presence of α and β transcripts of the expected size, as well as larger mRNAs, presumably resulting from aberrant initiation, splicing, or termination of transcripts.

Analysis of the transfectants using the fluorescence activated cell sorter (FACS) and a panel of mAbs confirmed that the cells were expressing HLA-DQ products as evidenced by their positive reactivity with mAb Tu22 (HLA-DQ specific) and negative reactivity with mAbs B7/21 (DP specific) and DA6.164 (DR specific) (see Figure 3 and Table 1 for mAb references). More specifically, the cells expressed DQ molecules of the DQw3 serotype, e.g., they were positive with the DQw3 specific mAb IVD12 and negative with the DQw1 specific mAb 4.1. Interestingly, the mAb Tu39, previously thought to recognize DP, DQ, and DR products, did not bind to the transfectants, emphasizing the value of transfection as a means of delineating unequivocally the specificities of mAbs.

The mAb 11.4.1 recognizes the L cell's own class I molecules and comparison of the FACS profile obtained with this mAb and those obtained with the anti-HLA class II mAbs allows a quantitative estimation of the level of cell surface expression of the human products. As can be seen from the figure, the FACS was set at a high sensitivity for this analysis, with the profile for mAb 11.4.1 being offscale. The level of expression of HLA-DQ is in fact on the order of one tenth of that of the mouse class I products. Various attempts were made to increase this level, including further cell sorting and single-cell cloning but no significant improvement in expression was observed. The fact that in these cosmid transfectants the DQα and β genes are being transcribed from their own promoters, which would be expected to be weak in cells of non-lymphoid/myeloid origin, is probably at least partly responsible for this low level of expression. These genomic DNA transfectants were useful for mAb characterization, but initial experiments attempting to stimulate T cells were negative. Given the importance for functional studies of achieving high levels of expression of

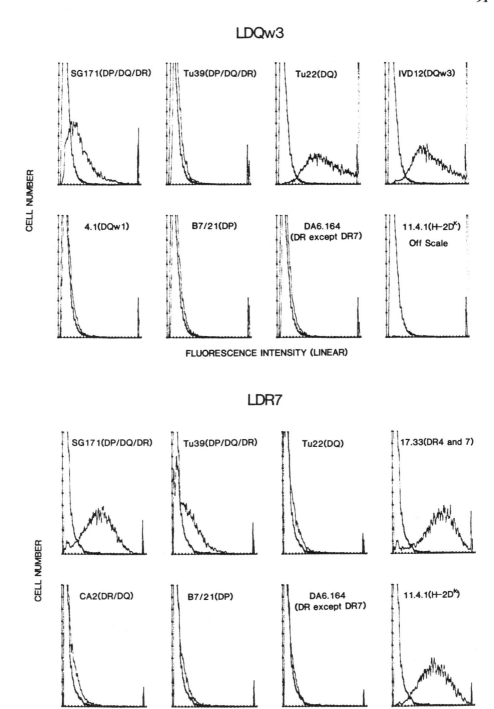

FIGURE 3. Flow microfluorimetric analysis of L cell transfectants expressing HLA-DQw3 (LDQw3) and HLA-DR7 (LDR7). Fluorescence intensity is represented on a linear scale for the LDQw3 transfectants and a logarithmic scale for the LDR7 transfectants. mAb specifities are shown in parentheses.

TABLE 1
Characterization of Monoclonal Antibody Specificities Using a Panel of HLA Class II Transfectants and Flow Microfluorimetric Analysis

		Transfectant cell lines and their specificities									
		DP			DQ	DR					
mAb	Reported specificity	L3.6.2 DPw2	L11.3 DPw4	L8a.5 DPw4	LDQw3 DQw3	LDR2a DR2a	LDR2b DR2b	R/RW-L DR4Dw15	R/3.2-L DR4Dw4	LDR7 DR7	R/MA-L DRw53
B7/21 (32)	DP	+	+	+							
ILR1 (33)	DPw2,3,5	+	−		−	−	−			−	
DP11.1 (34)	DPw2,4	+	+		−						
MHM4 (35)	DP?			−							
Tu22 (36)	DQ				+	−	−			−	
Leu10 (37)	DQ				+	−	−			−	
4.1 (38)	DQw1				−						
Genox 353 (39)	DQw1				−						
IVD12 (40)	DQw3				+						
2HB6 (41)	DQw3				+						
HU-18 (42)	DQw3				+						
TDR31.1 (43)	DR					+	+	+	+	+	+
Hok7 (44)	DR					+	+	+	−		+
Hu-30 (45)	DR1,2					+	+	−			+
DA6.164 (46)	DR except DR7			−	−	+	±	±	±	−	−
17.33B (47)	DR4,7					−	±	±	±	+	
SFR16-DR7M (48)	DR7					−	−			+	
SG171 (49)	Broadly reactive			+	+	+	+			+	
SG465 (50)	Broadly reactive				+	+	+	+	+	−	+
CA2 (51)	Broadly reactive					+	+	+	+	−	+
DA2 (52)	Broadly reactive			+	−					+	+
Tu39 (36)	Broadly reactive			+						+	

the transfected class II products, we decided to experiment with cDNA expression systems for producing transfectants.

2. Isolation of L-Cell Transfectants Expressing High Levels of HLA-DR7 Molecules Using cDNA Clones

The use of full length cDNA clones for transfection studies has the advantage that the sequences of interest can be subcloned into expression vectors containing promoters known to be highly active in the cell type to be used for transfection. We have experimented with a variety of expression vectors for transfection into L cells and have found the vector pJ4 to be particularly useful (Figure 2B). This vector contains the mouse Moloney leukemia virus long terminal repeat (MoMLV LTR) as promoter together with the SV40 small t intervening sequence (t IVS) and large T poly A site. In transient and long-term transfection experiments, the MoMLV LTR promoter is highly active in L cells, significantly more so than the SV40 early promoter. In our initial experiments, a full length DRα cDNA clone was subcloned into the SV40 based expression vector pcEXV3 and a DR7 cDNA clone subcloned into pJ4 and these two constructs cotransfected into L cells together with pSV2neo.[31] Transfectants were selected with G418, subjected to flow microfluorimetric sorting and then analyzed on the FACS using a panel of mAbs (Figure 3). In the FACS analysis shown, the instrument was set at a much lower sensitivity than for the analysis of the LDQw3 transfectants as illustrated by the fact that the profile for the anti-H-2K^k mAb 11.4.1 is on scale. The LDR7 transfectants in fact express tenfold more class II than the LDQw3 transfectant. Even higher levels of cell surface class II expression can be achieved when both transfected cDNA clones are expressed from the vector pJ4 (see below). The LDR7 transfectants displayed the expected pattern of reactivity with most of the mAbs tested, e.g., they bound the mAbs SG171 (DP/DQ/DR), Tu39 (DP/DQ/DR), and 17.33 (DR4 and DR7), but did not bind the mAbs DA6.164 (DR except DR7), Tu22 (DQ), or B7/21 (DP). Surprisingly, the cells also failed to bind the mAb CA2, a mAb previously thought to recognize all DR serotypes. This lack of reactivity with DR7 heterodimers is presumably obscured in DR7 B-cell lines, by the positive reactivity of this mAb with concomitantly expressed DP molecules. This observation emphasizes the caution that must be exerted when using mAbs as reagents in functional studies attempting to define the restriction pattern of human T cells.

3. Summary of Monoclonal Antibody Binding Profiles to HLA Class II Transfectants

Using the transfection strategies described, we have produced a panel of HLA class II expressing L-cell transfectants. A summary of the mAb-binding profiles of some of these transfectants together with the mAb references, is given in Table 1.[32-52] In addition to the specific comments made elsewhere in the text, there are several points to be made from these data. First, three mAbs were clearly DP specific. One of them, DP11.1, binds to the DPα chain but only on DPw2 and DPw4 cells, even though the DPα chain of DPw3 cells is identical in sequence.[68] The ILR1 mAb bound to the DPw2, but not the DPw4 transfectant, in confirmation of its published DPw2, 3 and DR5 specificity. The larger panel of DR transfectants permitted some further dissection of the available reagents. Surprisingly, 17.33B, originally thought to recognize DRw53 because of its reactivity with DR4 and 7 cells, in fact binds to both the DR4 and DR7 molecules and not DRw53.

Having outlined broadly the ultimate goals of, and the techniques involved in the transfection of HLA class II genes, these points will be illustrated more specifically by describing some of our work on the DRβ gene products of the DR2Dw2 haplotype.

```
                    +1

  DR2(Dw2) β a    GDTRPRFLQQDKYECHFFNGTERVRFLHRDIYNQEEDLRFDSDVGEYRAV

  DR2(Dw12)β a    -----------------------------G------NV-----------

  DR2(Dw2) β b    --------W-P-R-------------D-YF-----SV--------F---

  DR2(Dw12)β b    --------W-P-R-------------D-YF-----SV--------F---


                                                                 94

                  TELGRPDAEYWNSQKDFLEDRRAAVDTYCRHNYGVGESFTVQRR

                  -------------------------------------------

                  ----------------I--QA-------------V--------

                  ----------------I--QA---------------------
```

FIGURE 4.   β1 domain amino acid sequences of the cDNA clones DR2β a and DR2β b from the DR2Dw2 haplotype compared with the corresponding sequences from the DR2Dw12 haplotype. Sequences are shown using the one letter amino acid code.

III. SEROLOGICAL, BIOCHEMICAL, AND FUNCTIONAL ANALYSIS OF THE DRβ GENES OF THE HLA-DR2Dw2 HAPLOTYPE

Within haplotypes serologically defined as HLA-DR2, there are a number of cellularly defined subtypes, the most common of which are DR2Dw2 and DR2Dw12. We have used DNA mediated gene transfer to analyze serological, biochemical, and functional aspects of the DRβ genes of the HLA-DR2Dw2 haplotype, an HLA type of considerable interest due to its association with a number of diseases such as Goodpasture's syndrome, multiple sclerosis, narcolepsy, and tuberculoid leprosy.[53-56] There are two expressed DRβ genes in DR2 haplotypes and on the basis of cDNA sequencing and biochemical studies, one appears to vary markedly between the cellularly defined subtypes of DR2 (Dw2 and Dw12) while the other is more conserved.[57,58] These genes have been termed DR2β a (or β_1) and DR2β b (or β_2), respectively. The concomitant expression of these two genes, together with a lack of serological distinction between their encoded products, has prohibited a clear understanding of their relative contribution to HLA function and biochemistry. These genes were transfected into mouse L cells in order to clarify this situation.

A. ISOLATION AND TRANSFECTION OF DR2β a AND DR2β b cDNA CLONES INTO MOUSE L CELLS

Two full-length cDNA clones were isolated from a cDNA libraby made from a DR2Dw2/DR4 B-cell line which correspond exactly in sequence to the published DR2β a and DR2β b sequences for the DR2Dw2 homozygous cell line PGF (Figure 4). Both of these cDNA sequences were subcloned into the expression vector pJ4 and then separately transfected into mouse L cells together with a DRα/pJ4 construct. The transfected cell lines (designated LDR2a and LDR2b, respectively) were then subjected to four rounds of flow microfluorimetric sorting to obtain cells expressing high levels of surface expression (greater than 50% of an EBV transformed B-cell line, as measured by flow microfluorimetry) of the transfected gene products. Southern analysis confirmed the presence of both DR α and β gene sequences in genomic DNA from the transfectants and Northern analysis demonstrated the presence

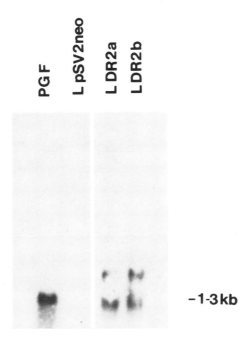

FIGURE 5. Northern analysis of the DR2Dw2 transfectants LDR2a and LDR2b. Total RNA samples isolated from the DR2Dw2 homozygous B-LCL PGF, the mock transfectant LpSV2neo and the DR2 transfectants LDR2a and LDR2b were electrophoresed on a formaldehyde containing agarose gel, blotted onto Hybond N and probed with a full length DRβ cDNA clone.

of high levels of RNA transcripts from both genes as shown for DRβ in Figure 5. The two different transcripts for the α and β genes in the transfectants probably represent polyadenylation from either the endogenous poly A site in the cDNA clones (approximately 1.3 kb) or from the SV40 poly A site in pJ4.

B. FLOW MICROFLUORIMETRIC ANALYSIS OF L-CELL TRANSFECTANTS EXPRESSING PRODUCTS OF THE HLA-DR2β a AND -DR2β b GENES

Analysis of the LDR2a and LDR2b transfectants using flow microfluorimetry and a panel of anti-HLA class II specific mAbs clearly shows them to be expressing HLA class II products (Figure 6). Both transfected cell lines bind the pan-HLA class II mAbs SG465 and CA2 and the DR-specific mAb DA6.164, but do not bind the HLA-DP-specific mAb B7/21 nor the HLA-DQ-specific mAb Leu10. The fact that mAb HU30 bound to both transfectants was surprising in the light of previous immunoprecipitation studies which suggested that this mAb is specific for the DRα/DR2β b heterodimer.[59] Of all the mAbs tested, TDR31.1 was the only mAb which discriminated between the transfectants, binding to both but having a higher affinity for the DRα/DR2β a heterodimer. Interestingly, this mAb has been shown to recognize determinants on the β1 domain.[69] These observations again serve to illustrate the present limitations in the human system of using mAbs as reagents for functional studies. Although many mAbs have been generated which are broadly reactive with HLA class II products, comparatively few allele specific mAbs are available, presumably because the polymorphic regions of human class II molecules are not the most immunogenic in the immunized mouse.

LDR2a

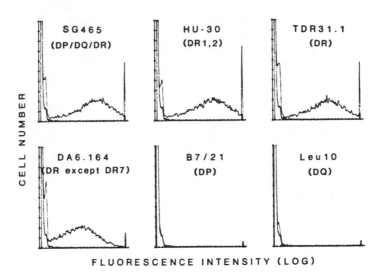

LDR2b

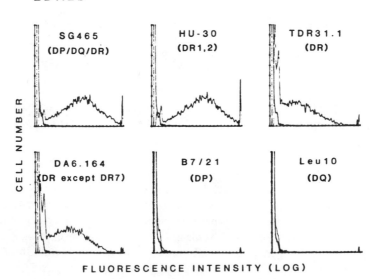

FIGURE 6.   Flow microfluorimetric analysis of the transfectant cell lines LDR2a
and LDR2b (which express cell surface DRα/DR2β a and DRa/DR2β b heterodimers,
respectively) using a panel of anti-HLA class II mAbs. The reported specificities of
the mAbs are shown in parentheses.

C. TWO-DIMENSIONAL PAGE ANALYSIS OF THE DR2β a AND DR2β b CHAINS

Although very similar serologically, the DRα/DR2β a and DRα/DR2β b heterodimers
are clearly distinguishable by two-dimensional polyacrylamide gel electrophoresis (2D-PAGE).
When biosynthetically labeled extracts from the DR2Dw2 homozygous B-cell line PGF are
immunoprecipitated with the DRα-specific mAb TAL 1B5 and analyzed by 2D-PAGE, two
sets of spots corresponding to the coprecipitated β chains DR2β a and DR2β b are clearly

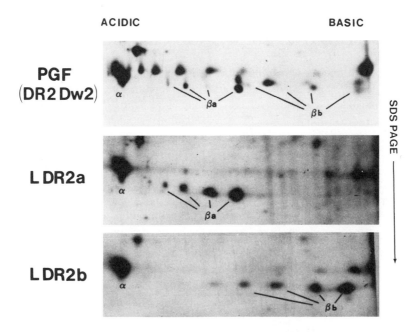

FIGURE 7. Two-dimensional gel analysis of the immunoprecipitated products from the biosynthetically labeled cell lines PGF (homozygous DR2Dw2 B-LCL), LDR2a and LDR2b using the anti-HLA-DRα mAb TAL 1B5. Isoelectric focusing was performed as the first dimension and SDS-PAGE as the second. The coprecipitated α and β spots corresponding to the β chain which is electrophoretically variable among DR2 subtypes are designated βa and those corresponding to the electrophoretically constant β chain designated βb.

seen (Figure 7). Comparison of the electrophoretic mobilities of these chains with those of the equivalent chains immunoprecipitated from a DR2Dw12 B-cell line shows that the more basic β chain has an indistinguishable electrophoretic mobility between Dw2 and Dw12 subtypes, whereas the more acidic chain has a variable mobility (data not shown). Analysis of DRβ cDNA clone sequences obtained from DR2Dw2 and DR2Dw12 cell lines has suggested that the DR2β a gene encodes the electrophoretically variable β chain and the DR2β b gene the electrophoretically constant chain, although both chains have sequence differences between the Dw2 and Dw12 subtypes.[57,58]

When biosynthetically labeled cell extracts from the LDR2a and LDR2b transfectants were immunoprecipitated with the DRα-specific mAb TAL 1B5 and subjected to 2D-PAGE, the DRα chain and a set of spots corresponding to the different glycosylated forms of the β chain were clearly visible (Figure 7). Comparison of the localization of the DRβ chains in PGF and in the transfectants unequivocally shows that of the β chains present in PGF, the more acidic, variable β chain is the only β chain present in the LDR2a transfectant and the more constant β chain the only one present in the LDR2b transfectant. The variable β chain can thus formally be defined as the product of the DR2β a sequence and the more constant chain as that of the DR2β b sequence. Significantly, the individual β chains in the transfectants have electrophoretic mobilities indistinguishable from the β chains in the human B cell, suggesting that there are no gross differences in post-translational modification of the human proteins when expressed in human and mouse cells. This observation is reassuring when using the transfectants for functional studies as it is conceivable that the pattern of glycosylation could affect recognition of HLA class II glycoproteins by mAbs and T cells.

D. FUNCTIONAL ANALYSIS OF THE LDR2a AND LDR2b TRANSFECTANTS USING T-CELL CLONES ISOLATED FROM HLA-DR2 LEPROSY PATIENTS

Leprosy is a disease in which infection of different individuals with the same bacillus leads to a wide range of clinical phenotypes, from lepromatous leprosy where there is a state of immunological unresponsiveness to tuberculoid leprosy, where there is a state of immunological hyper-responsiveness.[60] The observation that the type of pathology developed is at least partly dependent upon the HLA type of the individual makes leprosy an exciting model for looking at human Ir gene effects. The DR2 (and DR3) allele is over represented in tuberculoid leprosy patients and substantial progress has been made in the analysis of epitopes on the mycobacterium recognized by DR2-restricted T-cell clones.[61] However, precise mapping of the interaction between peptide, HLA class II molecule, and T-cell receptor has been hampered by the inability to define unambiguously which DR2β chain is being used by the *M.leprae*-specific T cells. Given this scenario, we were interested to see whether our DR2 transfectants could present antigen to human T cells and thereby define which DR2β chain is recognized by individual T-cell clones from HLA-DR2 leprosy patients.

1. Definition of the Restriction Element Used by *M.leprae*-Specific T Cell Clones from HLA-DR2 Leprosy Patients

T-cell clone 2F10, derived from a tuberculoid leprosy patient, is specific for a peptide, p17, derived from the sequence of the *M.leprae* 65-kDa antigen, and has been shown from mapping with homozygous B-cell lines and antibody blocking to be DR2 restricted.[62] The clone responds to antigen when presented by the autologous B-cell line REIZ (DR2Dw2, DR3), the cDNA donor B-cell line ROF (DR2Dw2, DR4Dw4) and the DR2Dw12 homozygous B-cell line TOK (data not shown). Given that from sequence data, both the electrophoretically variable DR2β a and electrophoretically constant DR2β b chains have amino acid sequence differences between the Dw2 and Dw12 subtypes (Figure 4) and that none of the mAbs we have tested discriminate clearly between these chains, it is impossible to say from these data which DR2β chain clone 2F10 is restricted by.

When the two transfectants were tested for their ability to present p17 to clone 2F10, LDR2a was shown to present antigen efficiently (Figure 8). The LDR2b transfectant, despite expressing a similar level of surface DR, could not present antigen to 2F10, generating a response as low as the mock transfectant. This inability to present was not overcome by increasing the peptide concentration tenfold to 0.1 μg/ml. Thus, DR2 transfected mouse L cells can present peptide to human T cells and clone 2F10 is restricted by the DRα/DR2β a heterodimer. This result is interesting given that both DR2Dw2 and DR2Dw12 cell lines present antigen to 2F10 and the DR2β a chain is the more variable of the two expressed DRβ chains between these subtypes. Comparison of β1 domain sequences of DRβ alleles shows that there are essentially three areas of hypervariability, these being in the regions of amino acid residues 4 to 14, 25 to 33 and 67 to 74. The Dw2β a and Dw12β b chains have similar sequences in the first and third hypervariable regions but differ markedly in the second (Figure 4), suggesting that the specific amino acid sequence of the second hypervariable region is not critically important to either binding of p17 or recognition by the T-cell receptor of 2F10. In this connection it is noteworthy that the DR2 B-cell line AZH, which has undergone a recombination event such that its β a chain is identical to the Dw2/Dw12β b chain in the third hypervariable region, is unable to present peptide to this clone, emphasizing the importance of this region of the β chain in presentation.[70]

Of two other *M.leprae*-specific clones which are known to proliferate in response to antigen presented by DR2 cell lines, both utilized the DR2β a chain when screened on the transfectants. Similarly, of four DR2-restricted T-cell clones specific for malarial antigens we have tested, all were restricted by the DRα/DR2β a heterodimer. Although T-cell clones

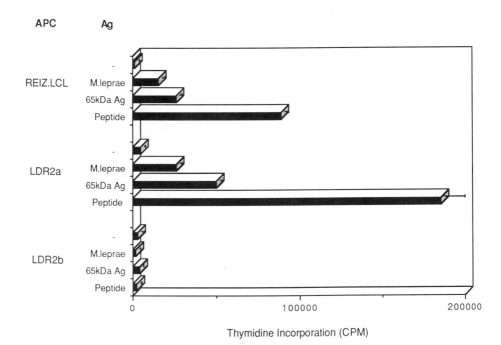

APC   Ag

FIGURE 8.   Antigen presentation by the LDR2a transfectant to the *M.leprae*-specific, HLA-DR2 restricted human T-cell clone 2F10. Cloned T cells (1×10^4) were added to the autologous B-LCL REIZ, or the transfectants LDR2a and LDR2b as a source of APCs (5×10^4 in each case). Cells were cultured in the absence of antigen or with *M.leprae* (20 μg/ml), recombinant 65 kDa mycobacterial antigen (10 μg/ml) or synthetic peptide (0.1 μg/ml). The bars shown mean ^{3}H-thymidine incorporation at 72 h ($\pm$ S.E.M.).

represent a highly selected population of T cells and one must be wary when attempting to relate the restriction patterns and specificities of clones back to peripheral blood lymphocytes, these observations strongly suggest that the DRα/DR2β a heterodimer is the dominant restriction element used by DR2-restricted T cells. The role of the DR2β b chain in DR2 restriction may thus be analogous to the DRw52 and DRw53 supertypic specificities of other HLA-DR haplotypes, which probably as a consequence of their low cell surface expression compared to the more variable DRβ chain products, rarely function as T-cell restricting elements.

2. Immunological Features of Antigen Presentation by L-Cell Transfectants

Having clearly demonstrated that the LDR2a transfectants are capable of presenting antigen to a number of T-cell clones, we were curious to assess more fully the antigen presentation function of these cells, particularly with regard to the role of accessory molecules and antigen processing.

Antigen presentation to clone 2F10 by LDR2a and the autologous B-cell line (REIZ) were compared with respect to susceptibility to mAb blocking (Figure 9). The first point to be made is that presentation by LDR2a is a bona fide class II-restricted response since the class II framework mAb CA2 almost totally blocked the response, whereas mAbs with irrelevant specificites such as W6/32 (anti-HLA-class I) and Genox 3/53 (anti-HLA-DQ) had no significant inhibitory effects.

A variety of cell surface molecules (e.g., CD4 and LFA-1) have been implicated as accessory molecules in antigen presentation, increasing the avidity of interaction between the responding T cell and the antigen-presenting cell in an antigen-independent fashion. MHC class II molecules have been proposed as the ligand for CD4 on the T cell, and as HLA-DR molecules are the only human products expressed by the mouse cell transfectant,

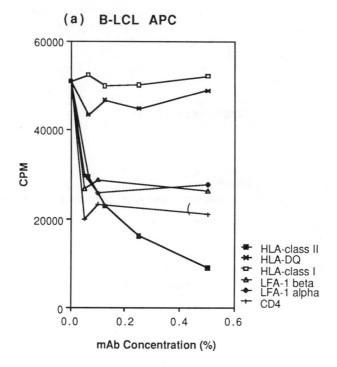

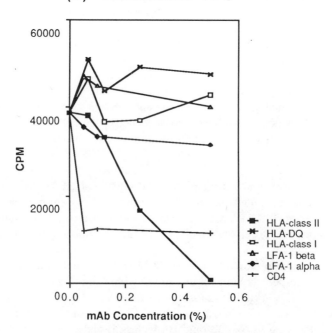

FIGURE 9. Antibody blocking of antigen presentation to T cell clone 2F10. The autologous B-LCL (REIZ) (panel a) or the LDR2a transfectant (panel b) were tested for presentation of peptide p17 (added to cultures at 0.01 μg/ml) in the presence of mAbs at the indicated concentrations. Results are given as ΔCPM, that is ³H thymidine incorporation in the presence of peptide minus incorporation by T cells plus APC in the absence of peptide.

we were interested to see whether interaction between CD4 and the transfected HLA-DR product was involved in the antigen presentation by LDR2a. As can be seen in Figure 9, the anti-CD4 mAb Leu3a inhibited antigen presentation by both the autologous B-cell line and the LDR2a transfectant by approximately 70%, suggesting that the transfected HLA-DR product is indeed the ligand for CD4.[63] In contrast, while presentation to 2F10 by the autologous B-cell line was significantly inhibited by mAbs against CD11a (LFA-1α) or CD18 (LFA-1β), these mAbs had no effect on presentation by the transfectants.[64] This is in line with the notion that the function of LFA-1 in T cell/antigen-presenting cell interactions represents a nonobligatory means of enhancing intercellular avidity and that the ligand for LFA-1, probably ICAM-1, is not expressed on the surface of the L cell.[65] These observations corroborate those made in studies of presentation by mouse class II transfectants to mouse T-cell clones. Curiously, some *M.leprae*-specific T-cell clones which, from panel studies on HLA-typed presenting cells, were expected to be restricted by one of the two DR products of the DR2Dw2 haplotype failed to proliferate to antigen when presented by either LDR2a or LDR2b. It is conceivable that this observation may be a reflection of heterogeneity among human T cells in their dependence upon accessory cell function. We are currently investigating this notion by introducing ICAM-1 into the transfected cell lines as a means of examining the requirements for T-cell activation in different T cells.

Experiments using L cells transfected with mouse MHC class II genes as antigen-presenting cells to murine T cells have not produced a clear consensus as to whether L cells can serve all the normal antigen-processing requirements of a conventional antigen-presenting cell. Some studies using whole protein antigens, e.g., keyhole limpet hemocyanin, have suggested that L cells can indeed process antigen normally while others using different antigens, e.g., hen eggwhite lysozyme, demonstrated that the L cells could not present the whole antigen but only synthetic peptides derived from its sequence.[66,67] Given these apparently contradictory observations, we investigated the ability of the LDR2a transfectants to present nonpeptidic antigens to clone 2F10. As can be seen in Figure 8, these transfectants were clearly capable of presenting both the 65 kDa recombinant *M.leprae* antigen and the whole mycobacterium to clone 2F10. Given that the antigen recognized by this clone has been defined as a peptidic fragment of the 65 kDa antigen, these data suggest that the L-cell transfectants are capable of processing the whole antigen. It is possible that most cell types have an antigen-processing capability, but subtle differences exist in the enzymatic degradation steps involved in particular cell types. If this is so, the seemingly contradictory observation that one cell type can process some but not other antigens could be explained if the particular enzymatic processing steps in that cell type led to a selective exposure or degradation of potential T-cell epitopes in the whole antigen.

IV. CONCLUDING REMARKS

We have used DNA-mediated gene transfer to produce transfected cell lines expressing single defined HLA class II products in order to address questions concerning the serology, biochemistry, and function of these cell surface glycoproteins. Such transfectants allow the unequivocal assignment of mAb and biochemical specificities to particular HLA class II products in lymphoid/myeloid cells. Using cDNA expression systems, we have produced L-cell transfectants expressing high levels of HLA class II glycoproteins which are functionally capable of presenting peptidic and native antigens to human T-cell clones. By producing transfectants which individually express the products of the two DRβ genes of the DR2Dw2 haplotype, we have been able to examine the properties of these chains in isolation. Despite the very similar patterns of reactivity of DRα/DR2β a and DRα/DR2β b heterodimers with existing mAbs, these products can clearly be distinguished biochemically and functionally. The DR2β a gene encodes the more acidic β chain which varies in electrophoretic mobility between the cellularly defined subtypes of DR2 while the DR2β b gene encodes the electrophoretically constant chain. From our studies using T-cell clones, the DRα/DR2β a heterodimer appears to be the dominant restriction element used by DR2-restricted T cells, whereas the DRα/DR2β b heterodimer is probably equivalent to the DRw52 and DRw53 supertypic specificities found in other haplotypes. Having established a system whereby transfectants expressing defined HLA class II products can present peptidic antigens to human T-cell clones, we are now in a position to map in fine detail the trimolecular interactions between HLA class II, antigen, and the T-cell receptor. This is being carried out by considering the effects of amino acid substitutions in the peptide antigen upon binding to the class II product and recognition by the T-cell receptor. Ultimately, given the recent publication of the three-dimensional structure of HLA-A2, and its implications for HLA class II structure, the precise role of individual class II amino acid residues in interactions with the antigen and T-cell receptor can be evaluated using site-directed mutagenesis.

ACKNOWLEDGMENTS

We are indebted to Rene De Vries, Alejandro Madrigal, Chris Lock, Jay Morgenstern, Jill Maddox, Judy Heyes, and Pat Miller for their invaluable contribution to the work presented here. J. G. Bodmer, A. Ziegler, S. F. Radka, S. M. Goyert, M. Aizawa, A. J. McMichael, and others generously donated mAbs and Kim Richardson provided excellent secretarial assistance.

REFERENCES

1. **Klein, J.,** *Natural History of the Major Histocompatibility Complex,* John Wiley & Sons, New York, 1986.
2. **Korman, A. J., Boss, J. M., Spies, T., Sorrentino, R., Okada, K., and Strominger, J. L.,** Genetic complexity and expression of human class II histocompatibility antigens, *Immunol. Rev.,* 85, 45, 1986.
3. **Trowsdale, J., Young, J. A. T., Kelly, A. P., Austin, P. J., Carson, S., Meunier, H., So, A., Erlich, H. A., Spielman, R. S., Bodmer, J., and Bodmer, W. F.,** Structure, sequence and polymorphism in the HLA-D region, *Immunol. Rev.,* 85, 4, 1985.
4. **Bohme, J., Andersson, M., Andersson, G., Möller, E., Peterson, P. A., and Rask, L. J.,** *Immunology,* 135, 2149, 1985.
5. **Spies, T., Sorrentino, R., Boss, J. M., Okada, K., and Strominger, J. L.,** Structural organization of the DR subregion of the human major histocompatibility complex, *Proc. Natl. Acad. Sci. U.S.A.,* 82, 5165, 1985.
6. **Rollini, P., Mach, B., and Gorski, J.,** Linkage map of three HLA-DRβ chain genes: evidence for a recent duplication event, *Proc. Natl. Acad. Sci. U.S.A.,* 82, 7197, 1985.
7. **Tonnelle, C., de Mars, R. and Long, E. O.,** DOβ: a new β chain gene in HLA-D with a distinct regulation of expression, *EMBO J.,* 4, 2839, 1985.
8. **Servenius, B., Gustafsson, K., Widmark, E., Emmoth, E., Andersson, G., Larhammar, D., Rask, L., and Peterson, P. A.,** Molecular map of the human HLA-SB (HLA-DP) region and sequence of an SBα (DPα) pseudogene, *EMBO J.,* 3, 3209, 1984.
9. **Ando, A., Inoko, H., Nakatsuji, T., Sato, T., Awataguchi, S., and Tsuji, K.,** Isolation and characterization of genomic clones for new HLA class II antigen light chains, in *HLA in Asia-Oceania 1986,* Aizawa, M., Ed., Hokkaido University Press, Sapporo, Japan, 1986, 859.
10. **Meunier, H. F., Carson, S., Bodmer, W. F., and Trowsdale, J.,** An isolated β exon next to the DRα gene in the HLA-D region, *Immunogenetics,* 23, 172, 1986.
11. **Trowsdale, J. and Kelly, A.,** The human HLA class II α chain gene DZα is distinct from genes in the DP, DQ and DR subregions, *EMBO J.,* 4, 2231, 1985.
12. **Auffray, C., Lillie, J. W., Arnot, D., Grossberger, D., Kappes, D., and Strominger, J. L.,** Isotypic and allotypic variation of the class II human histocompatibility antigen α chain genes. *Nature (London),* 308, 327, 1984.
13. **Charron, D. J., Lotteau, V., and Turmel, P.,** Hybrid HLA-DC antigens provide molecular evidence for gene *trans*-complementation. *Nature (London),* 312, 157, 1984.
14. **Lotteau, V., Teyton, L., Burroughs, D., and Charron, D.,** A novel HLA class II molecule (DRα-DQβ) created by mismatched isotype pairing, *Nature (London),* 329, 339, 1987.

15. **Germain, R. N. and Malissen, B.**, Analysis of the expression and function of class II major histocompatibility complex-encoded molecules by DNA-mediated gene transfer, *Annu. Rev. Immunol.*, 4, 281, 1986.

16. **Germain, R. N. and Quill, H.**, Unexpected expression of a unique mixed-isotype class II MHC molecule by transfected L-cells, *Nature (London)*, 320, 72, 1986.

17. **Malissen, B., Shastri, N., Pierres, M. and Hood, L.**, Cotransfer of the E_α^d and A_β^d genes into L cells results in the surface expression of a functional mixed-isotype Ia molecule, *Proc. Natl. Acad. Sci. U.S.A.*, 83, 3958, 1986.

18. **Shimonkevitz, R. J., Kappler, J. W., Marrack, P., and Grey, H.**, Antigen recognition by H-2 restricted T cells. I. Cell free antigen processing, *J. Exp. Med.*, 158, 303, 1983.

19. **Babbitt, B. P., Allen, P. M., Matsueda, O., Haber, E., and Unanue, E. R.**, Binding of immunogenic peptides to Ia histocompatibility molecules, *Nature (London)*, 317, 359, 1985.

20. **Buus, S., Sette, A., Colon, S. M., Miles, C., and Grey, H. M.**, The relationship between major histocompatibility complex (MHC) restriction and the capacity of Ia to bind immunogenic peptides, *Science*, 235, 1353, 1987.

21. **Bjorkman, P. J., Saper, M. A., Samraoui, B., Bennett, W. S., Strominger, J. L., and Wiley, D. C.**, Structure of the human class I histocompatibility antigen, HLA-A2, *Nature (London)*, 329, 506, 1987.

22. **Germain, R. N. and Malissen, B.**, Analysis of the expression and function of the class II major histocompatibility complex-encoded molecules by DNA-mediated gene transfer, *Annu. Rev. Immunol.*, 4, 281, 1986.

23. **Austin, P., Trowsdale, J., Rudd, C., Bodmer, W. F., Feldmann, M., and Lamb, J.**, Functional expression of HLA-DP genes transfected into mouse fibroblasts, *Nature (London)*, 313, 61, 1985.

24. **Moen, T., de Preval, C., Rabourdin-Combe, C., Mach, B., Gaudernack, G., Bondervik, E., and Thorsby, E.**, Mouse L cells expressing human HLA-DR antigens after transfection with class II genes do not stimulate human T lymphocytes, in *Histocompatibility Testing 1984*, Albert, E. D., et al., Eds., Springer-Verlag, New York, 1984, 595.

25. **Wilkinson, D., de Vries, R. R. P., Madrigal, J. A., Lock, C. B., Morgenstern, J. P., Trowsdale, J., and Altmann, D. M.**, Analysis of HLA-DR glycoproteins by DNA-mediated gene transfer: definition of DR2β gene products and antigen presentation to T cell clones from leprosy patients, *J. Exp. Med.*, 167, 1442, 1988.

26. **Qvigstad, E., Moen, T., and Thorsby, E.**, T cell clones with similar antigen specificity may be restricted by DR, MT (DC) or SB class II HLA molecules, *Immunogenetics* 19, 455, 1984.

27. **Cresswell, P.**, Regulation of HLA class I and class II antigen expression, *Br. Med. Bull.*, 43, 66, 1987.

28. **Matis, L. A., Glimcher, L. H., Paul, W. E., and Schwartz, R. H.**, Magnitude of response of histocompatibility-restricted T cell clones is a function of the product of the concentrations of antigens and Ia molecules, *Proc. Natl. Acad. Sci. U.S.A.*, 80, 6019, 1983.

29. **Grosveld, F. G., Lund, T., Murray, E. J., Mellor, A. L., Dahl, H. M., and Flavell, R. A.**, The construction of cosmid libraries which can be used to transform eukaryotic cells, *Nucleic Acids Res.*, 10, 6715, 1982.

30. **Wigler, M., Pellicer, A., Silverstein, S., and Axel, R.**, Biochemical transfer of single copy eukaryotic genes using total cellular DNA as donor, *Cell*, 14, 725, 1978.

31. **Young, J. A. T., Wilkinson, D., Bodmer, W. F., and Trowsdale, J.**, Sequence and evolution of HLA-DR7 and -DRw53 associated β-chain genes, *Proc. Natl. Acad. Sci. U.S.A.*, 84, 4929, 1987.

32. **Watson, A. J., DeMars, R., Trowbridge, I. S., and Bach, F. H.**, Detection of a novel human class II HLA antigen, *Nature (London)*, 304, 358, 1983.

33. **Nadler, L. M., Stashenko, P., Hardy, R., Tomaselli, K. J., Yunis, E. J., Schlossman, S. F., and Pesando, J. M.**, Monoclonal antibody identifies a new I-A like polymorphic system linked to the HLA-D/DR region, *Nature (London)*, 290, 591, 1981.

34. **Heyes, J., Austin, P., Bodmer, J., Madrigal, A., Mazzilli, M. C., and Trowsdale, J.**, Monoclonal antibodies to HLA-DP transfected mouse L cells, *Proc. Natl. Acad. Sci. U.S.A.*, 83, 3417, 1986.

35. **Makgoba, M. W., Hildreth, J. E. K., and McMichael, A. J.**, Identification of a human I-A antigen that is different from HLA-DR and DC antigens, *Immunogenetics*, 17, 623, 1983.

36. **Pawelec, G. P., Shaw, S., Ziegler, A., Muller, C., and Wernet, P.**, Differential inhibition of HLA-D-directed or SB-directed lympho-proliferative responses with monoclonal antibodies detecting human I-A-like determinants, *J. Immunol.*, 129, 1070, 1982.

37. **Becton Dickinson Ltd.**, Mountain View, California.

38. **Bodmer, J., Heyes, J., and Lindsay, J.**, Study of monoclonal antibodies to the HLA-D region products DQw1 and DRw52, in *Histocompatibility Testing 1984*, Springer-Verlag, New York, 1984, 432.

39. **Brodsky, F. M., Parham, P., and Bodmer, W. F.**, Monoclonal antibodies to HLA-DRw determinants, *Tissue Antigens*, 16, 30, 1980.

40. **Giles, R. C., Nunez, G., Hurley, C. K., Nunezroldan, A., Winchester, R., Stastny, P., and Capra, J. D.**, Structural-analysis of a human I-A homolog using a monoclonal-antibody that recognizes an MB3-like specificity, *J. Exp. Med.*, 157, 1461, 1983.

41. **Shannon, A. D., Rudd, C. E., Bodmer, J.G., Bodmer, W. F., and Crumpton, M. J.,** Characterization of the HLA-D region DQw3 specificity using the monoclonal antibodies 2HB6 and IVD12, in *Histocompatibility Testing 1984*, Springer-Verlag, New York, 1984, 217.

42. **Kasahara, M., Takenouchi, T., Ogasaware, K., Ikeda, H., Okuyama, T., Ishikawa, N., Moriuchi, J., Wakisaka, A., Kikuchi, Y., Aizawa, M., Kaneto, T., Kashiwagi, N., Nishimura, Y., and Sasazuki, T.,** Serologic dissection of HLA-D specificities by the use of monoclonal antibodies, *Immunogenetics*, 17, 485, 1983.

43. **de Kretser, T. A., Crumpton, M. J., Bodmer, J. G., and Bodmer, W. F.,** Demonstration of two distinct light chains in HLA-DR associated antigens by two-dimensional gel electrophoresis, *Eur. J. Immunol.*, 12, 214, 1982.

44. **Kasahara, M., Ikeda, H., Ogasawara, K., Ishikawa, N., Okuyama, T., Fukasawa, Y., Kojima, H., Kumikane, H., Hawkin, S., Ohashi, T., Natori, T., Wakisaka, A., Kikuchi, Y., and Aizawa, M.,** Inhibition of autologous mixed lymphocyte reaction by monoclonal antibodies specific for the β chain of HLA-DR antigens, *Immunology*, 53, 79, 1984.

45. **Kasahara, M., Takenouchi, T., Ogasawara, K., Ikeda, H., Okuyama, T., Ishikawa, N., Wakisaka, A., Kikuchi, Y., and Aizawa, M.,** A monoclonal antibody that detects a polymorphic determinant common to HLA-DR1 and 2, *Tissue Antigens*, 21, 105, 1983.

46. **van Heyningen, v., Guy, K., Newman, R., and Steel, C. M.,** Human MHC class II molecules as differentiation markers, *Immunogenetics*, 16, 459, 1982.

47. **Ozato, K., Mayer, N., and Sachs, D. H.,** Hybridoma cell lines secreting monoclonal antibodies to mouse H-2 and I-A antigens, *J. Immunol.*, 124, 533, 1980.

48. **Radka, S. F., Amos, D. B., Quackenbush, L. J., and Cresswell, P.,** HLA-DR7-specific monoclonal antibodies and a chimpanzee anti-DR7 serum detect different epitopes on the same molecules, *Immunogenetics*, 19, 63, 1984.

49. **Goyert, S. M. and Silver, J.,** Isolation of I-A subregion-like molecules from subhuman primates and man, *Nature (London)*, 294, 266, 1981.

50. **Goyert, S. M. and Silver, J.,** Further characterization of HLA-DS molecules — implications for studies assessing the role of human Ia molecules in cell-interactions and disease susceptibility, *Proc. Natl. Acad. Sci. U.S.A.*, 80, 5719, 1983.

51. **Eckels, D. D., Woody, J. N., and Hartzman, R. J.,** Monoclonal and xenoantibodies specific for HLA-DR inhibit primary responses to HLA-D but fail to inhibit secondary proliferative (PLT) responses to allogeneic cells, *Hum. Immunol.*, 3, 133, 1981.

52. **Brodsky, F. M., Parham, P., Barnstable, C. J., Crumpton, M. J., and Bodmer, W. F.,** Monoclonal antibodies for analysis of the HLA system, *Immunol. Rev.*, 47, 3, 1979.

53. **Rees, A. J., Peters, D. K., Compston, D. A. S., and Batchelor, J. R.,** Strong association between HLA-DRw2 and antibody mediated Goodpasture's syndrome, *Lancet*, 1, 966, 1978.

54. **Tiwari, J. L. and Terasaki, P. I.,** Neurology, in *HLA and Disease Associations*, Springer-Verlag, New York, 1985, chap. 8.

55. **Langdon, N., Welsh, K., van Dam, M., Vaughan, R. M., and Parkes, J. D.,** Genetic markers in narcolepsy, *Lancet*, 2, 1178, 1984.

56. **van Eden, W., de Vries, R. R. P., Mehra, N. K., Vaidya, M. C., D'Armaro, J., and van Rood, J. J.,** HLA segregation of tuberculoid leprosy: confirmation of the DR2 marker, *J. Infect. Dis.*, 141, 693, 1985.

57. **Wu, S., Yabe, T., Madden, M., Saunders, T. L., and Bach, F. H.,** cDNA cloning and sequencing shows that the electrophoretically constant DRβ2 molecules, as well as the variable DRβ1 molecules from HLA-DR2 subtypes have different amino acid sequences including a hypervariable regions for a functionally important epitope, *J. Immunol.*, 138, 2953, 1987.

58. **Lee, B. S. M., Rust, N. A., McMichael, A. J., and McDevitt, H. O.,** HLA-DR2 subytpes form an additional supertypic family of DRβ alleles, *Proc. Natl. Acad. Sci. U.S.A.*, 84, 4591, 1987.

59. **Takenouchi, T., Kasahara, M., Ogasawara, K., Ikeda, H., Ishikawa, N., Hawkin, S., Waksiaka, A., and Aizawa, M.,** Electrophoretic analysis of HLA-DR2 molecules isolated from HLA-Dw2 and HLA-Dw12 cell lines, *J. Immunogenet.*, 12, 65, 1985.

60. **Bloom, B. R. and Godal, T.,** Selective primary health care: strategies for control of disease in the developing world, *Rev. Infect. Dis.*, 5, 765, 1983.

61. **Ottenhoff, T. H. M., Klatser, P. R., Ivanyi, J., Elferink, D. G., de Wit, M. Y. L., and de Vries, R. R. P.,** Mycobacterium leprae specific protein antigens defined by cloned human helper cells, *Nature (London)*, 319, 66, 1986.

62. **Ottenhoff, T. H. M., Neuteboom, S., Elferink, D. G., and de Vries, R. R. P.,** Molecular localization and polymorphism of HLA class II restriction determinants defined by *M.leprae*-reactive helper T cell clones from leprosy patients, *J. Exp. Med.*, 164, 1923, 1986.

63. **Doyle, C. and Strominger, J. L.,** Interaction between CD4 and class II MHC molecules mediates cell adhesion, *Nature (London)*, 330, 256, 1987.

64. **Hildreth, J. E. K., Gotch, F. M., Hildreth, P. D. K., and McMichael, A. J.,** A human lymphocyte associated antigen involved in cell mediated lympholysis, *Eur. J. Immunol.,* 13, 202, 1983.
65. **Rothlein, R., Dustin, M. L., Marlin, S. D., and Springer, T. A.,** A human intercellular adhesion molecule (ICAM-1) distinct from LFA-1, *J. Immunol.,* 137, 1270, 1986.
66. **Lechler, R. I., Norcross, M. A., and Germain, R. N.,** Qualitative and quantitative studies of antigen presenting cell function using I-A expressing L cells, *J. Immunol.,* 135, 2914, 1985.
67. **Shastri, N., Malissen, B., and Hood, L.,** Ia transfected L cell fibroblasts present a lysozyme peptide but not the native protein to lysozyme-specific T cells, *Proc. Natl. Acad. Sci. U.S.A.,* 82, 5885, 1985.
68. **Young, J. A. T., et al.,** in preparation.
69. **Maddox, J.,** personal communication.
70. **de Vries, R., Elferink, D., Reinsmoen, N.,** personal communication.

Chapter 6

MOLECULAR DIVERSITY OF HLA CLASS II GENES AND HAPLOTYPES: IMPLICATIONS FOR DISEASE SUSCEPTIBILITY

Peter K. Gregersen and Jack Silver

TABLE OF CONTENTS

I. INTRODUCTION

Human leucocyte antigen (HLA) class II molecules, encoded within the human major histocompatibility complex (MHC), play a central role in the immune response by presenting processed peptide antigens to CD4-positive T lymphocytes. Since their original discovery by serologic methods, the extraordinary allelic diversity of the HLA class II glycoproteins has occupied the attention of a wide variety of scientists. It is now apparent that allelic polymorphisms of class II molecules play a direct role in regulating patterns of immune responsiveness in mammalian organisms.[1,2] This realization has provoked a major effort to define the nature and extent of this polymorphism at the molecular level in the human population. The contributors to this volume have made major contributions to this endeavor; the results have provided considerable insight into the complexity of the class II gene family and have stimulated new approaches to understanding the relationship between HLA polymorphism and autoimmune disease.

In keeping with the intent of the editor, we will focus this chapter on those areas which have been of particular interest to this laboratory. In general, we have attempted to analyze allelic polymorphisms in the context of the various haplotype family groups which exist in the population. A comparison of DNA sequences of various alleles at one particular class II locus, DRβ, leads to the conclusion that part of this gene, the third hypervariable region (HV3), is under the influence of special genetic mechanisms which result in the sharing of HV3s among many different DRβ chain alleles. In addition, sequence comparisons of alleles encoded within the DR and DQ subregions among certain haplotypes indicates that recombination has been a major mechanism by which modern day class II haplotypes have evolved. Finally, by comparing a closely related group of haplotypes which encode the serologic determinant HLA-DR4, we have gained insight into specific polymorphisms which strongly influence T cell recognition and susceptibility to autoimmune diseases such as rheumatoid arthritis.

II. GENETIC ORGANIZATION OF THE HLA CLASS II REGION

Briefly, the class II region of the human MHC contains a large cluster of genes located on chromosome 6 in a region of approximately 1 million base pairs.[3] As shown in Figure 1, this region has been subdivided into three major subregions designated DR, DQ and DP.[4] Each subregion encodes at least one set of functional α and β chains which pair noncovalently to form membrane heterodimers.[5] The assignment of a class II gene to a particular subregion is based on both structural and functional criteria. Genes located within the same subregion are more closely related to each other than are genes from different subregions. For example, the three β genes located in the DR subregion are more than 90% identical as are the two DQβ genes, whereas the DR and DQ β genes are approximately 70% identical to each other.[6,7] These sequence differences limit the association of α and β chains in the formation of functional heterodimers to those which are encoded by the same subregion. Thus, for example, within the DR subregion the α gene product can, in general, only interact with either the DRβI or DRβIII gene product to form functional DR complexes. Similarly, the DQα chain associates only with the DQβ chain, and the DPα chain associates only with the DPβ chain, to form functional DQ and DP complexes, respectively. Exceptions to this rule have been noted; for example, heterodimers consisting of DRα and DQβ chains (mixed isotypes) have been detected in some B cell lines.[8] However, the presence of such mixed isotypes appears to be a consequence of abnormal overexpression of one chain, DRα in this case, in relation to its cognate β chain (DRβ). The existence of mixed isotypes under normal physiologic conditions has been difficult to demonstrate.

As illustrated in Figure 1, not all of the α and β genes express functional products.

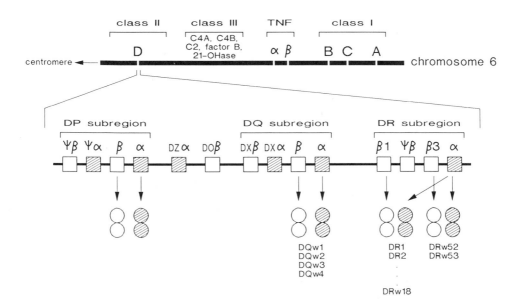

FIGURE 1. Genetic map of the HLA class II region on chromosome 6. Also shown are the locations of the HLA class I loci (HLA-A, -B and -C), the tumor necrosis factor (TNF) α and β genes, and the class III region which encodes complement components C4A, C4B, C2 and factor B.

Both the DR and DP subregions contain nonfunctional pseudogenes designated DRβII and SXα, SXβ.[4] Furthermore, the DQ subregion contains a set of genes designated DXα and DXβ[9] which are not pseudogenes[10] but nevertheless are not expressed. In addition, the DZα[4] and DOβ[11,12] genes located between the DQ and DP subregions do not appear to form a functional heterodimeric pair. Thus, despite the large number of α and β genes, the class II region encodes only four functional α-β heterodimers: DRβI, DRβIII, DQ and DP.

III. POLYMORPHISM OF HLA CLASS II GENES

The polymorphism of the HLA class II system has classically been defined by serologic methods. Using alloantisera from multiparous women, three major allelic series of class II alloantigenic determinants have been defined.[13] These are the DR1-DRw18 specificities, the DRw52 and DRw53 specificities, and the DQw1-DQw4 specificities. A variety of biochemical and molecular approaches has made it possible to assign these specificities to particular loci within the class II region. Two of these groups of specificities are encoded by products of the DR subregion. The DR1-DRw18 specificities are encoded by the highly polymorphic DRβI locus, while the DRw52 and DRw53 specificities are encoded by the less polymorphic DRβIII locus. The DRα gene is, with minor exceptions, invariant.[4] The DQw1-DQw4 specificities are expressed by HLA-DQ molecules. Although both DQ α and β chains are polymorphic, the DQβ chain appears to be primarily responsible for the serologically defined DQ determinants (see below).

The polymorphism of HLA class II proteins detectable at the serologic level is a consequence of sequence variation among class II alleles. The sequencing of a large number of class II genes has provided enormous insight into the nature of this structural variation. Both the α and β chains of class II molecules consist of two domains of approximately 90 to 95 amino acids each, designated α1 and α2, and β1 and β2.[14] The membrane proximal, or second, domains are homologous to immunoglobulin (Ig) constant region domains,[15] and include an internal disulfide loop similar to that found in Ig. Like Ig constant regions, the second domains of class II chains show little variability between alleles. In contrast the N-

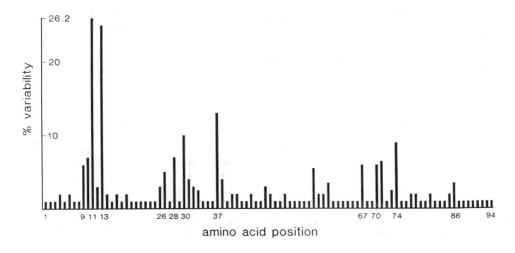

FIGURE 2. Variability plot of DRβ chain diversity.

terminal, or first, domain is the major site of variation between the different class II alleles. More than 3/4 of the variation is found in the region.[6] Furthermore, the relatively high ratio of productive: silent substitutions in the variable region (4:1) rather than the expected ratio of ~2.5:1 suggests that the fixation of mutations by strong selection pressures plays a major role in generating this diversity.[6] In contrast, the second domains or constant regions have a productive:silent substitution ratio of 2:3 suggesting that strong selection pressures act to conserve the structure of this part of the molecule.[6]

A detailed structure analysis of the polymorphism at one particular locus, DRβI, reveals several other striking similarities to immunoglobulin genes. Within the first domain, DRβ1 alleles may differ by multiple amino acids with some alleles differing by as many as twenty amino acids. (A compilation of these sequences with references may be found in the Appendix of this volume.) In addition, these amino acid differences are concentrated within three regions referred to as hypervariable (HV) regions. A variability plot of diversity at the DRβI locus (Figure 2) shows that the three HV regions are located at amino acid positions 9 to 13, 26 to 37, and 67 to 78. Additional sites of variability are also observed at positions 57, 60, and 86. It is intriguing to note the periodicity of the polymorphic residues within each of the hypervariable regions; in the first and second hypervariable regions the polymorphic residues (positions 11, 13, 28, and 30) alternate with nonpolymorphic residues (positions 10, 12, 29, and 31), while in the third hypervariable region polymorphic residues are observed at every 3 to 4 amino acids (positions 67, 70, 71, and 74). These patterns of variability are highly relevant in that they relate to the function and three-dimensional structure of class II molecules. As we discuss below, the diversity of DRβI genes appears to have arisen as a consequence of two general mechanisms, point mutation/selection and gene conversion, with the latter operating primarily on the third hypervariable region (HV3).

IV. HLA CLASS II HAPLOTYPE FAMILIES AND THEIR RELATION TO DRβ CHAIN SEQUENCE POLYMORPHISM

An analysis of sequence similarities between various DRβ chain alleles clearly indicates the existence of a close evolutionary relationship between certain class II haplotypes (Figure 3). This is particularly obvious among haplotypes which encode the DR3, DR5, and DRw6 serologic determinants. When alleles of the DRβI locus from these haplotypes are compared, striking similarities are observed, even within the first, or variable, domain of the DRβI molecule. Thus, the EYST sequence at positions 9 to 12 of the first domain (the first

Block 1 (residues 1–50)

```
                      10           20           30           40          50
             G D T R P R F L W Q  L K F E C H F F N G  T E R V R L L E R C  I Y N Q E E S V R F  D S D V G E Y R A V
DR1βI        - - - - - - - - - -  - - - - - - - - - -  - - - - - - - - - -  - - - - - - - - - -  - - - - - - - - - -
DR2 Dw2βI    - - - - - - - - Q -  - D - Y - - - - - -  - - - - - F - H - D  - - - - - - - - - -  - D L - - - - F - -

DR3βI        - - - - - - - - E Y  S T S - - - - - - -  - - - - - F - H - D  - - Y - - - - N - -  - - - - - - - F - -
DR5 Dw5βI    - - - - - - - - E Y  S T S - - - - - - -  - - - - - F - D - Y  F - - - - - - Y - -  - - - - - - - F - -
DR5 JVM      - - - - - - - - E Y  S T S - - - - - - -  - - - - - F - D - Y  F - - - - - - Y - -  - - - - - - - F - -
DRw6aβI      - - - - - - - - E Y  S T S - - - - - - -  - - - - - F - D - Y  F H - - - - - Y - -  - - - - - - - F - -
DRw6bβI      - - - - - - - - E Y  S T S - - - - - - -  - - - - - F - D - Y  F H - - - - - Y - -  - - - - - - - F - -
DRw8βI       - - - - - - - - E Y  S T G - Y - - - - -  - - - - - F - D - Y  F - - - - - - Y - -  - - - - - - - - - -
DRw12        - - - - - - - - E Y  S T G - Y - - - - -  - - - - - - - - - H  - - - - - - - L L -  - - - - - - - - - -

DR4 Dw10βI   - - - - - - - - E -  - V - H - - - - - -  - - - - - F - D - Y  F - H - - - - Y - -  - - - - - - - - - -
DR4 Dw14βI   - - - - - - - - E -  - V - H - - - - - -  - - - - - F - D - Y  F - H - - - - Y - -  - - - - - - - - - -
DR7βI        - - Q - - - - - - -  - G - Y K - - - - -  - - - - - Q F - - L  F - - - - - - F - -  - - - - - - - - - -
DRw9βI       - - Q - - - - - K -  - D - - - - - - - -  - - - - - - Y - H - G  - - - - - - - N - -  - - - - - - - - - -
```

Block 2 (residues 51–94)

```
                      60           70           80           90
             T E L G R P D A E Y  W N S Q K D L L E Q  R R A A V D T Y C R  H N Y G V G E S F T  V Q R R
DR1βI        - - - - - - - - - -  - - - - - - - - - -  - - - - - - - - - -  - - - - - - - - - -  - - - -       [11]
DR2 Dw2βI    - - - - - - - - - -  - F - - - - - - - D  - - - - - - - - - -  - - - - - - - - - -  - - - -       [79]

DR3βI        - - - - - - - - - -  - - - - - - - - - -  K - G R - - - N - -  - - - - - V - - - -  - - - -       [6]
DR5 Dw5βI    - - - - - E - - - -  - F - - - - - - - D  E - - - - - - - - -  - - - - - V - - - -  - - - -       [19]
DR5 JVM      - - - - - E - - - -  - I - - - - - - - D  E - - - - - - - - -  - - - - - V - - - -  - - - -       [24]
DRw6aβI      - - - - - - - - A H  - I - - - - - - - R  - - E - - - - - - -  - - - - - V - - - -  - - - -       [17]
DRw6bβI      - - - - - S - - - -  - F - - - - - - - D  - - L - - - - - - -  - - - - - V - - - -  - - - -       [80]
DRw8βI       - - - - - - - - - -  - H - - - - - - - D  E - - - - - - - - -  - - A V - - - - - -  - - - -       [24]
DRw12        - - - - - V - - - -  - I - - - - - - - D  E - - - - - - - - -  - - - - - - - - - -  - - - -       [21]

DR4 Dw10βI   - - - - - - - - - -  - - - - - - - - - D  E - - - - - - - - -  - - - - - V - - - -  - - - -       [27]
DR4 Dw14βI   - - - - - - - - - -  - - - - - - - - - -  - - G Q - - - - - -  - - - - - V - - - -  - - - -       [29]
DR7βI        - - - V - - - - - S  - I - - - - - - - D  - - - - - V - - - -  - - - - - - - - - -  - - - -       [22]
DRw9βI       - - - V - - - - - S  - F - - - - - - - R  E - - - - V - - - -  - - - - - - - - - -  - - - -       [24]
```

FIGURE 3. Comparison of first domain amino acid sequences of a representative group of DRβ chains. The DRβ1 alleles are grouped to reflect their family relationships. The DR3, -5, -6, -8, and -12 alleles belong to the DRw52 family of haplotypes and share a common sequence at positions 9 to 12 of the first hypervariable region (boxed region). In contrast, simple inspection reveals examples of sequences within the third hypervariable region which are shared among DRβ chains from different family groups. Such sharing of third hypervariable region sequences between different DRβ chains may be a result of gene conversion events acting on this region of the DRβ genes (see text). References for these sequences are given in brackets. A complete compilation of DRβ chain alleles with original references is given in the Appendix.

TABLE 1
Nucleotide Differences in the 3′ Untranslated Region of DRβI Chains

	DRw53 family			DRw52 family					
	DR4	DR7	DR9	DR3	DR5	DRw6	DRw12	DR2	DR1
DR4		2.7	3.7	15.3	14.8	13.8	11.8	12.5	12.4
DR7			1.8	13.1	12.1	12.8	11.3	13.7	11.3
DR9				16.1	15.9	14.3	13.4	13.2	13.9
DR3					0.3	0.7	2.5	11.7	7.5
DR5						0.3	2.2	12.1	7.1
DRw6							2.2	12.3	6.5
DRw12								10.8	5.5
DR2									10.1

Note: Numbers represent percent difference. Sequence data is from the following sources: DR4,
DR7, and DR9 (Reference 22), DR1 (Reference 11), DR2 (Reference 78), DR3 (Reference
6), DR5 (Reference 19), DRw6 (Reference 20), DRw12 (Reference 21).

hypervariable region) shared in common among these alleles is a reflection of their common
ancestry. This family relationship among DR3, -5 and -w6 haplotypes has been noted
previously on the basis of their very similar genomic organization.[16] In addition, a particularly
striking feature of this group of haplotypes is the fact that the DR3, DR5 and DRw6 DRβI
genes are linked to structurally similar DRβIII genes, of which three variant forms have
recently been described and are designated DRw52a, DRw52b and DRw52c.[17,18] Therefore,
haplotypes which type as DR3, -5, and -w6 may be viewed as part of the DRw52 family
of haplotypes.

The close family relationship between DRβI alleles encoded on haplotypes of the DRw52
family is further confirmed by an analysis of sequence similarities in the 3′ untranslated
regions of these genes. As shown in Table 1, the 3′ut regions of the DR3, DR5, and DRw6
DRβI alleles differ by only 0.3 to 0.7% from each other,[6,19,20] whereas much greater sequence
divergence is seen when comparisons are made with DRβI alleles outside the DRw52 family
(6.5 to 16.1%). Haplotypes encoding the DRw8 and DRw12 serologic specificities may also
belong to the DRw52 family. We have recently determined the DNA sequence of a DRβI
cDNA clone obtained from the DRw12 homozygous cell line HERLUF.[21] This DRw12
haplotype types serologically as DRw52 and expresses the DRw52b allele at the DRβIII
locus. Sequence comparisons in the 3′ut region confirm that the DRw12 allele is a member
of the DRw52 family. This allele differs by 2.2 to 2.5% from the DR3, -5, and -w6 alleles,
compared to the much larger divergence seen when comparisons are made with other DRβI
alleles (5.5 to 13.4%). Interestingly, the DRw12 gene is identical to the DRw8 allele
throughout the first hypervariable region (Figure 3). Both of these alleles carry the EYST
sequence at positions 9 to 12 characteristic of the DRw52 group, but in addition have distinct
amino acid substitutions at positions 13 and 16. Overall, the data indicate that the DRw8
and DRw12 haplotypes may form a subgroup within the DRw52 family which may have
diverged slightly earlier than the DR3, -5, and -w6 haplotypes. It will be of interest to see
if these relationships are supported by the DNA sequence analysis of the 3′ untranslated
regions of DRw8 DRβ chain alleles, when they become available.

Haplotypes encoding the DR4, -7, and -9 serologic determinants constitute a second
closely related family of haplotypes.[22] The DR4, DR7, and DR9 βI genes are all linked to
a second expressed DRβ gene (designated DRβIII) which encodes an identical allele (designated DRw53) in all three haplotypes.[22-24] However, unlike the DR3, -5, and -6 family
of haplotypes, sequence comparisons between the DR4, -7, and -9 DRβI chains do not

reveal an obvious family relationship in the variable first domain. Nevertheless, when the 3' untranslated sequences are compared, DRβI genes from these haplotypes are seen to be closely related. Thus, as shown in Table 1, DR4, DR7, and DR9 DRβI chains differ from one another by only 1.8 to 3.7% in the 3' untranslated region, whereas they differ to a much greater extent from other DRβI alleles (11.3 to 16.1%) in this region. RFLP analysis also supports a close genetic relationship between DRw53 haplotypes.[16]

We have previously suggested that haplotypes encoding the DR1 and DR2 serologic determinants may constitute a third broad family of haplotypes.[25] This idea rests primarily on the observation that these haplotypes have a similar genomic organization based on sharing of restriction fragments on Southern blot analysis.[6] In addition, the recent finding of a DR"X" sequence which is common to DR1 and DR2 haplotypes lends support to the concept of a DR1/2 family of haplotypes.[26] However, an analysis of DRβ chain 3' untranslated sequences does not reveal striking similarities among DRβ alleles derived from DR1 and DR2 haplotypes (see Table 1). Therefore, if DR1 and DR2 haplotypes do derive from a common ancestral haplotype, they nevertheless appear to be less closely related to one another than haplotypes within the DRw52 or DRw53 haplotype families. A definitive resolution of this question will clearly require more detailed study of the genomic organization of these haplotypes.

V. THE ROLE OF GENE CONVERSION IN GENERATING DRβ DIVERSITY

One striking feature of DRβ chain polymorphism is the sharing of HV3s among DRβ chains belonging to different haplotype families, as illustrated in Figure 3. For example, the DRw6a and DR4 Dw10 DRβI genes are members of quite distinct haplotype families, yet they share an identical third hypervariable region sequence, with nucleotide sequence identity extending from codon 37 through the end of the first domain.[17,27] A more recently described example[20] is shown in Figure 4, and involves the DRw6 DRβI allele found in cell line AMALA. The AMALA haplotype belongs to the DRw52 haplotype family, yet the DRβI chain of this haplotype is identical in sequence to a DR1 DRβ chain allele in the third hypervariable region. The DR4 Dw14βI gene also shares this sequence.[27,28] This phenomenon of sequence identity between different DRβ genes has been ascribed to the much discussed mechanism of gene conversion which may occur during the evolution of class II haplotypes.[17] Examples of gene conversion have been described in both murine class II[29] and murine and human class I[30,31] genes, as well as in other gene families.[32,33] Alternatively, shared third hypervariable region sequences among DRβ chains may reflect strong selection pressures for particular sets of sequences in this portion of the molecule. However, if this were the case one might not expect to see the high degree of conservation of nucleotide as well as protein sequences among HV3s shared by different DRβI alleles. The point we wish to make is that the pattern of third hypervariable region sequence sharing between DRβ chain alleles must be due to an active genetic mechanism, because when viewed in the context of the various haplotype family groups, such sharing of HV3 sequences obviously cannot be ascribed to simple divergence from a common ancestral DRβ gene.

VI. POLYMORPHISM OF DQα GENES

In the case of DQα, DNA sequence analysis indicates that they can also be placed into three major groups based on similarities in sequence. A representative sample of these DQα sequences is shown in Figure 5. These family groupings are further supported in many cases by the finding of similar RFLP patterns for alleles within each group.[34] Interestingly, these groups often follow the haplotype family groups discussed above for the DR subregion. For

```
                                              10                                              20
CB6B    GGG GAC ACC AGA CCA CGT TTC TTG GAG TAC TCT ACG TCT GAG TGT CAT TTC TTC AAT GGG
AMALA   --- --- --- --- --- --- --- --- --- --- --- --- --- --- --- --- --- --- --- ---
DR1     --- --- --- C-- --- --- --- --- TG- C-G CT- -A- -T- --A --- --- --- --- --- ---

CB6B    Gly Asp Thr Arg Pro Arg Phe Leu Glu Tyr Ser Thr Ser Glu Cys His Phe Phe Asn Gly
AMALA    -   -   -   -   -   -   -   -   -   -   -   -   -   -   -   -   -   -   -   -
```

```
                                              30                                              40
CB6B    ACG GAG CGG GYG CGG TTC CTG GAC AGA TAC TTC CAT AAC CAG GAG GAG AAC GTG CGC TTC
AMALA   --- --- --- --- --- --- --- --G --- --- --- --- --- --- --- --- --- --- --- ---
DR1     --- --- --- --- --- --G --- --A --- -G- A-- T-- --- --A --- --- TC- --- --- ---

CB6B    Thr Glu Arg Val Arg Phe Leu Asp Arg Tyr Phe His Asn Gln Glu Glu Asn Val Arg Phe
AMALA    -   -   -   -   -   -   -  Glu  -   -   -   -   -   -   -   -   -   -   -   -
```

```
                                              50                                              60
CB6B    GAC AGC GAC GTG GGG GAG TTC CGG GCG GTG ACG GAG CTG GGG CGG CCT GAT GCC GAG TAC
AMALA   --- --- --- --- --- --- -A- --- --- --- --- --- --- --- --- --- --- --- --- ---
DR1     --- --- --- --- --- --- -A- --- --- --- --- --- --- --- --- --- --- --- --- ---

CB6B    Asp Ser Asp Val Gly Glu Phe Arg Ala Val Thr Glu Leu Gly Arg Pro Asp Ala Glu Tyr
AMALA    -   -   -   -   -   -  Tyr  -   -   -   -   -   -   -   -   -   -   -   -   -
```

```
                                              70                                              80
CB6B    TGG AAC AGC CAG AAG GAC ATC CTG GAA GAC GAG CGG GCC GCG GTG GAC ACC TAC TGC AGA
AMALA   --- --- --- --- --- --- C-- --- --G C-G AG- --- --- --- --- --- --- --- --- ---
DR1     --- --- --- --- --- --- C-- --- --G C-G AG- --- --- --- --- --- --- --- --- ---

CB6B    Trp Asn Ser Gln Lys Asp Ile Leu Glu Asp Glu Arg Ala Ala Val Asp Thr Tyr Cys Arg
AMALA    -   -   -   -   -   -  Leu  -   -  Gln Arg  -   -   -   -   -   -   -   -   -
```

```
                                              90
CB6B    CAC AAC TAC GGG GTT GTG GAG AGC TTC ACA GTG CAG CGG CGA
AMALA   --- --- --- --- --- -GT --- --- --- --- --- --- --- ---
DR1     --- --- --- --- --- -GT --- --- --- --- --- --- --- ---

CB6B    His Asn Tyr Gly Val Val Glu Ser Phe Thr Val Gln Arg Arg
AMALA    -   -   -   -   -  Gly  -   -   -   -   -   -   -   -
```

FIGURE 4. Comparison of the DNA and predicted protein sequences of DRβ1 chains derived from two DRw6 cell lines, AMALA (now designated DRw14) and CB6B (now designated DRw13).[20] Also shown is the nucleotide sequence of a DR1 DRβ chain[11] which is identical to the AMALA sequence from positions 38 to 94 (including the third hypervariable region). Thus, the AMALA DRw14 allele may have arisen by means of a gene conversion event with the DR1 DRβ1 allele acting as donor. (From Kao, H. T., Gregerson, P. K., Tang, J. C., et al., *J. Immunol.*, 142, 1743, 1989. With permission.)

example, DR4, -7, and -9 haplotypes belong to the DRw53 family and all carry similar DQα gene alleles. Likewise, many haplotypes within the DRw52 family carry highly similar DQα chains such as those found in DR3 DQw2, DR3 DQw4, DR5 DQw3, and DRw6 DQw3. The DQα genes found in DR1 DQw1 and DR2 DQw1 haplotypes form a third group of related alleles. The demonstration that DQα genes can be subdivided into similar family groupings as the DRβI and DRβIII genes suggests that all three genes, DRβI, DRβIII and DQα, evolved together as a single unit and in general remained linked to each other throughout evolution.

VII. POLYMORPHISM OF DQβ GENES

The DQβ genes may be divided into four major groups, DQw1-DQw4, according to the serologic specificities with which they are associated (Figure 6). The number of alleles within each group is quite variable. Only one sequence has been found for DQw2.[35] Three variants of the DQw3 DQβ chain have been reported: DQ3.1, DQ3.2, and DQ3.3.[36-38] They differ by one to four amino acids. Two minor variants of the DQw4 DQβ chain have also

```
                      20          30          40          50          60          70          80
                 YGVNLIQSYG  PSGQYSHEFD  GDEEFYVDLE  RKETVWQLPL  FRRFRRFDPQ  FALTNIAVLK  HNLNIVIKRS  [26]
DRw53    DR4,DQw3  ----------  ----FT----  ----------  ------K---  --H-L-----  ----------  ------L----  [81]
family   DR7,DQw2  ----------  ----------  ----------  ----------  ----------  ----------  -----------  [81]
         DR9,DQw3

DRw52    DR3,DQw2  ----------  ----T-----  ---Q-----G  ------C--V  L-Q- R----  ----------  ----SL----   [82]
family   DR3,DQw4  ----------  ----T-----  ---Q-----G  ------C--V  L-Q- R----  ------T---  -----L----   [83]
         DR5,DQw3  ----------  ----T-----  ---Q-----G  ------C--V  L-Q- R----  ----------  ----SL----   [84]
         DR6,DQw3  ----------  ----T-----  ---Q-----G  ------C--V  L-Q- R----  ----------  ----SL----   [20]

DR1/2    DR1,DQw1  C-------F--  ----T-----  ----------  ----A-RW-E  -SK-GG----  G--R-M--A-  -----M---Y   [36]
family   DR2,DQw1  C-------F--  ----T-----  ---Q------  ----A-RW-E  -SK-GG----  G--R-M--A-  -----M---Y   [36]
         DR5,DQw1  C-------F--  ----FT----  ---Q------  K---A-RW-E  -SK-GG----  G--R-M--A-  -----M---Y   [20]
         DR6,DQw1  C-------F--  ---FT-----  ---Q------  K---A-RW-E  -SK-GG----  G--R-M--A-  -----M---Y   [20]
```

FIGURE 5. Amino acid sequences of DQα chains (first domain). The sequences are grouped in three families (DR1/2, DRw52, and DRw53) reflecting sequence similarities within each group. These DQα alleles are generally found in linkage disequilibrium with DR alleles from the same family group, except where recombination disrupts this relationship, as in DR5 DQw1 and DRw6 DQw1 haplotypes (see text). References for the sequences are given in brackets.

been reported.[27,39] Based on sequence analysis of DQα and DQβ genes from haplotypes of different DQ type, it is clear that the DQβ chains are primarily responsible for the DQw2, DQw3, and DQw4 serologic specificities. However, assignment of the DQw1 specificity to either the DQα or β chain has been a matter of controversy. The observations by several groups that DQw1 α chains (Figure 5) are less variable than DQw1β chains (Figure 6) has prompted the suggestion that the DQw1 serologic specificity is encoded by the DQw1α chain.[36] However, recent data from our laboratory indicate that the DQw1 specificity may also be determined by the β chain.[40] A close inspection of DQβ gene sequences reveals a patchwork pattern similar to that observed in the DRβ genes. For example, the DQw2-DQw4β genes have an identical sequence between amino acids 84 and 90 which differs greatly from that observed in this region in the DQw1β gene. Indeed, since it is relatively conserved among DQw1 haplotypes, this region of variability between positions 84 and 90 may be involved in forming DQw1 serologic determinants. As in DRβ, these sequence similarities between different alleles may also reflect gene conversion events during the course of evolution of the DQβ genes.

VIII. RECOMBINATION MAY GENERATE CLASS II HAPLOTYPE DIVERSITY

Extensive serologic analysis of the HLA class II region in the Caucasian population has led to the observation that characteristic patterns of linkage disequilibrium exist between alleles of the DR and DQ subregion.[13] For example, most Caucasians that type as DR4 also type as DQw3 and similarly, most DR7 Caucasians type as DQw2. However, exceptions to these patterns of linkage disequilibrium can be found in a minority of Caucasian haplotypes or in haplotypes derived from non-Caucasian racial groups. We have analyzed the sequence polymorphism of DR and DQ subregion alleles of five such "exceptional" haplotypes in order to define the molecular basis for these differences in linkage disequilibrium.[25] These are listed in Table 2. These haplotypes were compared in pairwise fashion to the more common Caucasian haplotypes of identical DR type. Thus, the DR4 DQw4 haplotype found in the Japanese population was compared to the Caucasian DR4 DQw3 haplotype.[27] Likewise the DR7 DQw3 haplotype[38] (found in a minority of Caucasians) and the DR3 DQw4 haplotype[39] present in American blacks were compared with the DR7 DQw2 and DR3 DQw2 haplotypes, respectively. In addition, a DR5 DQw1 haplotype found in the Ashkenazi Jewish population[41] and a DRw6 DQw3 haplotype found in the Warao Indian population of Venezuela[42] were compared to the Caucasian DR5 DQw3 and DRw6 DQw1 haplotypes, respectively.

Although a number of possible mechanisms such as point mutation, gene conversion or recombination might be invoked to explain these differences in linkage disequilibrium patterns, we have suggested that recombination is the most likely explanation for the DQ differences in haplotypes of identical DR type.[25] With some haplotypes the recombination event appears to have occurred between the DQα and DQβ genes. For example, sequence analysis of class II genes from the DR4 DQw3 and the DR4 DQw4 haplotypes reveals that they share identical DQα chains yet differ at the DQβ locus. Likewise, the DR7 DQw2 and DR7 DQw3 haplotypes share identical DQα alleles, but have distinct DQβ chain alleles. These sequence relationships are shown schematically in Figure 7. Another example of recombination between DQα and DQβ is indicated by the comparison of the DR3 DQw4 haplotype found in the American black population[39] with the common Caucasian haplotype DR3 DQw2. These two DR3 haplotypes express highly similar DQα chains, as shown in Figure 5, yet have quite distinct DQβ chain alleles. Thus, for these three pairs of haplotypes recombination between DQα and DQβ is the best explanation for the DQ differences.

The recombination events that appear to have taken place in these haplotypes between the DQα and DQβ genes are very likely important in that they provide an additional means

FIGURE 6. Comparison of first domain sequences of DQβ chains. The sequences are grouped according to the serological specificities with which they are associated (DQw1—DQw4). The DR haplotypes on which these DQβ alleles have been commonly found are given in parenthesis. The bracketed numbers refer to original references for these sequences.

```
                                   10                  20                  30                  40                  50
                    R D S P E D F V Y Q   F K G L C Y F T N G   T E R V R G V T R H   I Y N R E E Y V R F   D S D V G V Y R A V
DQw1.1 (DR1;DRw10)  - - - - - - - - - -   - - - - - - - - - -   - - - - - - - - - -   - - - - - - - - - -   - - - - - - - - - -
DQw1.2  (DR2)       - - - - - - - - F -   - - M - - - - - - -   - - - - - - L - Y -   - - - - - - - A - -   - - - - - - - - - -
DQw1.12 (DR2)       - - P - - - - - L -   - A M - - - - - - -   - - - - - - Y - Y -   - - - - - - - D - -   - - - - - - - - - -
DQw1.AZH (DR2)      - - - - - - - - - -   - - - - - - - - - -   - - - - - - - - - -   - - - - - - - A - -   - - - - - - - - - -
DQw1.9  (DRw6)      - - - - - - - - - -   - - M - - - - - - -   - - - - - - L - Y -   - - - - - - - A - -   - - - - - - - - - -
DQw1.18 (DRw6)      - - - - - - - - - -   - - M - - - - - - -   - - - - - - L - Y -   - - - - - - - A - -   - - - - - - - - - -
DQw1.19 (DRw6)      - - - - - - - - - -   - - - - - - - - - -   - - - - - - - - - -   - - - - - - - - - -   - - - - - - - - - -

DQw2 (DR3;DR7)      - - - - - - - - - -   - - M - - - - - - -   - - - - - L - - S S   - - - - - - - I - -   - - - - E F - - - -

DQw3.1 (DR4;DR5)    - - - - - - - - - -   - A M - - - - - - -   - - - - - Y - - Y -   - - - - - - - A - -   - - - - E - - - - -
DQw3.2 (DR4)        - - - - - - - - - -   - - M - - - - - - -   - - - - - L - - Y -   - - - - - - - A - -   - - - - - - - - - -
DQw3.3 (DR7;DR9)    - - - - - - - - - -   - - M - - - - - - -   - - - - - L - - Y -   - - - - - - - A - -   - - - - - - - - - -

DQw4.1 (DR4)        - - - - - - - - F -   - - M - - - - - - -   - - - L - - - - Y -   - - - - - - - A - -   - - - - - - - - - -
DQw4.2 (DRw8;DR3)   - - - - - - - - F -   - - M - - - - - - -   - - - - - - - - Y -   - - - - - - - A - -   - - - - - - - - - -
```

```
                                   60                  70                  80                  90
                    T P Q G R P V A E Y   W N S Q K E V L E G   A R A S V D R V C R   H N Y E V A Y R G I   L Q R R
DQw1.1 (DR1;DRw10)  - - - - - - - - - -   - - - - - - - - - -   - - - - - - - - - -   - - - - - - - - - -   - - - -   [36,83]
DQw1.2  (DR2)       - - - - - D - - - -   - - - - - - - - - -   T - E L - T - - - -   - - - - - - - F - -   - - - -   [79]
DQw1.12 (DR2)       - - - - - D - - - -   - - - D I - - R - -   T - E L - T - - - -   - - - - - - - F - -   - - - -   [79]
DQw1.AZH (DR2)      - - - - - S - - - -   - - - - - - - - - -   - - - - - - - - - -   - - - - - - - - - -   - - - -   [79]
DQw1.9  (DRw6)      - - - - - D - - - -   - - - - - - - - - -   - - - - - - - - - -   - - - - - - - - - -   - - - -   [36]
DQw1.18 (DRw6)      - - - - - D - - - -   - - - - - - - - - -   T - E L - T - - - -   - - - - - - - F - -   - - - -   [20,36]
DQw1.19 (DRw6)      - - - - - - - - - -   - - - D I - - R - -   T - E L - T - - - -   - - - - - - - G - -   - - - -   [36]

DQw2 (DR3;DR7)      - L - L - A - - - -   - - - - - - - - - -   K - - A - - - - - -   - Q L E L - - - - -   - - - -   [35]

DQw3.1 (DR4;DR5)    - - L - P - D - - -   - - - D I - - R - -   T - E L - T - - - -   - Q L E L - T T - -   - - - -   [36]
DQw3.2 (DR4)        - - L - P - A - - -   - - - D I - - R - -   T - E L - T - - - -   - Q L E L - T T - -   - - - -   [37]
DQw3.3 (DR7;DR9)    - - L - P - D - - -   - - - D I - - R - -   T - E L - T - - - -   - Q L E L - T T - -   - - - -   [36,38]

DQw4.1 (DR4)        - - L - - L D - - -   - - - D I - - E - -   D - - - - T - - - -   - Q L E L - T T - -   - - - -   [27]
DQw4.2 (DRw8;DR3)   - - L - - L D - - -   - - - D I - - E - -   D - - - - T - - - -   - Q L E L - T T - -   - - - -   [36,84]
```

TABLE 2
Variation in Linkage Disequilibrium of DR
and DQ Alleles among Haplotypes of
Different Ethnic Groups

Common Caucasian haplotypes	Uncommon or non-Caucasian haplotypes
DR4 DQw3	DR4 DQw4 (Japanese)
DR7 DQw2	DR7 DQw3
DR5 DQw3	DR5 DQw1 (Ashkenazi Jewish)
DR3 DQw2	DR3 DQw4 (American black)
DRw6 DQw1	DRw6 DQw3 (Venezuelan Indian)

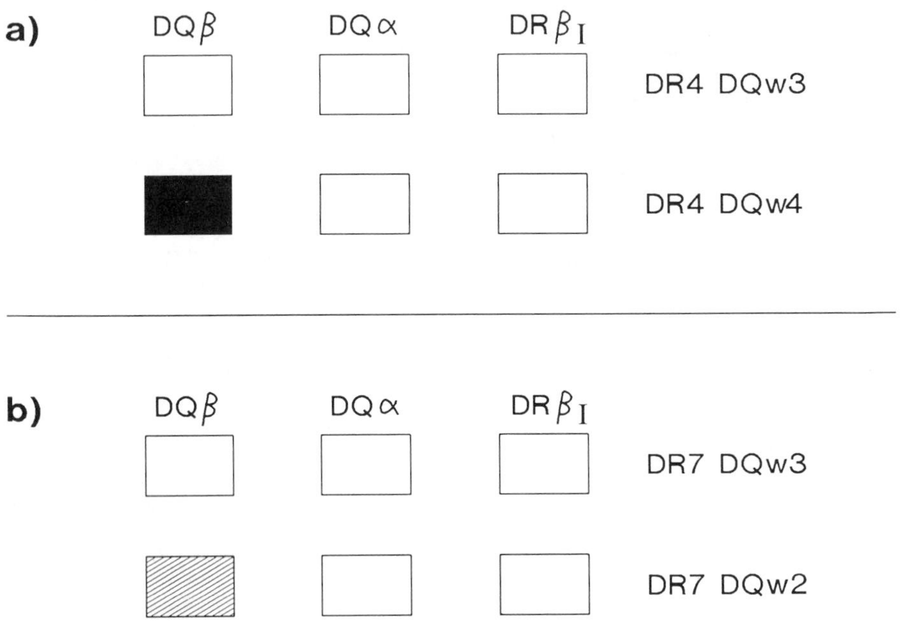

FIGURE 7. (a) Schematic comparison of DR4 DQw3 and DR4 DQw4 haplotypes. (b) Schematic comparison of DR7 DQw2 and DR7 DQw3 haplotypes. Boxes of different character (white, black, shaded) signify alleles which differ in amino acid sequence by more than 10% in the first (variable) domain at the indicated locus. Recombination between the DQα and DQβ genes appears to be the best explanation for the differences observed between these two sets of haplotypes. The sequences for the DQα and DQβ alleles from these haplotypes are shown in Figures 5 and 6.

of generating HLA-DQ diversity: such events can lead to the expression of HLA-DQ molecules consisting of the same β chain in association with different α chains. Furthermore, if as suggested by these recombinant haplotypes, DQ α and β chains can freely associate to form HLA-DQ heterodimers then clearly HLA heterozygous individuals would in many cases be able to express four rather than two DQ molecules at the cell surface. The two additional DQ molecules would result from *trans*-association of DQ α and β chains as shown schematically in Figure 8. Indeed, the occurrence of such "hybrid" DQ molecules formed by the association of DQα and DQβ chains in a *trans*-configuration has been reported by several investigators.[43,44] The formation of such "hybrid" DQ molecules by heterozygous individuals has important functional implications in that heterozygous individuals may have unique HLA-DQ molecules not expressed by either of the parental haplotypes. The ability

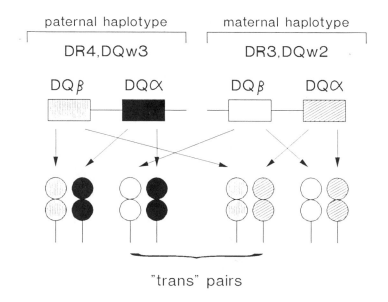

"trans" pairs

FIGURE 8.   Formation of ''hybrid'' DQ molecules as a result of *trans* pairing of DQα and DQβ chains in a heterozygous individual. As discussed in the text, recombination between the DQα and DQβ genes appears to have generated many modern day class haplotypes. This may constitute a selection mechanism for relatively unrestricted pairing among DQ α and β chains. Thus, *trans*-association of DQ α and β chains may be a common occurrence in heterozygous individuals.

of DQ heterozygotes to form DQ molecules by *trans*-complementation may influence not only their ability to respond to foreign pathogens, but also their ability to respond to autoantigens and thus affect their susceptibility to autoimmune disease.

The fact that recombination between the DQα and DQβ genes has generated many of the human class II haplotypes is in distinct contrast to the observations in mice, where recombination between the Aα and Aβ genes (the murine homologs of DQα and DQβ) rarely occurs.[45] Interestingly, restrictions on pairing of particular Aα and Aβ chain alleles have been reported in the mouse,[46] whereas no such restrictions have as yet been reported in humans. We suggest that the apparent ability of the various human DQ α and β chain alleles to pair freely with one another is a direct result of selection pressures brought about by recombination between the DQ α and β loci. In contrast, restrictions on Aα and Aβ pairing in the mouse may be a consequence of the absence of recombination between these two loci, with no concomitant selection for *trans*-pairing of α and β chains.

In contrast to the above described haplotypes where recombination has occurred between DQα and DQβ, other haplotypes appear to have arisen as a consequence of recombination between the DRβI and DQα genes. Examples of these are the DR5 DQw1 haplotype found in the Ashkenazi Jewish population,[41] and the DRw6 DQw1 haplotype found in high frequence in the Caucasian population. In these haplotypes recombination between the DRβI and DQα genes disrupts the usual linkage relationships between DRβ and DQα genes; the DQα chain found in these haplotypes is similar to DQα chains found commonly in Caucasian DR1 and DR2 haplotypes, rather than those found in DR5 DQw3 and DRw6 DQw3 as shown in Figure 5. These relationships for DR5 haplotypes are outlined schematically in Figure 9. A similar recombinational event may have generated the DR8 DQw1 haplotype, as described by Amar et al.[47] More recently, Liu et al. have reported on an unusual DR2 DQw3 haplotype found in a South American Indian population.[48] This haplotype contains a DQα chain which is identical to that found commonly in DRw52 haplotypes such as DR5 DQw3 and DR3 DQw2. Thus, comparisons with Caucasian DR2 DQw1 haplotypes suggest

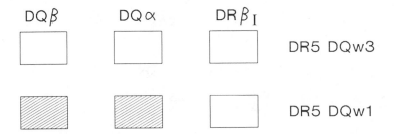

FIGURE 9.   Schematic comparison of the DR5 DQw3 haplotype and the DR5 DQw1
haplotype. Shaded or white boxes signify alleles which differ in amino acid sequence by
more than 10% (in the variable domain) at the locus indicated. In this case, recombination
between DRβ1 and DQα appears to be the best explanation for the differences between
these two haplotypes. The sequences for the DQα and DQβ alleles from these haplotypes
are shown in Figures 5 and 6.

that the DR2 DQw3 haplotype may also have arisen by means of recombination between
DRβI and DQα.

A consideration of these patterns of recombination in the context of the various haplotype
families has suggested to us that certain haplotype groups may have a tendency to undergo
recombination at particular locations. Although relatively few haplotypes have been examined
at the sequence level, it is striking that no haplotype suggestive of recombination between
DRβ and DQα has been found in the DRw53 family. This may indicate a restriction on the
ability of DRw53 haplotypes to recombine in this region. We have hypothesized that this
may be a general feature of this group of haplotypes and that in general the possible sites
of recombination in class II haplotypes may be haplotype dependent.[25] Haplotype-dependent
patterns of recombination have been observed in the mouse MHC.[45,49,50] The site and frequency of recombination in the murine class II region varies depending on the particular
mouse strains involved.[45] It has been postulated that these results might be explained by
recombinational "hotspots" containing sequences similar to the chi or minisatellite sequences
that appear to facilitate genetic exchange. Hotspots with recombination frequencies on the
order of 1% (similar to the frequencies described in the mouse) would be unlikely in the
DR and DQ subregions of the human MHC. If such high recombination frequencies occurred
in this region, the striking degree of linkage disequilibrium between DR and DQ in the
outbred human population would be difficult to explain. Nevertheless, a differing molecular
organization among the different human class II haplotype families may impose constraints
on where homologous cross-over events can occur to generate recombinant haplotypes. For
example, using pulse gel electrophoresis, Inoko and colleagues have shown that the distance
between DRβ and DQα is different in the DRw53 family compared with other haplotypes
(Reference 51 and Chapter 10 of this volume). Such differences would support the concept
of haplotype dependent recombination sites.

The hypothesis that recombination sites in the human class II region are haplotype
dependent rests on an admittedly limited sample of haplotypes from the human population.
Recently, Merryman et al.[40] have made the unexpected observation that DRw10 haplotypes,
which type as DQw1 and presumably belong to the DR1/DR2 family of haplotypes, have
a DQα gene which is identical in sequence to that found in DR4 haplotypes. Two separate
DRw10 haplotypes, one found in the Raji cell line and the other in an individual with
rheumatoid arthritis, both contain this DQα chain allele. This observation might suggest
that DRw10 haplotypes are in fact distant members of the DRw53 family, or alternatively,
that recombination can occur between DRβ and DQα in some DRw53 haplotypes. Clearly,
more information on class II haplotype organization will be required to address these possibilities. Parenthetically, it should be noted that the DRw10 haplotypes express DQβ chains

TABLE 3
**A Summary of the Nucleotide and Amino Acid Sequence Differences
among DR4 Subtypes**

aa Position	37	57	67	69	70	71	74	86	Ref.
Dw4	Tyr	Asp	Leu	Glu	Gln	Lys	Ala	Gly	27
	TAC	GAT	CTC	GAG	CAG	AAG	GCG	GGT	
Dw10	—	—	Ile	—	Asp	Glu	—	Val	27
			A	A	G C	G		TG	
Dw13	—	—	—	—	—	Arg	Glu	Val	28
						G	A	TG	
Dw14	—	—	—	—	—	Arg	—	Val	27,28
						G		TG	
Dw15	—	Ser	—	—	—	Arg	—	—	27
		AGC				G			
KT2	Ser	—	—	—	—	Arg	Glu	Val	54
	C					G	A	TG	
JHA	—	—	—	—	—	Arg	Glu	—	55
						G	A		

Note: Only those residues that differ in sequence are shown.

which are identical in amino acid sequence to DQβ chains found commonly on DR1 DQw1 haplotypes. This strongly suggests that the DQw1 serologic specificity in this instance is determined by the DQβ chain and is not affected by the pairing of a DQw1β chain with a DQα chain normally found in DR4 DQw3 haplotypes.

IX. HAPLOTYPE COMPARISONS WITHIN THE DR4 FAMILY: IMPLICATIONS FOR T-CELL RECOGNITION

Before the availability of extensive sequence data on class II genes it was apparent that classical serologic typing methods concealed considerable heterogeneity among certain groups of related haplotypes. Among these were haplotypes which encoded the serologic determinant HLA-DR4. Largely through the work of Bach, Reinsmoen, and colleagues (also contributors to this volume), it was shown that these haplotypes could be distinguished by means of the mixed lymphocyte culture (MLC) assay.[52] Thus, serologically indistinguishable DR4 individuals could be distinguished on the basis of the ability of their lymphocytes to cross-stimulate in an MLC. In this manner, a number of MLC subtypes of DR4 were defined which have been designated Dw4, Dw10, Dw13, Dw14, and Dw15.[53]

Because the DR and DQ subregions exhibit strong linkage disequilibrium in the human population, we reasoned that the phenotypic differences which account for stimulation in the MLC assay might be encoded by any of the polymorphic genes in these two subregions. We therefore sequenced and compared cDNA clones for DQα, DQβ, DRβI, and DRβIII from four of the major DR4 subtypes: Dw4, Dw10, Dw14, and Dw15.[27] The DRβI sequence from a and a Dw13 haplotype had been previously reported.[28] The results were striking. In general, differences among these haplotypes were restricted to the DRβI locus. Furthermore, most of the amino acid sequence differences between the DRβI chains were localized to the third hypervariable region of the molecule. A summary of these amino acid differences is given in Table 3. Note that in some instances only a single amino acid difference in this region, as between Dw13 and Dw14, accounts for the T-cell stimulation in MLC between these two haplotypes. In other cases, such as the Dw10 haplotype, multiple amino acid substitutions are seen, many of a nonconservative character. The only exception to this pattern of polymorphism among the DR4 subtypes is the Dw15 haplotype found commonly in the Japanese population. In this haplotype, in addition to the changes at the DRβI locus, the DQβ gene is quite distinct from those in the other DR4 haplotypes.[27] This DQβ gene encodes the DQw4 serologic specificity. Thus, stimulation by the Dw15 haplotype in the MLC may be due to differences at either the DQβ or DRβI loci, or both, when non-Dw15 DR4 individuals are used as responders.

We have further investigated the molecular heterogeneity among the Dw13 group of DR4 haplotypes.[54] A Dw13 haplotype found in the Yemenite Jewish population (the TAS haplotype) was shown to be identical in sequence to that found in a North American Caucasian Dw13 haplotype. In contrast, another Japanese DR4 haplotype which had been difficult to distinguish from Dw13 by the MLC assay[53] was shown to differ from the Caucasian Dw13 allele by a single amino acid substitution at position 37 of DRβI. This is the KT2 DRβ chain listed in Table 3. It is of interest that the KT2 allele is identical to the other Dw13 haplotypes in the third hypervariable region; this may account for the difficulty in distinguishing this haplotype by MLC. This further emphasizes the special importance of third hypervariable region sequence differences in allospecific T-cell recognition.

Additional heterogeneity among DR4 haplotypes may be detected with the use of T-cell clones. In the course of studying a group of T-cell clones directed against Dw14 haplotypes, a T-cell clone (designated E 38) was identified[55] which responded to the Dw13 haplotypes mentioned above, but which was not stimulated by the DR4 haplotype found in cell line JHa, another DR4 haplotype which had been placed in the Dw13 group.[56] Monoclonal antibody blocking studies indicated that these differences were due to polymorphism in the DRβ subregion. Sequence analysis of the DRβI chain indicates that a gly for val change at position 86 accounts for this difference in recognition by T-cell clone E38, as shown in Table 3. This result confirmed oligonucleotide analysis suggesting a position 86 difference in this haplotype.[84] Position 86 of the DRβI chain is of interest because this limited polymorphism (gly or val) is common among many DRβ chain alleles. The fact that allospecific T-cell recognition is influenced by this conservative amino acid substitution suggests that position 86 polymorphisms may also affect immune response patterns or disease susceptibility.

X. HLA-DR4 HAPLOTYPE DIVERSITY: IMPLICATIONS FOR GENETIC SUSCEPTIBILITY TO RHEUMATOID ARTHRITIS

A reconsideration of the previously defined HLA associations with rheumatoid arthritis (RA) in light of the patterns of DRβ polymorphism outlined above has led to the surprising conclusion that the third hypervariable region (HV3) may be of central importance in conferring disease risk. The association of RA with the DR4 specificity was first described more than 10 years ago[57] and has been confirmed in numerous different ethnic groups including North American and Northern European Caucasians, Japanese, Hispanics, and North American blacks.[58] When the incidence of RA was analyzed in the context of the various DR4 subtypes an association was found with three of them, Dw4, Dw14, and Dw15.[59-61] In contrast, a recent study, although somewhat limited in size, suggests a lack of association with the Dw10 subtype; individuals with the Dw10 subtype are underrepresented in the RA population when compared to the normal population.[62] This finding not only suggests that the Dw10 subtype does not confer risk for RA but also provides a satisfying explanation for the paradoxical observation that in the Israeli population RA is not significantly associated with DR4;[63] in contrast to other ethnic groups where Dw10 represents a relatively infrequent DR4 subtype, a majority of DR4 Israeli Jews are Dw10.[64] In this context note that the Dw10 sequence is conspicuously different from the other DR4 subtypes in the third diversity region, having two negatively charged residues at positions 70 and 71, whereas the susceptible alleles have neutral and positive residues, respectively, at these positions (Table 3).

HLA-DR1 has also been associated with RA; indeed DR1 is the predominant risk conferring haplotype in some populations, such as Asian Indians[65] and Israeli Jews.[63] Weaker associations with DR1 have been described in populations where DR4 is the predominant RA-associated haplotype.[66] Furthermore, a serologic specificity designated MC1 which cross reacts with both DR1 and DR4 has been strongly associated with risk for rheumatoid arthritis.[67] We would emphasize that DR1 and DR4 haplotypes belong to quite distinct family groups, yet they seem to share some feature which confers risk for RA. A likely candidate for such a feature is the sequence similarity between DR1 and the RA associated DR4 alleles in the third hypervariable region of the DRβI chain. Indeed, as shown in Figure 3, DR1 and DR4 Dw14 have identical third hypervariable region sequences.

On the basis of these observations, we have proposed that susceptibility to RA may in fact be due to a group of similar HV3s which share certain conformational and functional attributes.[68] These disease-associated sequences may be found on a number of different DRβ chain alleles, such as DR1 and several of the DR4 subtypes (Dw4, Dw14, and Dw15). The functional relevance of such shared sequences in the third hypervariable region of DRβ chains is supported by the existence of both allospecific[69] and antigen-specific[70] T-cell clones which recognize both DR4 Dw14 and DR1 haplotypes. A recent study emphasizes the similarity of DR1 and DR4 alleles for presentation of well-defined peptide fragments to T-cells.[71] In these instances, antigen-specific T-cell recognition events appear to depend on conformational determinants encoded within the third hypervariable region of the DRβ chain, even when such determinants are present on quite distinct DRβ chain alleles.

It remains a question whether other DRβ chains which contain this third hypervariable region sequence are also associated with risk for RA. For example, the DRw6 allele found in the AMALA haplotype is present in some native Indian populations of North and South America.[42] There is currently no information concerning the association of this haplotype with RA. However, it is of interest that T-cell clones have been described which recognize the DRw6 allele found in cell line AMALA, as well as DR1 and DR4 Dw14.[69] As shown in Figure 4, the AMALA DRβI chain has a third hypervariable region sequence identical to DR1, perhaps as a result of a gene conversion event. Here again, allospecific T-cell recognition appears to follow the presence of a particular third hypervariable region sequence, indicating a degree of functional equivalence of this sequence when it is present in quite distinct DRβ chain alleles.

A related question is whether subtypes of DR1 vary in their association with rheumatoid arthritis. At least three different DRβ chain alleles may be found on DR1 haplotypes (see the Appendix). One of these, found on DR1 Dw1 haplotypes,[11] is seen commonly in the Northern European population.[72] This DR1 haplotype is probably the one which associates with RA in these populations.[66] A second DR1 allele which differs from Dw1 DRB chain in the third hypervariable region has been described by Erlich.[26] Its role in RA is unknown at present. A third DR1 allele, designated NASC DR1, was first reported in a Jewish individual with RA.[73] This DR1 allele differs from the DR1 Dw1 allele by two conservative amino acid substitutions at positions 85 and 86. These amino acid differences may well account for the Dw typing differences between these haplotypes,[74] indicating that these position 85/86 differences are important for T-cell recognition. The NASC DR1 allele has since been found in some black and hispanic individuals, a number of whom have rheumatoid arthritis.[85] However, it has not yet been shown that this DRβ chain is associated with risk for RA in population studies. Only weak associations of RA with DR1 have been reported in the black population.[75] The role of the NASC DR1 allele in conferring risk for RA in the black population is under active study in this laboratory.

XI. RECONCILING STRUCTURE WITH FUNCTION

It is clear from the studies described above that the third hypervariable region plays a fundamental role in the recognition of class II molecules by T-cells. An understanding of

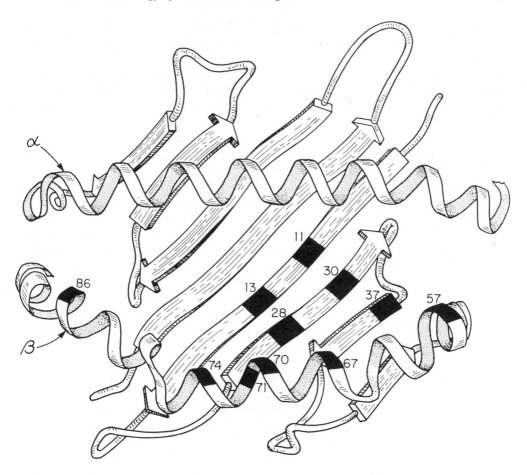

FIGURE 10.   Predicted model of the peptide binding cleft of an HLA class II molecule.[76,77] Numbers signify amino acid positions on the α and β chains of the DR molecule.

the three-dimensional structure of class II molecules provides insight into this relationship. The recent determination of the three-dimensional structure of a class I molecule[76] has allowed the modeling of a hypothetical 3-D structure for class II molecules[77] (Figure 10). This model predicts a molecule with an antigen-binding "pocket" formed by the first, or variable, domains of the class II molecule. In this model the three hypervariable regions play a central role in forming the cleft in which the antigen sits. The base of the cleft consists of β-strands formed by amino acids 1 to 50 while the rim of the cleft consists of α-helices formed by amino acids 55 to 90. The β-strands have a conformation in which alternating residues point up into the groove; thus, the polymorphic residues in the first and second hypervariable regions, notably amino acids 11, 13, 26, 28, and 30 point up towards the antigen allowing them to form specific interactions with the peptide fragment that sits in the cleft. In contrast, the α-helices which are formed by the third hypervariable regions and require 3.4 residues/ turn consist of two kinds of polymorphic residues; some point inwards into the cleft allowing them to form bonds with the antigenic fragment, while others, notably amino acids 67, 70, and 74 point upwards, away from the cleft. These presumably are involved in interacting with the T-cell receptor. With such a model it is possible to envision why the third hypervariable region plays such a prominent role in T-cell recognition.

The role of the third hypervariable region in conferring susceptibility to rheumatoid arthritis is somewhat more difficult to explain. A number of models have been proposed to explain the association of HLA class II genes with autoimmune diseases. These include: (1)

determinant selection, wherein the association is based on the ability of some class II alleles to present autoantigen to the immune system; (2) effects on the T-cell receptor repertoire, wherein the expression of some class II alleles actively selects either for or against certain T-cell receptors which may be involved in the immune response to "self"; and (3) molecular mimicry, wherein cross-reactivity between some pathogen and class II alleles leads to a break in tolerance to self class II molecules. The fact that susceptibility to RA is associated with differences in the HV3 of DRβI genes would suggest that the association is based on the ability of class II molecules to influence the T-cell repertoire. However, since some of the polymorphic residues in the HV3 are also apparently involved in binding antigen, the determinant selection model also remains a viable possibility. Indeed, it is possible that a combination of several mechanisms plays a role in determining susceptibility to RA. The delineation of the relative contributions of each of these mechanisms to the pathogenesis of autoimmune disease in general and RA in particular remains one of the unsolved puzzles in human immunology.

ACKNOWLEDGMENTS

This work has been supported by NIH grant #AI22005 (JS) and by a Pfizer Faculty Scholar Award (PKG, 1988). The authors wish to thank Ann Rupel for assistance in preparation of the manuscript and figures.

REFERENCES

1. **Benacerraf, B.,** Role of MHC gene products in immune regulation, *Science,* 212, 1229, 1981.
2. **Schwartz, R. H.,** Immune response (Ir) genes of the murine major histocompatibility complex, *Adv. Immunol.,* 38, 31, 1986.
3. **Hardy, D. A., Bell, J. I., Long, E. O., Lindsten, T., and McDevitt, H. O.,** Mapping of the class II region of the human major histocompatibility complex by pulsed-field gel electrophoresis, *Nature (London),* 323, 453, 1986.
4. **Trowsdale, J., Young, J. A. T., Kelly, A. P., Austin, P. J., Carson, S., Meunier, H., So, A., Erlich, H. A., Spielman,R. S., Bodmer, J., and Bodmer, W. F.,** Structure, sequence and polymorphism in the HLA-D region, *Immunol. Rev.,* 85, 5, 1985.
5. **Kaufman, J. F., Auffray, C., Korman, A. J., Shackelford, D. A., and Strominger, J. L.,** The class II molecules of the human and murine major histocompatibility complex, *Cell,* 36, 1, 1984.
6. **Gustafsson, K., Wiman, K., Emmoth, E., Larhammar, D.,Bohme, J., Hyldig-Nielsen, J. J., Ronne, H., Peterson, P., and Rask, L.,** Mutations and selection in the generation of class II histocompatibility antigen polymorphism, *EMBO J.,* 3, 1655, 1984.
7. **Curtsinger, J. M., Hilden, J. M., Cairns, J. S., and Bach, F. H.,** Evolutionary and genetic implications of sequence variation in two nonallelic HLA-DR β chain cDNA sequences, *Proc. Natl. Acad. Sci. U.S.A.,* 84, 209, 1987.
8. **Lotteau, V., Teyton, L., Burroughs, D., and Charron, D. J.,** A novel HLA class II molecule (DR alpha-DQ beta) created by mismatched isotype pairing, *Nature (London),* 329, 339, 1987.
9. **Okada, K., Boss, J., Prentice, H., Spies, T., Mengler, R., Auffray, C., Lillie, J., Grossberger, D., and Strominger, J. L.,** Gene organization of DQ and DX subregions of the human major histocompatibility complex, *Proc. Natl. Acad. Sci. U.S.A.,* 82, 3410, 1985.
10. **Jonsson, A.-K., Hyldig-Nielsen, J. J., Servenius, B., Larhammar, D., Andersson, G., Jörgensen, F., Peterson, P. A., and Rask, L.,** Class II genes of the human major histocompatibility complex. Comparisons of the DQ and DX α and β genes, *J. Biol. Chem.,* 262, 8767, 1987.
11. **Tonelle, C., DeMars, R., and Long, E. O.,** DOβ: a new β chain gene in HLA-D with a distinct regulation of expression, *EMBO J.,* 4, 2839, 1985.
12. **Servenius, B., Rask, L., and Peterson, P. A.,** Class II genes of the human major histocompatibility complex. The DOβ gene is a divergent member of the class II β gene family, *J. Biol. Chem.,* 262, 8759, 1987.
13. **Bodmer, W. F.,** The HLA system, 1984, in *Histocompatibility Testing 1984,* Albert, E. D., Baur, M. P., and Mayr, W. R., Eds., Springer-Verlag, Berlin, 1984.

14. **Korman, A. J., Boss, J. M., Spies, T., Sorrentino, R., Okada, K., and Strominger, J. L.,** Genetic complexity and expression of human class II histocompatibility antigens, *Immunol. Rev.,* 85, 45, 1985.

15. **Williams, A. F.,** A year in the life of the immunoglobulin superfamily, *Immunol. Today,* 8, 298, 1987.

16. **Böhme, J., Andersson, M., Andersson, G., Möller, E., Peterson, P., and Rask, L.,** HLA-DR β genes vary in number between different DR specificities, whereas the number of DQ β genes is constant, *J. Immunol.,* 135, 2149, 1985.

17. **Gorski, J., and Mach, B.,** Polymorphism of human Ia antigens: gene conversion between two DRβ loci results in a HLA-D/DR specificity, *Nature (London),* 322, 67, 1986.

18. **Gorski, J., Eckels, D. D., Thiercy, J. M., Ucla, C., and Mach, B.,** Sequence analysis of the DRw13 β chain gene: the Dw19 specificity may be encoded by the DRβIII locus, in *Immunobiology of HLA, Vol. 2, Immunogenetics and Histocompatibility,* Dupont, B., Ed., Springer-Verlag, New York, 1989, 220.

19. **Tieber, V. L., Abruzzini, L. F., Didier, D. K., Schwartz, B. D., and Rotwein, P.,** Complete characterization and sequence of an HLA class II β chain from the DR5 haplotype, *J. Biol. Chem.,* 261, 2738, 1986.

20. **Kao, H. T., Gregersen, P. K., Tang, J. C., Takahashi, T., Wang, C. Y., and Silver, J.,** Molecular analysis of the HLA class II genes in two DRw6-related haplotypes, DRw13 DQw1 and DRw14 DQw3, *J. Immunol.,* 142, 1743, 1989.

21. **Navarrete, C., Seki, T., Miranda, A., Winchester, R., and Gregersen, P. K.,** DNA sequence analysis of the HLA-DRw12 allele, *Hum. Immunol.,* 25, 51, 1989.

22. **Gregersen, P. K., Moriuchi, T., Karr, R. W., Obata, F., Moriuchi, J., Maccari, J., Goldberg, D., Winchester, R. J., and Silver, J.,** Polymorphisms of HLA-DR β chains in DR4, -7, and -9 haplotypes: implications for the mechanisms of allelic variation, *Proc. Natl. Acad. Sci. U.S.A.,* 83, 9149, 1986.

23. **Spies, T., Sorrentino, R., Boss, J. M., Okada, K., and Strominger, J. L.,** Structural organization of the DR subregion of the human major histocompatibility complex, *Proc. Natl. Acad. Sci. U.S.A.,* 82, 5165, 1985.

24. **Bell, J. I., Denney, D., Foster, Belt, T., Todd, J. A., and McDevitt, H. O.,** Allelic variation in the DR subregion of the human major histocompatibility complex, *Proc. Natl. Acad. Sci. U.S.A.,* 84, 6234, 1987.

25. **Gregersen, P. K., Kao, H. T., Nunez-Roldan, A., Hurley, C. K., Karr, R. W., and Silver, J.,** Recombination sites in the HLA class II region are haplotype dependent, *J. Immunol.,* 141, 1365, 1988.

26. **Erlich, H. A., Scharf, S., Long, C., and Horn, G.,** Analysis of isotypic and allotypic sequence variation in the HLA-DR beta region using in vitro enzymatic amplification of specific DNA segments, in *Immunobiology of HLA, Vol. 2., Immunogenetics and Histocompatibility,* Dupont, B., Ed., Springer-Verlag, New York, in press.

27. **Gregersen, P. K., Shen, M., Song, Q., Merryman, P., Degar, S., Seki, T., Maccari, J., Goldberg, D., Murphy, H., Schwenzer, J., Wang, C. Y., Winchester, R. J., Nepom, G. T., and Silver, J.,** Molecular diversity of HLA-DR4 haplotypes, *Proc. Natl. Acad. Sci. U.S.A.,* 83, 2642, 1986.

28. **Cairns, J. S., Curtsinger, J. M., Dahl, C. D., Freeman, S., Alter, B. J., and Bach, F.,** Sequence polymorphism of HLA-DRβ1 alleles relating to T cell recognized determinants, *Nature (London),* 317, 166, 1985.

29. **Mengle-Law, L., Conner, S., McDevitt, H. O., and Fathman, C. G.,** Gene conversion between murine class II major histocompatibility complex loci: functional and molecular evidence from the bm12 mutant, *J. Exp. Med.,* 160, 1184, 1984.

30. **Nathenson, S. G., Geliebter, J., Pfaffenbach, G. M., and Zeff, R. A.,** Murine major histocompatibility complex class-I mutants: molecular analysis and structure-function implications, *Annu. Rev. Immunol.,* 4, 471, 1986.

31. **Parham, P., Lomen, C. E., and Lawlor, D. A.,** Nature of polymorphism in HLQ-A, -B and -C molecules, *Proc. Natl. Acad. Sci. U.S.A.,* 85, 4005, 1988.

32. **Bentley, D. L. and Rabbits, T. H.,** Evolution of immunoglobulin V genes: evidence indicating that recently duplicated V_k sequences have diverged by gene conversion, *Cell,* 32, 181, 1983.

33. **Hill, A. V. S., Nicholls, R. D., Thein, S. L., and Higgs, D. R.,** Recombination within the human embryonic zeta-globin locus: a common zeta-zeta chromosome produced by gene conversion of the pseudo zeta gene, *Cell,* 42, 809, 1985.

34. **Spielman, R. S., Lee, J., Bodmer, W. F., Bodmer, J. G., and Trowsdale, J.,** Six HLA-D region α-chain genes on human chromosome 6: polymorphisms and associations of DCα-related sequences with DR types, *Proc. Natl. Acad. Sci. U.S.A.,* 81, 3461, 1984.

35. **Boss, J. M. and Strominger, J. L.,** Cloning and sequence analysis of the human major histocompatibility complex gene DC-3β, *Proc. Natl. Acad. Sci. U.S.A.,* 81, 5199, 1984.

36. **Todd, J. A., Bell, J. I., and McDevitt, H. O.,** HLA-DQβ gene contributes to susceptibility and resistance to insulin dependent diabetes mellitus, *Nature (London),* 329, 599, 1987.

37. **Larhammar, D., Hyldig-Nielsen, J. J., Servenius, B., Andersson, G., Rask, L., and Peterson, P.,** Exon-intron organization and complete nucleotide sequence of a human major histocompatibility antigen DCβ gene, *Proc. Natl. Acad. Sci. U.S.A.,* 84, 209, 1987.

38. **Song, Q. L., Gregersen, P. K., Karr, R. W., and Silver, J.,** Recombination between DQα and DQβ genes generates human histocompatibility leukocyte antigen class II haplotype diversity, *J. Immunol.,* 139, 2993, 1987.

39. **Hurley, C. K., Gregersen, P. K., Steiner, N., Bell. J., Hartzman, R., Nepom, G., Silver, J., and Johnson, A. H.,** Polymorphism of the HLA-D region in American blacks, *J. Immunol.,* 140, 885, 1988.

40. **Merryman, P., Silver, J., Gregersen P. K., Solomon, G., and Winchester, R.,** A novel association of DQα and DQβ genes in the DRw10 haplotype: determination of a DQw1 specificity by the DQβ chain, *J. Immunol.,* 143, 2068, 1989.

41. **Navarrete, C., Jaraquemada, D., Hui, K., Awad, J., Okoye, R., and Festenstein, H.,** Different functions and associations of HLA-DR and HLA-DQ (DC) antigens shown by serological, cellular and DNA assays, *Tissue Antigens,* 25, 130, 1985.

42. **Layrisse, A., Heinene, H. D., and Simonney, N.,** HLA-D typing with homozygous cells identified in an American indigenous isolate. II. Family studies and D/DR relationship, *Tissue Antigens,* 20, 86, 1982.

43. **Charron, D., Lotteau, V., and Tumel, P.,** Hybrid HLA DC antigens provide molecular evidence for gene transcomplementation, *Nature (London),* 312, 157, 1984.

44. **Nepom, B. S., Schwartz, D., Palmer, J., and Nepom, G. T.,** Transcomplementation of HLA genes in IDDM. HLA DQa and b chains produce hybrid molecules in DR3/DR4 heterozygotes, *Diabetes,* 36, 114, 1987.

45. **Steinmetz, M. D., Stephan, D., and Lindahl, K. F.,** Gene organization and recombinational hotspots in the murine major histocompatibility complex, *Cell,* 44, 895, 1986.

46. **Braunstein, N. S. and Germain, R. N.,** Allele-specific control of Ia molecule surface expression and conformation: Implications for a general model of Ia structure-function relationships, *Proc. Natl. Acad. Sci. U.S.A.,* 84, 2921, 1987.

47. **Amar, A., Radka, S. F., Holbeck, S. L., Kim, S. J., Nepom, B. S., Nelson, K., and Nepom, G. T.,** Characterization of specific HLA-DQα specificities by genomic, biochemical, and serologic analysis, *J. Immunol.,* 138, 3986, 1987.

48. **Liu, C.-P., Bach, F. H., and Wu, S.,** Molecular studies of a rare DR2/LD-5a/DQw3 HLA class II haplotype, *J. Immunol.,* 140, 3631, 1988.

49. **Kobori, J. A., Strauss, E., Minard, K., and Hood, L. E.,** Molecular analysis of the hotspot of recombination in the murine major histocompatibility complex, *Science,* 234, 173, 1986.

50. **Lafuse, W. P. and David, C. S.,** Recombination hot spots within the I region of the mouse H-2 complex map to the Eβ and Eα genes, *Immunogenetics,* 24, 352, 1986.

51. **Inoko, H., Tsuji, K., Groves, V., and Trowsdale, J.,** Mapping of HLA class II genes by pulsed-field gel electrophoresis and size polymorphism, in *Immunobiology of HLA,* Vol. 2, *Immunogenetics and Histocompatibility,* Dupont, B., Ed., Springer Verlag, New York, 1989, 83.

52. **Reinsmoen, N. L., and Bach, F.,** Five HLA-D clusters associated with HLA-DR4, *Hum. Immunol.,* 4, 249, 1982.

53. **Jaraquemada, D., Reinsmoen, N. R., Ollier, W., Okoye, R., Bach, F. H., and Festenstein, H.,** First level testing of HLA-DR4 associated new HLA-D specificities: Dw13(DB3), Dw14(LD40), Dw15(DYT) and DKT2, in *Histocompatibility Testing 1984: Report on the Ninth International Histocompatibility Workshop and Conference,* Albert, E. D., Baur, M. P. and Mayr, W. R., Eds., Springer-Verlag, Heidelberg, 1984.

54. **Gregersen, P. K., Goyert, S. M., Song, Q. L., and Silver, J.,** Microheterogeneity of HLA-DR4 haplotypes: DNA sequence analysis of LD"KT2" and LD"TAS" haplotypes, *Hum. Immunol.,* 19, 287, 1987.

55. **Lang, B., Navarrete, C., Winchester, R., Silver, J. and Gregersen, P. K.,** in preparation.

56. **Mickelson, E., Brautbar, C., Nisperos, B., Cohen, N., Amar, A., Kim, S. J., Lanier, A., and Hansen, J.,** HLA-DR2 and DR4 further defined by two new HLA-D specificities (HTC) derived from Israeli Jewish donors: comparative study in Caucasian, Korean, Eskimo and Israeli populations, *Tissue Antigens,* 24, 197, 1984.

57. **Stastny, P.,** Association of the B-cell alloantigen DRw4 with rheumatoid arthritis, *New Engl. J. Med.,* 298, 869, 1978.

58. **Tiwari, J. L. and Terasaki, P. I.,** *HLA and Disease Associations,* Springer Verlag, New York, 1985, 55.

59. **Stastny, P.,** Mixed lymphocyte cultures in rheumatoid arthritis, *J. Clin. Invest.,* 57, 1148, 1976.

60. **Nepom, G. T., Seyfried, C. E., Holbeck, S. L., Wilske, K. R., and Nepom, B. S.,** Identification of HLA-Dw14 genes in DR4 + rheumatoid arthritis, *Lancet,* 2, 1002, 1987.

61. **Ohta, N., Nishimura, Y. K., Tanimoto, K., Horiuchi, Y., Abe, C., Shiokawa, Y., Abe, T., Katagiri, M., Yoshiki, T., and Sasazuki, T.,** Association between HLA and Japanese patients with rheumatoid arthritis, *Hum. Immunol.,* 5, 123, 1982.

62. **Zoschke, D. and Segall, M.,** Dw subtypes of DR4 in rheumatoid arthritis: evidence for a preferential association with Dw4, *Hum. Immunol.,* 15, 118, 1986.

63. **Schiff, B., Mizrachi, Y., Orgad, S., and Gazit, E.,** Association of HLA-Aw31 and HLA-DR1 with adult rheumatoid arthritis, *Ann. Rheum. Dis.,* 41, 403, 1982.
64. **Amar, A., Oksenberg, J., Cohen, N., Cohen, I., and Brautbar, C.,** HLA-D locus in Israel. Characterization of 14 local HTC's and a population study, *Tissue Antigens,* 20, 198, 1982.
65. **Woodrow, J. C., Nichol, F. E., and Zaphiropoulos, G.,** DR antigens and rheumatoid arthritis: a study of two populations, *Br. Med. J.,* 283, 1287, 1981.
66. **Legrande, L., Lathrop, G. M., Marcelli-Barge, A., Dryll, A., Bardin, T., Debeyre, N., Poirier, J. C., Schmid, M., Ryckewaert, A., and Dausset, J.,** HLA-DR genotype risks in seropositive rheumatoid arthritis, *Am. J. Hum. Genet.,* 36, 690, 1984.
67. **Duquesnoy, R. J., Marrari, M., Hackbarth, S., and Zeevi, A.,** Serological and cellular definition of a new HLA-DR associated determinant, MC1, and its association with rheumatoid arthritis, *Hum. Immunol.,* 10, 165, 1984.
68. **Gregersen, P. K., Silver, J., and Winchester, R. J.,** The shared epitope hypothesis: an approach to understanding the molecular genetics of susceptibility to rheumatoid arthritis, *Arthritis Rheum.,* 30, 1205, 1987.
69. **Seyfried, C. E., Mickelson, E., Hansen, J. A., and Nepom, G. T.,** A specific nucleotide sequence defines a functional T cell recognition epitope shared by diverse HLA-DR specificities, *Hum. Immunol.,* 21, 289, 1988.
70. **Weyand, C. and Goronzy, J.,** Shared conformational T cell epitopes on DR-molecules of the HLA-DR1 and DR4 haplotypes associated with rheumatoid arthritis, *Arthritis Rheum. (Suppl.),* 30, S25, 1987.
71. **Lamb, J. R., Rees, A. D. M., Bal, V., Ikeda, H., Wilkinson, D., De Vries, R. R. P., and Rothbard, J. B.,** Prediction and identification of an HLA-Dr-restricted T cell determinant in the 19-kDa protein of *Mycobacterium tuberculosis, Eur. J. Immunol.,* 18, 973, 1988.
72. **Jaraquemada, D., Okoye, R., Ollier, W., Asad, J., and Festenstein, H.,** HLA-D and DR in different populations, in *Histocompatibility Testing 1984,* Albert, E. D., Baur, M. P., and Mayr, W. R., Eds., Springer Verlag, New York, 1984.
73. **Merryman, P., Gregersen, P. K., Lee, S., Silver, J., Nunez-Roldan, A., Crapper, R., and Winchester, R. J.,** Nucleotide sequence of a DRw10 β chain cDNA clone, *J. Immunol.,* 140, 2447, 1988.
74. **Hurley, C. K., Ziff, B. L., Silver, J., Gregersen, P. K., Hartzman, R., and Johnson, A. H.,** Polymorphism of the HLA-DR1 haplotype in American blacks, *J. Immunol.,* 140, 4019, 1988.
75. **Alarcon, G. S., Koopman, W. J., Acton, R. T., and Barger, B. O.,** DR antigen distribution in blacks with rheumatoid arthritis, *J. Rheumatol.,* 10, 579, 1983.
76. **Bjorkman, P. J., Saper, M. A., Samraoui, B., Bennett, W. S., Strominger, J. L., and Wiley, D. C.,** Structure of the human class I histocompatibility antigen, HLA-A2, *Nature (London),* 329, 506, 1987.
77. **Brown, J. H., Jardetzky, T., Saper, M., Samraoui, B., Bjorkman, P. J., and Wiley, D. C.,** A hypothetical model of the foreign antigen binding site of class II histocompatibility molecules, *Nature (London),* 332, 845, 1988.
78. **Wu, S., Yabe, R., Madden, M., Saunders, T. L., and Bach, F. H.,** cDNA cloning and sequencing reveals that the electrophoretically constant DR β_2 molecules, as well as the variable DR β_1 molecules, from HLA-DR2 subtypes have different amino acid sequences including a hypervariable region for a functionally important epitope, *J. Immunol.,* 138, 2953, 1987.
79. **Lee, B. S. M., Rust, N. A., McMichael, A. J., and McDevitt, H. O.,** HLA-DR2 subtypes for an additional supertypic family of DRB alleles, *Proc. Natl. Acad. Sci. U.S.A.,* 84, 4591, 1987.
80. **Gorski, J.,** First domain sequence of the HLA-DRβ1 chain from two HLA-DRw14 homozygous typing cell lines: TEM (Dw9) and AMALA (Dw16), *Hum. Immunol.,* 24, 145, 1989.
81. **Moriuchi, J., Moriuchi, T., and Silver, J.,** Nucleotide sequence of an HLA-DQ α chain derived form a DRw9 cell line: genetic and evolutionary implications, *Proc. Natl. Acad. Sci. U.S.A.,* 82, 3420, 1985.
82. **Schenning, L., Larhammar, D., Bill, P., Wiman, K., Jonsson, A.-C., Rask, L. and Peterson, P.,** Both α and β chain of HLA-DC class II histocompatibility antigens display extensive polymorphism in their amino terminal domains, *EMBO J.,* 3, 447, 1984.
83. **Hurley, C. K., Gregersen, P. K., Steiner, N., Bell, J., Hartzman, R., Nepom, G., Silver, J., and Johnson, A. H.,** Polymorphism of the HLA-D region in American blacks, *J. Immunol.,* 140, 885, 1988.
84. **Schiffenbauer, J. D., Didier, K., Klearman, M., Rice, K., Shuman, S., Tiber, V. L., Kittlesen, D. J., and Schwartz, B. D.,** Complete sequence of the HLA DQα and DQβ cDNA from a DR5/DQw3 cell line, *J. Immunol.,* 139, 228, 1987.
85. **Nepom, G. T.,** personal communication.
86. **Gregersen, P. K. and Winchester, R. J.,** unpublished observations.

Chapter 7

HLA CLASS II SEQUENCE VARIATION AND DISEASE SUSCEPTIBILITY

Henry A. Erlich, Glenn Horn, Stephen Scharf, and Teodorica Bugawan

TABLE OF CONTENTS

I. INTRODUCTION

A variety of autoimmune diseases have been associated with serologically defined variants of the HLA class II antigens.[1] These cell surface glycoproteins consist of an α chain ($\sim$32 kDa) and a β chain ($\sim$29 kDa) encoded by loci on the short arm of chromosome 6. Three related class II antigens have been identified and are designated HLA-DR, HLA-DQ, and HLA-DP. With the exception of the DRα and DPα chains, these proteins are highly polymorphic with the variability being localized to the NH_2-terminal outer domain encoded by the second exon (reviewed in References 2, 3). T cell recognition of foreign antigen results from an interaction between the T cell receptor, antigen peptide fragments, and the polymorphic residues of the class II products.[4,5] In the mouse, immune responsiveness to defined antigens has been mapped to specific class II alleles.[6] The HLA association of human autoimmune diseases may reflect a similar relationship between a specific immune response and polymorphic class II residues. Alternatively, the class II variants may simply be genetic markers in linkage disequilibrium with other putative disease susceptibility alleles. The analysis of class II haplotypes by cellular, biochemical, and molecular techniques has revealed extensive genetic heterogeneity *within* the serologically defined types (reviewed in References 2,3,7). The study of specific class II DNA-defined alleles among patients and controls may elucidate the nature of HLA disease associations and should provide more informative markers for genetic susceptibility.

In this chapter, the general issue of HLA-linked disease susceptibility will be explored by focusing on the autoimmune dermatologic disease *pemphigus vulgaris* (PV) and the autoimmune endocrine disease insulin-dependent diabetes mellitus (IDDM). For these, as well as for other HLA-associated diseases, our experimental approach has been to examine the distribution of HLA class II allelic sequences in patients and in HLA-matched controls. This rather formidable task has been greatly facilitated by the use of the polymerase chain reaction (PCR) method of specific *in vitro* DNA amplification.[8,9] The PCR amplified HLA class II sequences can be directly cloned into M13 vectors for sequence analysis[10] or sequenced directly.[11] To generate the data set required for statistical evaluation, the frequency of the class II alleles defined by sequencing can be measured in patients and controls using allele-specific oligonucleotide (ASO) probes to analyze PCR amplified DNA in a simple dot-blot format.[12] By comparing the sequences of several class II loci from a variety of patient and control haplotypes, one can begin to address a fundamental genetic question: Is HLA-associated disease susceptibility conferred by (1) specific *epitopes*, short stretches of amino acids found in several different allelic frameworks, (2) specific class II *alleles*, or (3) by specific combinations of class II alleles or epitopes in *haplotypes* or *genotypes*?

II. INSULIN-DEPENDENT DIABETES MELLITUS (IDDM)

Insulin-dependent diabetes mellitus (IDDM or Type I Diabetes) is an autoimmune disease in which dysfunctional regulation of glucose metabolism results from the immunologically mediated destruction of the insulin-producing islet cells of the pancreas.[13] The 50% concordance rate of monozygotic twins indicates a significant genetic component for IDDM and suggests, as well, that an ''environmental trigger'' (e.g., viral infection) may be required to elicit the clinical disease in genetically susceptible individuals.

Loci in the HLA region of chromosome 6 contribute a major portion of the genetic predisposition based on the concordance rates (25%) of HLA identical sibs, patterns of haplotype sharing among affected sib pairs, and linkage analysis in families.[1] In addition, population analysis has revealed that the frequency of certain serologically defined variants of the HLA class II antigens is different among patients and controls. These studies found that the serologic types DR3 and DR4 were positively associated with IDDM while DR2 was negatively associated.[1]

Recently, we have described our analysis of the sequence polymorphism in the HLA-DQα and DQβ loci, correlating the pattern of allelic variation with the major serologic specificities (DQw1, DQw2, DQw3, and DQ-blank) and with susceptibility to IDDM.[14]

A. DQ ALLELIC SEQUENCE VARIATION

Allelic variation in the HLA-DQ region has previously been defined by the serological DQw1, DQw2, and DQw3 specificities, and by DQ-blank.[15] Thus far, eight allelic DNA sequence variants have been identified at the DQα locus (Figure 1) and thirteen at DQβ (Figure 2)[14] We have compared these sequence variations with conventional HLA classifications and propose a new nomenclature that has its basis in the correspondence between current DQw serological specificities and sequence patterns. Specifically, the DNA-defined types DQB1, DQB2, and DQB3 designate β chain sequences derived from the serologically defined DQw1, DQw2, and DQw3 haplotypes, respectively, while DQB4 designates those sequences derived from DQ(blank) haplotypes, an apparently homogeneous type.[14] In the suggested nomenclature for these DNA-defined alleles, we have adopted the convention of using greek letters for the genetic loci and for the protein products (e.g., the DQβ locus encoding the DQβ chain), and have designated the specific allelic sequence variants with a number preceded by a Roman capital letter (e.g., the DQB2 allele). Sequence variants that subdivide these types are designated by a subtype number (e.g., DQB1.2 or DQA1.3). The designation of DQα allelic variants, however, does not always correspond to the DQw specificity; for example, the DQA4 type is associated with both the DQw2 and DQw3 haplotypes. This is because the DQw2 and DQw3 specificities appear to be determined by polymorphic epitopes on the β-chain, independent of allelic variation on the α chain.[14] This nomenclature, based on the DQw specificities, differs from our previous designations for the DQα alleles[12] which referred to the associated DR serotype. In this system, new sequences can be assigned a new number (e.g., DQA5) or a new subtype number (e.g., DQA1.4), depending on their homology with other alleles.

B. SUSCEPTIBILITY TO AUTOIMMUNE DISEASE

The availability of a large number of DQβ allelic sequences (Figure 2), including 20 sequences from diabetic patients, allows the tentative localization of epitopes including putative disease susceptibility epitopes. DR4-bearing haplotypes occur at increased frequency in IDDM patients.[1] However, the DRβ variants, which correlate with the cellular Dw typing,[16] are only weakly associated with IDDM.[17]* By contrast, the two major allelic variants found in DQβ, (DQB3.1 and DQB3.2[17-19]) have been shown by serologic and RFLP studies to correlate well with the disease, with DQB3.2 accounting for almost all of the DR4-associated susceptibility.[20-23] The DQB3.1 and DQB3.2 variants are distinguished by their reactivity to the monoclonal antibody TA10[24] as well as by their susceptibility to IDDM. The DQB3.1 β chains react with this antibody (TA10+) while the more frequent DQB3.2 allele encodes an antigen which fails to bind TA10 (TA10−).

From Figure 2, it can be seen that the DQB3.1 and DQB3.2 alleles differ by only four amino acids in the outer domain. It is likely that their differential IDDM susceptibility and TA10 reactivity involves one or more of these polymorphic residues. The cell line KOZ (DQB3.3) differs from DQB3.2 only at position 57. KOZ (Asp-57) is TA10−, whereas other sequences with the same asparate substitution at 57 (DQB3.1) are TA10+. Thus, the TA10 epitope does not map to position 57 and, based on the pattern of allelic DQβ allelic variation, can be tentatively assigned to the region around the glutamate residue at position 45 since this is the only residue that correlates with TA10 reactivity and the DQB3.1 allele.

It is the charge of the polymorphic residue at position 57, however, that is associated

---

* In combination with the DQB3.2 allele, however, specific DRβI subtypes of DR4 (e.g., Dw4 and Dw10) *do* contribute to IDDM susceptibility (see below).

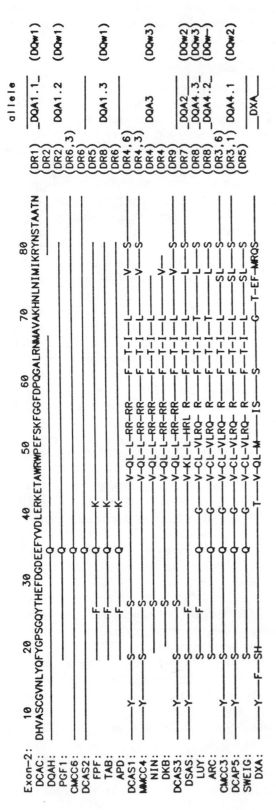

FIGURE 1. Alignment of HLA-DQα protein sequences. The DNA sequence of the DQα alleles was translated to the standard one-letter amino acid code and aligned to show patterns of homology. A dash indicates homology with the equivalent amino acid in the prototypic DQA1.1 allele. A blank indicates that the sequence was not determined, except at position 55 in the DQA2 alleles, where a space was included for alignment purposes. Note that the PCR amplification procedure only determines the sequence between the oligonucleotide primers.[14] Positions within the mature DQα protein are shown at the top, and the locations of the PCR amplification primers are shown on the bottom. The source of each sequence is designated on the left of each line, and its DR serologic type is shown on the right. An asterisk after the DR type indicates that the allele was determined from a patient with IDDM. On the far right is the designation of the allele, corresponding with the DQw typing, when available, of the haplotype. Sequences LG2, CMCC, and MMCC are from Reference 41; the other sequences are reported in or referenced in Reference 14.

133

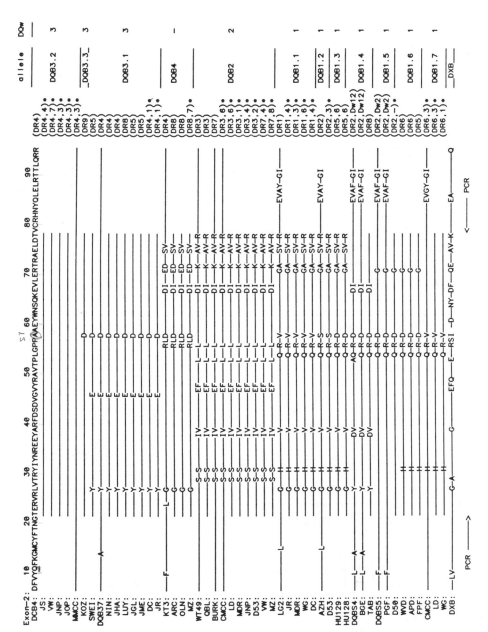

FIGURE 2. Alignment of HLA-DQβ protein sequences. The DNA sequence of the DQβ alleles was translated to protein as in Figure 1, and aligned to show patterns of homology. See the legend to Figure 1 for the explanation of the labels and symbols. Most of the sequences are reported in or referenced in Reference 14. D53 (DR2,3) and D50 (DR2⁻) are both DR2⁺ IDDM patients. HU128 (DR5,6) and HU129 (DR5,6) are both PV patients. The asterisk indicates that the sequence was determined from an IDDM patient.

TABLE 1
Distribution of DQβ Alleles in DRw6, DQw1
Haplotypes

	Controls	IDDM
√ Val-57		
DQB1.7 (Dw19)	9/26 (34.6%)	14/20 (70%)
DQB1.1	2/26 (7.7%)	2/20 (10%)
Ɖ Asp-57		
DQB1.6 (Dw18)	12/26 (46.2%)	4/20 (20%)
DQB1.3 (Dw9)	3/26 (11.5%)	0/20 (0%)

with IDDM susceptibility (see below). As with the DR4 haplotype, DQβ allelic variation defines subtypes of the DRw6 haplotype. Four alleles have been identified to date, (DQB1.1, DQB1.3, DQB 1.6, and DQB1.7).[25] The DQβ sequence derived from three DRw6 IDDM patients contained valine at position 57 (DQB1.7) while the DQβ sequence from two DRw6 HTC's contained aspartate (DQB1.6) (Figure 2). The analysis of PCR amplified DNA with sequence-specific oligonucleotide probes[12,25] revealed that the DQB1.7 allele (Val-57), which is associated with the DRw6, Dw19 subtype, was present in most of the DRw6 patients (70%) and in only 35% of DRw6 controls (Table 1). Most DRw6 controls (58%) had the DQB1.6 allele (Asp-57) associated with the Dw18 subtype or the DQB1.3 allele (Asp-57) associated with the Dw9 subtypes (Table 1).

The correlation between residue 57 in the DQβ chain and IDDM susceptibility extends beyond the DR4 and DRw6 subtypes, discussed above, to the general pattern of haplotypic disease association. DR2 has been negatively associated with IDDM, an observation interpreted to mean that this haplotype confers "resistance" or "protection." DQβ allelic variation subdivides the DR2 haplotype into the common alleles DQB1.4 and DQB1.5. These correspond to the MLC-defined subtypes, Dw12 and Dw2, respectively, and both contain Asp-57. DQβ RFLP analysis of DR2 IDDM patients has revealed that most of these DR2 haplotypes are of the rare "AZH" (also known as "MN2") type.[17] Unlike Dw2 and Dw12 which have an aspartate at position 57, the DQβ allele from the AZH HTC (termed DQB1.2 in Figure 2) contains a serine at position 57.[26] Recently, we have determined the sequence of two DR2 IDDM patients (Figure 2); one DR2 haplotype contained the rare DQβ allele (DRB1.2, Ser-57) found in AZH and one contained the DQB1.5 allele (Dw2) (Asp-57) found in the majority (~90%) of Caucasian DR2 controls. A positive association for DR1 and IDDM has been recently reported[27] as well as a negative association for DR5.[28] Consistent with the codon 57 pattern described above, the DR1 DQβ allele (DQB1.1) contains Val-57 and the DQβ allele from the DR5 haplotype (DQB3.1) contains Asp-57. Like the DR4 haplotype, the DQβ allele (DQB2) from the DR3 haplotype, which has been positively associated with IDDM, contains Ala at position 57.

Further support for the hypothesis that position 57 in DQβ is relevant to diabetes is found in the sequences of the analogous murine I-Aβ alleles. Whereas all of the "normal" Aβ alleles contain aspartate at position 57, the allele in the diabetes-susceptible NOD mouse contains serine[29] (Table 2). This variation in the NOD strain involves the same substitution of a hydrophobic or neutral amino acid residue for the charged aspartate as is seen in the human alleles. This overall pattern of homology suggests that this region of the DQβ protein, and in particular the charge of residue 57, plays an important role in the development or suppression of autoimmune responses. Similar results have been reported recently by Todd et al.[30] This hypothesis can account in part for the complex pattern of DR haplotypes positively (e.g., DR3, DR4, DR1) and negatively (e.g., DR2, DR5) associated with IDDM. Moreover, it suggests that it is the class II antigens themselves and not some putative

TABLE 2

**Correlation of IDDM Susceptibility with Residue 57 of
Class II β Chains**

Locus	Haplotypes	Sequence	IDDM assocation
	Human	⇓	
DQβ	DR2, DRw6, (DQB1.6)	GRP D AEY	−
	DRw8	GRL D AEY	No
	DR4 (DQB3.1), DR5	GPP D AEY	−
	DR1, DRw6 (DQB1.7)	GRP V AEY	+
	DR4 (DQB3.2)	GPP A AEY	+
	DR3	GLP A AEY	+
	DR7	GLP A AEY	No
	DR2 "AZH"	GRP S AEY	+
DRβIII	DRw52b (DR3,5,w6)	GRP D AEY	No
	DRw52a,c (DR3,w6)	GRP V AES	+
	Mouse		
I-Aβ	d,k,b,u,s,q	GRP D AEY	−
	f	GRS D AEY	−
	NOD	GRH S AEY	+

Note: Shown are residues 54 to 60 of class II β chains. The arrow indicates the position of residue 57. A " + " indicates a positive association with IDDM (a higher frequency of the marker in patients than in controls). A " − " indicates a negative association (a lower frequency of the marker in patients than in controls), and "no" indicates equal frequencies of the marker in patients and controls). See text for discussion. The IDDM association of the DR serotypes are reported in References 1, 27, and 28. The association of the sequence defined subtypes of DR2, DR4, and Drw6 are reported in References 7, 14, and 30. The association of DRw52a and 52c is reported in Reference 7 and is based on unpublished results.

susceptibility gene in linkage disequilibrium with the serologic markers that is responsible for HLA-linked IDDM predisposition. Whether the aspartate at position 57 confers "resistance" or "protection" by eliciting specific immunosuppression for a diabetes-related antigen or by eliminating specific T cell clones by cross-tolerance or whether the alanine (and valine or serine) preferentially stimulates an autoimmune response remains to be established.

The sequence around position 57, as well as the Asp to Ala (or Val) polymorphism, is generally conserved in other class II β-chains (Table 2). The DR3 DRβIII variant with Val-57 (DRw52a) appears to be associated with higher IDDM susceptibility than the DR3βIII variant with Asp-57 (DRw52b).[19] Susceptibility to other autoimmune diseases such as PV (discussed below) is also related to DQβ codon 57 polymorphism.[25]

For IDDM and DQβ sequence variation, the disease association of a haplotype can generally be predicted by the charge of the polymorphic residue at position 57, in that only uncharged residues appear to be associated with susceptibility. However, not all DQβ alleles with the charged Asp-57 residue are "equally protective," since DR2 haplotypes are rarer among patients than are DR5 haplotypes. Similarly, some haplotypes with a neutral residue at position 57 of the DQβ chain appear to confer more risk than do others. One specific exception to the position 57 pattern, discussed above, is the DR7 haplotype which, like DR3, contains the DQB2 allele with Ala-57 but, unlike DR3, is *not* increased among patients.

TABLE 3
DR4 DR-beta I Subtypes Defined by Oligonucleotide
Probes

		GH78 "DW10"	GH101 "Dw4"	GH100 "Dw14"	CRX01 "Dw13"
I.	IDDM	7/66	42/66	13/66	2/40
	% Positive	11	64	20	5
II.	Controls	1/50	21/50	23/50	2/46
	% Positive	2	42	46	4

Thus, either the DR3 haplotype contains additional susceptibility alleles absent from the DR7 haplotype or, conceivably, the DR7 DQα-chain (DQA2) modifies the conformation of the β-chain. Also, the increased risk for DR3/DR4 individuals relative to DR3/DR3 and DR4/DR4[1] suggests that the susceptibility conferring sequences on these two haplotypes are *different* as both DQβ alleles contain Ala at position 57. In general, it is likely to be the DQβ *allele* and not simply the residue at position 57 that confers susceptibility, as demonstrated in a recent study of *Pemphigus vulgaris*.[25]

Another exception is the susceptible genotype DR1/DR4 where, for the DR4 haplotype, the DRβI Dw4 allele appears to be increased in patients but *not* the DQB3.2 allele (Ala-57).[54] In general, the Dw4 and Dw10 DRβI alleles are both increased in DR4[+] IDDM patients relative to DR4[+] controls and the Dw14 allele decreased (Table 3), suggesting a role for *both* the DQβ and DRβI products in IDDM predisposition. The analysis of DQβ and DRβI alleles using oligonucleotide probes as well as cellular and serological typing indicates that DR4 susceptibility is correlated with a specific combination of DQβ (3.2) and DRβI (Dw4 and Dw10) epitopes.[32,33] The polymorphic residues in the third hypervariable region of DRβI that distinguish Dw10 from the other DR4 subtypes may be involved in IDDM susceptibility since the same polymorphic epitope was recently identified on a DR1 encoded DRβI chain and a DR5 chain from IDDM patients (Figure 3). This sequence is also found on most DRw6 DRβI alleles in both patients and controls.

A role of HLA linked sequences other than DQβ in IDDM susceptibility is also suggested by the distribution of HLA markers among Japanese patients. Here, the DR4 and DRw9 serotypes are increased and the DR2 serotype decreased.[34] In Japanese, the DR4 haplotype carries the DQB4 allele (Asp-57), the DRw9 haplotype, the DQB3.3 allele (Asp-57), the DR2 haplotype, and the DQB1.4 allele (Asp-57) (Figure 2). Although the reduced incidence of IDDM in Japan may reflect in part the nature of the amino acid at position 57, this DQβ residue cannot account for the observed DR4, DRw9, and DR2 associations within the Japanese population. Thus, although the pattern of DQβ allelic variation clearly implicates position 57 in IDDM predisposition, this region does not appear to be the *only* class II epitope on susceptibility-conferring haplotypes which contributes to autoimmune diabetes.

C. VIRUSES AND IDDM

An environmental "triggering" agent, such as viral infection, may also be required for disease to develop in genetically susceptible individuals. A variety of viruses have been associated with autoimmune diabetes in humans, including cytomegalovirus (CMV) and rubella (reviewed in Reference 35). Congenital infection with rubella dramatically increases the risk for IDDM.[36] Unlike Coxsackie B, which has also been implicated in IDDM, CMV and rubella have been associated with the production of autoantibodies such as islet cell antibody or cytotoxic beta cell surface antibody.[37] In two recent studies, the presence of the CMV genome was also shown to be increased in IDDM patients[37] and was correlated with the HLA class II genotype defined by RFLP analysis.[38] If the region around position 57 of the DQβ protein is involved with antigen binding or T cell recognition (see below), then it

Alignment of HLA DR-beta-I Protein Sequences

```
                    10         20         30         40         50         60         70         80         90
CONSENSUS: RFLEQ*K*ECHFFNGTERVRFLDRYFY*QEEYVRFDSDVGEYRAVTELGRPDAEYWNSQKDLLEQ*RAAVDTYCRHNYGVGESFTVQRR
DR1:       ——W—L—F—————————————L—E—CI—N——S———————————————————————————————R————————————————————————————
DR1:       ——W—L—F—————————————L—E—CI—N——S—————————————————————————————I——DE—————————————————————————*

DR4-Dw4:   ——————V—H———————————————————————————————————————————————————————K——————————————————————————
DR4-Dw10:  ——————V—H————————————————H——————————————————————————————————I——DE———————————————————V——————*
DR4-Dw13:  ——————V—H————————————————H———————————————————————————————————R——E——————————————————V——————
DR4-DwKT2: ——————V—H————————————————H————————S——————————————————————————R——E——————————————————V——————
DR4-Dw14:  ——————V—H————————————————H————————————————————————————————————R——————————————————————V——————
DR4-Dw15:  ——————V—H————————————————H——————————————————————S————————————R——————————————————————V——————

DR5:       ————————————————————YSTS———————N————————————————————E————————F——DR——————————————————————————
DR5:       ————————————————————YSTS———————N————————————————————E————————I——DE———————————————————V——————*
```

FIGURE 3. Alignment of selected DRβI amino acid sequences from the variable amino terminal domain. Allelic sequences sharing the same polymorphic residues (ILEDE) are marked by the asterisk. They have been found in a DR1/1 IDDM patient, a DR4/5 IDDM patient, and in many DR4+ IDDM patients and controls. These DR5 and DR1 alleles are very rare among controls.

TABLE 4
Homologies between IDDM-Associated HLA-DQβ Alleles and EBV

DQβ Allele	HLA Epitope	EBV Homology	Position	Phase	ORF Size	ORF Name
		**Alleles Positively Associated with IDDM**				
DQB2	GLPAAEY	GLPAA	792	R1	134	
(DR3)			67134	R3	188	
			80004	R1	30	
			118884	R3	57	
			134232	F1	72	
		PAAEY	73713	R3	1374	BOLF1
DQB3.2	GPPAAEY	GPPAA	12404[a]	F1	129	BWRF1
(DR4)			61311	R1	20	
			100137[b]	F3	872	BERF4
			100257	F3	872	BERF4
		PPAAEY	73713	R3	1374	BOLFI
		**Allele Negatively Associated with IDDM**				
DQB1.4	GRPDAEY	RPDAE	167112	R3	101	BNLF26
(DR2)						

Note: The segment of HLA epitope (column 2) centered on position 57 (underlined) was used to search for exact matches of 5 or more residues (column 3) in the translations of the entire EBV genome. The position, translation phase, and open reading frame (ORF) for each match was then determined.

[a] Repeated 12 times as part of the 3072-bp "IR1" repeat.
[b] Directly repeated 6 times.

could conceivably also be a target of molecular mimicry by pathogens to avoid host immune defenses. Thus, individuals whose class II molecules share epitopes with a virus might fail to respond to that specific viral epitope. Alternatively, shared epitopes could serve as cross-reactive target auto-antigens for the immune response to the pathogen. Since the entire genome of the Epstein-Barr virus (EBV) is known, we searched the sequence for shared pentapeptides with DQB and DRβI sequences. Peptides and potential epitopes which include position 57 of the DQB2 (DR3) and DQB3.2 (DR4) alleles are more homologous to this persistent viral pathogen (EBV) than are peptides from the DR2 allele (Table 4), and many of the DR4-associated peptides are found in repeated segments of the EBV genome.[39] Particularly striking is a segment in the BERF4 open reading frame near position 100137 of EBV, which encodes a sixfold direct repeat of the "GPPAA" epitope found at position 57 in the DQB3.2 allele. Repeated structures of this type are known to be more immunogenic than isolated epitopes.[40] Although the relationship of this viral homology to HLA-associated disease susceptibility is not clear, it seemed interesting that the E1 envelope protein of rubella,[41] a virus implicated in the pathogenesis of IDDM,[36] also contains the GPPAA peptide at position 261. A shared peptide sequence has also been reported for the CMV IE2 protein (PLGRPD at residues 81 to 88) and residues 52 to 57 of the DRβI chain (ELGRPD) and of some DQβ chains (PQGRPD)*.[42] In addition, a peptide in the third hypervariable region of DRβI (DR4, Dw4; DR3; DR4, Dw14) is also shared by the gp110 protein of EBV (Figure

---

* This peptide is encoded by the DRB1.4, 1.5, and 1.6 alleles (see Figure 2).

```
808
--  E  Q  K  R  A  A  --    gp110


70
--  E  Q  K  R  A  A  --    DR4, Dw4

--  E  Q  K  R  A  G  --    DR3


--  E  Q  R  R  A  A  --    DR4, Dw14; DR1; others
```

FIGURE 4.   Shared epitopes for DRβI and EBV gp110. The amino acid residues from position 807 to 812 of the EBV glycoprotein gp110 and from position 69 to 74 of the DRβI chain are shown.

4). It has recently been suggested that these DRβI shared epitopes could be involved in predisposition to rheumatoid arthritis.[43] Such shared structures could serve as a cross-reactive target autoantigen for an immune response to the viral pathogen. Alternatively, individuals with the shared epitope might be unable to respond to the viral epitope due to cross-tolerance. While these correlations are intriguing, the functional relationship between HLA class II epitopes, viral antigen mimicry, and aberrant immunological responses remains to be determined.

III. *PEMPHIGUS VULGARIS*

Pemphigus vulgaris is an autoimmune dermatologic disease mediated by autoantibodies to an, as yet, uncharacterized epidermal membrane protein.[44] In population studies, it is strongly associated with the serotypes HLA-DR4 and HLA-DRw6[45,46] with less than 5% of the patients possessing neither marker. Disease associations with two different HLA haplotypes can be interpreted to mean that either these two haplotypes share a comon allele or epitope or, alternatively, that different alleles on the two haplotypes are capable of conferring disease susceptibility.

As noted above, molecular analysis of serologically defined class II haplotypes (e.g., DR4 and DRw6) has revealed significant variation at the DR and DQ loci. Sequence analysis of class II genes from various haplotypes demonstrated that a DRβI epitope at position 68 to 72 (termed here "IDE") was shared between a subset of DR4 haplotypes ("Dw10")[11] and a subset of DRw6 haplotypes (DRw13).[47,48] This observation suggests the possibility that these shared polymorphic residues of the DRβI chain, which determines the DR specificity, may contribute to PV autoimmunity and account for both the DR4 and DRw6 associated susceptibility.

A. DR4-ASSOCIATED SUSCEPTIBILITY

The sequence analysis of DRβI, DRβIII, and DQβ alleles from three U.S. DR4+ (DR4/4, DR4/5, and DR4/5) PV patients revealed that all the patients contained the "Dw10" DRβI allele[49] found in only 10% of U.S. DR4+ controls.[50] Furthermore, three of the four DQβ alleles from the DR4 haplotypes were the DQB3.2 allele present on 60 to 80% of control DR4 haplotypes. These sequence data, although obtained from only three patients, suggested that, for the DR4 susceptibility to PV, the DRβI Dw10 allele may contribute to autoimmunity but that the DQβ allele does not.

The analysis of PCR-amplified DNA from a panel of PV patients with a set of DRβI sequence specific oligonucleotide probes, revealed that virtually all DR4+ haplotypes con-

TABLE 5

Oligonucleotide Probe Analysis of HLA-DRβI Subtype Frequencies

	Probes					
	GH78 "Dw10"		GH101 "Dw4"		GH100 "Dw14"	
DR Haplotype	PV	Controls	PV	Controls	PV	Controls
Israeli DR4	24/24	15/25	0/24	1/25	6/24	10/25
% Positive	100	60	0	4	25	40
		($p = 0.001$)				
Israeli DR6	4/14	8/13	—	—	—	—
% Positive	29	62	—	—	—	—
		($p = 0.035$)				
Non-Israeli DR4	10/14	1/19	1/14	6/19	3/14	10/19
% Positive	71	5	7	30	21	55
		($p = 0.001$)				
Non-Israeli DR6	2/10	7/7	—	—	—	—
% Positive	20	100	—	—	—	—
		($p = 0.02$)				

TABLE 6

Oligonucleotide Probe Analysis of HLA DQβ 3.1 and 3.2 Subtype Frequencies Analysis

	Probes			
	DQ3.1 (GH92)		DQ3.2 (GH74)	
DR Haplotype	PV	Controls	PV	Controls
Israeli DR4	2/24	1/25	22/24	23/25
% Positive	8	4	92	92
Non-Israeli DR4	2/14	8/20	12/14	12/20
% Positive	14	40	86	60

tained the "Dw10" DRβI allele as defined by the GH78 ("IDE") probe[25] (Table 5). The frequency of the Dw10 subtype defined by MLC typing, among U.S. DR4[+] controls, is estimated to be about 10%.[50] The frequency determined by oligonucleotide probe analysis in our control panel was 5% for non-Israeli DR4[+] controls and 60% for Israeli DR4[+] controls[25] (Table 5). Sequence analysis of the DQβ alleles in three DR4[+] PV patients[49] as well as the oligonucleotide probe analysis of a large panel of patients and controls[25] (Table 6) revealed that most (92%) DR4[+] haplotypes from patients contained the DQB3.2 allele, as did most (92%) DR4[+] control haplotypes. The DQα and the DRβIV loci on DR4 haplotypes are invariant. Thus, the observed distribution of DRβI and DQβ alleles among patients and controls suggests that it is the DRβI allele ("Dw10") on the DR4 haplotype that is disease-associated and that may confer susceptibility to PV. Moreover, the association of DR4, Dw10 with PV reported here is consistent with the ethnic distribution of this disease since the frequencies of PV and of the DR4, Dw10 haplotype are both increased among Ashkenazi Jews[25,45,46] (Tables 3 and 5).

The DR4, Dw10, haplotype differs from the other DR4 haplotypes (Dw4, Dw14, Dw13) by only three amino acid residues ("IDE") at positions 68 to 72 (also known as HVR3 or the third hypervariable region) of the DRβI chain.[16] Thus, the differential PV susceptibility of the DR4, Dw10 haplotype appears to be determined by the residues in the third hypervariable region. These polymorphic residues must constitute an epitope recognized by the

T cell receptor since they distinguish the various MLC-defined Dw subtypes of DR4. Polymorphic residues in this region have also been implicated in the DR4 associated susceptibility to insulin-dependent diabetes[14,33] (Table 3, Figure 3) and to rheumatoid arthritis.[51]

B. DRw6 Susceptibility

The DRw6 susceptibility to PV, however, cannot be accounted for by the sequence or shared "epitope" at position 68 to 72 of the DRβI chain since only 29% of DRw6[+] patient haplotypes and 62% of control DRw6 haplotype contain this "IDE" sequence, as determined by oligonucleotide probe analysis (Table 5). In the case of DR4/DRw6 samples, specific hybridization of the GH78 probe ("IDE") to the PCR amplified DNA could be attributed to either haplotype. However, the unambiguous assignment of the probe-defined epitope to a specific haplotype could be accomplished by restriction endonuclease digestion of the PCR-amplified DNA with an enzyme that cleaves one of the alleles and Southern blot analysis with the GH78 probe. *Ddel* cleaves the DRβI allele of the DRw6 haplotype but not the DRβI allele from the DR4 haplotype, allowing identification of the particular DRβI allele which hybridizes to the probe.[25]

Since the shared "IDE" epitope on the DRβI chain cannot account for the DRw6 susceptibility, the nucleotide sequence of DQβ alleles from two DR5/DRw6 PV patients was determined. Both patient DRw6 haplotypes contained a DQβ allele (DQB1.3) (Figure 2) which we have designated DQB1.3[14,25] and is associated with the Dw9 subtype of DRw6, based on analysis of homozygous typing cells (HTCs).[54] It differs from the DR1 DQβ allele, DQB1.1, by only a Val to Asp substitution at amino acid position 57. At the nucleotide level, it is identical in this region to the DQB1.4 allele found on DR2, Dw12 and some DR8 haplotypes.[14]

The frequency of the DQB1.3 allele among patient and control DRw6 haplotypes, was determined by using a panel of 4 oligonucleotide probes[25] (Figure 5, Table 7) to distinguish the DR1-like DQβ "framework" sequence (in DQβ1.1, 1.2 and 1.3) from the DR6-like framework sequence (in DQβ1.6 and 1.7) and to identify the sequences at codon 57. These four oligonucleotide probes generate a hybridization pattern corresponding to the four DQβ alleles found on DRw6, DQw1 haplotypes (Figure 5, Table 7). Two of the Austrian DRw6 patients failed to hybridize to either the CRX03 ("QRD") or the CRX02 ("QRV") probe. When their DQβ sequences were determined, the nucleotide sequence in this region differed from the DQB1.3 sequence by one nucleotide but still encoded the same amino acid sequence. This observation is consistent with the hypothesis that this region of the DQβ chain is directly involved in autoimmunity and is not simply a genetic marker linked to some other susceptibility gene. Only 1/12 DRw6[+] controls has the probe hybridization pattern associated with the DQB1.3 allele. This allele (DQB1.3) was also found in two DRw6 HTCs ABO and EK/OH with the rare MLC-defined type, Dw9 and in 0/10 of the more frequent Dw18 and Dw19 DRw6 HTCs. These results indicate that the DRw6 susceptibility is strongly associated with a rare DQβ allele (DQB1.3) that differs from a common DQβ allele (DQβ1.1) by only a Val to Asp substitution at position 57. Similar results have been obtained recently by Sinha et al.[45] The relative risk for the DQB1.3 allele in the Israeli population is estimated to be >100 vs. 2.6 for DRw6.

The DQα allele determined by sequence analysis from one DRw6 patient was the DQA1.1 allele previously found on only DR1 haplotypes. Using oligonucleotide probe analysis, this DQα allele was found on most (10/13) of Israeli PV DRw6 haplotypes but only in (2/15) control haplotypes[25] (Table 7). This observation is consistent with the idea that this unusual PV-associated haplotype arose from a gene conversion that introduced new sequences around codon 57 into the DQβ locus on a DR1 haplotype. However, in two Dw9 DRw6 HTCs (ABO and EK/OH) and in 3 DRw6 haplotypes from PV patients, the rare DQB1.3 is linked to a DQA1.2 allele. Thus, in the contemporary population, this DQβ allele can be found on at least two different DRw6 haplotypes.

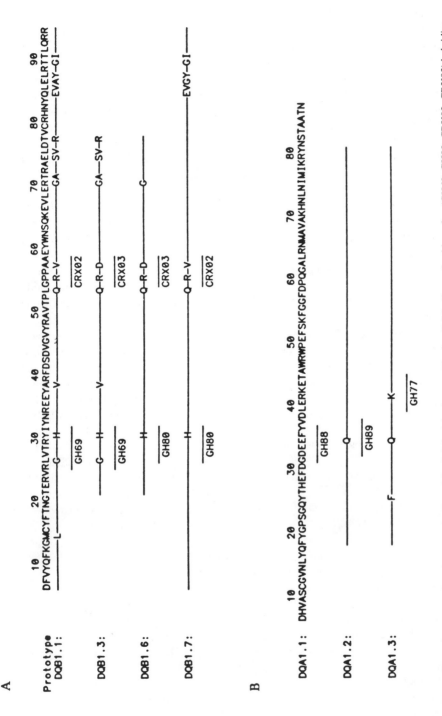

FIGURE 5. HLA DQβ and DQα alleles from DRw6, DRw1 haplotypes. The four DQβ oligonucleotide probes (GH69, GH80, CRX02, CRX03) hybridize to polymorphic nucleotide sequences encoding the amino acid residues underscored in the sequence alignment. The three DQα oligonucleotide probes GH88, GH89, and GH77 hybridize to nucleotide sequences encoding the underscored amino acid residues. The sequences and hybridization conditions for the oligonucleotide probes are described in Reference 25.

TABLE 7
Distribution of HLA DQβ and DQα Alleles and Haplotypes among DRw6 DQw1 Israeli and Austrian *pemphigus vulgaris* Patients and HLA Matched Controls

	Israelis		Austrians	
	PV	Controls	PV	Controls
DQβ				
DQB1.1	0/11	2/12	0/6	0/3
DQB1.3	11/11	1/12	3/6	0/3
DQB1.6	0/11	4/12	1/6	1/3
DQB1.7	0/11	3/12	2/6	2/3
DQα				
DQA1.1	10/13	2/15	5/9	0/3
DQA1.2	3/13	6/15	0/9	3/3
DQA1.3	0/13	7/15	4/9	0/3
DQβ 1.3 Haplotypes				
DQB1.3, DQA1.1	10/11	1/12	3/6	0/3
DQB1.3, DQA1.2	3/11	0/12	0/6	0/3

Note: The unusual DQB1.3 allele appears in 11/11 DRw6 PV patients and only 1/12 controls, $p = .001$. The frequency of the DQA1.1 allele, usually found only on DR1 haplotypes, is also increased among PV patients and the frequency of the DQA1.3 allele is reduced relative to controls.

It seems likely that it is this DQβ allele which confers susceptibility rather than the DQα allele because the DQ antigen encoded by the PV associated DRw6 haplotype differs from the one encoded by normal DR1 haplotype by *only* a Val to Arg substitution at residue 57 of the β chain. This observation clearly implicates position 57 as critical in generating the specific autoimmune response to the basement membrane autoantigen. A crucial role for the position 57 residue was suggested by a similar methodologic approach to the analysis of IDDM susceptibility[14,29] which also found that the charge of this polymorphic residue of the DQβ chain was generally correlated with genetic predisposition. In the case of PV, however, it is clear that it is the *allele*, not simply the *epitope* around position 57, that is disease associated since the most common DRw6 DQβ allele (DQB1.6) contains the same sequence around position 57 but in a different framework. Thus, this novel DQβ allele appears to account for the DRw6 susceptibility PV, consistent with the implications of the DQβ RFLP analysis.[46] To evaluate the potential contribution of other class II alleles on this DRw6 haplotype, DRβ and DPβ sequences were also determined. Recently, we have identified a new DRβI allele on some DRw6 haplotypes in DRw6+, DR4− PV patients.[54] It is present, however, only in some DRw6+ patients and may, therefore, contribute to genetic risk but cannot account for DRw6 susceptibility in general.

In summary, the DR4 association with PV can be accounted for by the DRβI Dw10 allele, implicating the polymorphic residues of the third hypervariable region as a critical ''susceptibility epitope'' involved in a putative interaction with T-cell receptor or antigen. This sequence is contained in most control DRw6 DRβI alleles, but cannot account for the DRw6 associations with PV. The DRw6 susceptibility is most strongly associated with a specific DQβ allele, DQB1.3, implicating the charge of the polymorphic residue at position 57.

IV. GENERAL CONCLUSIONS

The correlation between class II sequence variation and IDDM and PV discussed in this chapter support the general view that class II polymorphic residues are not simply linkage markers, but are involved directly in susceptibility to certain autoimmune diseases. Polymorphisms in the genes encoding T cell receptors and immunoglobulins as well as in other loci are also likely to contribute to predisposition. Here, we have focused only on the HLA-linked contribution.

For PV, DRβI polymorphic residues around position 70 (the Dw10 allele) were implicated in the DR4 susceptibility. The DRw6 susceptibility is associated with a rare DQβ allele which differs from a "normal" allele by only a Val to Asp substitution at position 57. The analysis of IDDM susceptibility associated with DQβ allelic variation on the DR4, DRw6, and DR2 haplotypes has also implicated the charge of the polymorphic residue at position 57 with an Asp residue present in the DQβ chain encoded by negatively associated haplotypes (e.g., DR2, Dw2) and an Ala, Val, or Ser residue by positively associated haplotypes (e.g., DR4, DQβ, 3.2). The DR3 susceptibility to IDDM may be conferred by class II sequences (e.g., DPβ) in addition to those in the DQβ gene. Recent studies of DPβ polymorphism have identified a specific DPβ allele associated with IDDM susceptibility,[54] as well as another DPβ allele correlated with the DR3-associated autoimmune disease, celiac disease.[31] DPβ polymorphisms also appear to be associated with myesthenia gravis[55] and pauciarticular juvenile rheumatoid arthritis.[56] The sequence pattern of the predisposing alleles implicates position 69 of the DPβ chain as a critical residue in genetic predisposition.[31] It may be relevant to potential mechanisms for autoimmune diseases that both the DQβ sequence around position 57 and the DRβI sequence around position 70 exhibit shared peptides with various viral pathogens[14,42,43] (Tables 4 and 5). Conceivably, these shared sequences could serve as a cross-reactive target autoantigen for the immune response to the virus. Alternatively, individuals whose HLA class II antigens contain a shared peptide might fail to respond to this viral epitope by cross-tolerance.

The association of polymorphic residues in the third hypervariable region of DRβI and at position 57 of the DQβ chain with susceptibility to both IDDM and PV suggest a critical role for these residues in the generation of the immune response. According to a recently reported structural model for the HLA class II heterodimers,[53] these residues are located on the α-helical domain and point in toward the putative antigen-binding cleft formed by the α-helices of the class II β and α chains. These polymorphic residues might serve either as antigen-binding sites or restriction elements for T-cell recognition. Alternatively, they could function to determine the T-cell repertoire by influencing the expression of T-cell receptors during maturation in the thymus. These mechanisms are, of course, not mutually exclusive and specific *haplotype* and *genotype* associated disease susceptibility might reflect their joint involvement in generating a specific autoimmune response. In fact, most DR disease associations are strongest with specific combinations of class II alleles, such as particular DR4 haplotypes (e.g., DRβ3.2, Dw4 for IDDM) or particular genotypes (e.g., DR3/DR4 for IDDM). In general, disease susceptibility appears to be associated with specific combinations of class II epitopes (e.g., alleles, haplotypes, or genotypes) rather than specific individual residues or epitopes.

In summary, molecular analysis has revealed the identity of specific class II sequences that are highly associated with autoimmune disease susceptibility. To move beyond this correlation toward an understanding of their role in autoimmunity, structural analysis of the class II molecules as well as *in vitro* and *in vivo* functional studies of interactions with putative autoantigens and T cell receptors will be required. In the meantime, DNA typing offers the potential for identifying individuals at high risk for specific autoimmune diseases.

ACKNOWLEDGMENTS

We thank Ann Begovich for helpful discussions and Linda Rutkowski for preparation of this manuscript.

REFERENCES

1. **Svejgaard, A., Platz, P., and Ryder, L. P.,** *Immunol. Rev.,* 70, 193, 1983.
2. **Giles, R. C. and Capra, J. D.,** *Adv. Immunol.,* 37, 1, 1985.
3. **Trowsdale, J., Young, J. A. T., Kelly, A. P., Austin, P. J., Carson, S., Meunier, H., So, A., Erlich, H. A., Spielman, R., Bodmer, J., and Bodmer, W. F.,** *Immunol. Rev.,* 85, 5, 1985.
4. **Bus, S., Sette, A., Colon, S. M., Miles, C., and Grey, H. M.,** *Science,* 235, 1353, 1987.
5. **Sette, A., Buss, S., Colon, S., Smith, J. A., Miles, C., and Grey, H. M.,** *Nature (London),* 328, 395, 1987.
6. **Mengel-Gaw, L., Conner, S., McDevitt, H. O., and Fathman, C. G.,** *J. Exp. Med.,* 160, 1184, 1984.
7. **Erlich, H. A., Horn, G. T., Bugawan, T. L., and Scharf, S.,** in *Molecular and Cellular Biology of Histocompatibility Antigens,* Schacter, B. et al., Eds., ASHI, New York, 1986, 93.
8. **Saiki, R. K., Scharf, S., Faloona, F., Mullis, K. G., Horn, G. T., Erlich, H. A., and Arnheim, N.,** *Science,* 230, 1350, 1985.
9. **Mullis, K. G. and Faloona, F.,** *Methods Enzymol.,* 155, 335, 1987.
10. **Scharf, S. J., Horn, G. T., and Erlich, H. A.,** *Science,* 233, 1076, 1986.
11. **Gyllensten, U. and Erlich, H. A.,** *Proc. Natl. Acad. Sci. U.S.A.,* 85, 7652, 1988.
12. **Saiki, R. K., Bugawan, T. L., Horn, G. H., Mullis, K. B., and Erlich, H. A.,** *Nature (London),* 324, 163, 1986.
13. **Eisenbarth, G. S.,** *N. Engl. J. Med.,* 314, 1360, 1986.
14. **Horn, G. T., Bugawan, T., Long, C. M., and Erlich, H. A.,** *Proc. Natl. Acad. Sci. U.S.A.,* 85, 6012, 1988.
15. **Schreuder, G. M. T., Doxiadis, I., Parlevliet, J., and Grosse-Wilde, H.,** *Histocompatility Testing 1984,* Albert, E. D. et al., Eds., Springer-Verlag, New York, 1984, 243.
16. **Gregerson, P. K., Shen, M., Song, Q.-L., Merryman, P., Degar, S., Seki, T., Maccari, J., Goldberg, D., Murphy, H., Schwenzer, J., Wang, Y. W., Winchester, R. J., Nepom, G. T., and Silver, J.,** *Proc. Natl. Acad. Sci. U.S.A.,* 83, 2642
17. **Bach, F. H., Rich, S. S., Barbosa, J., and Segall, M.,** *Hum. Immunol.,* 12, 59, 1985.
18. **Kim, S. J., Holbeck, S. L., Nisperos, B., Hansen, J. A., Maeda, H., and Nepom, G. T.,** *Proc. Natl. Acad. Sci., U.S.A.,* 82, 8139, 1985.
19. **Erlich, H. A., Bugawan, T. L., Scharf, S., Nepom, G. T., Tait, B. P., and Griffith, R.,** *Diabetes,* in press.
20. **Tait, B. D. and Boyle, A. J.,** *Tissue Antigens,* 28, 65, 1986.
21. **Arnheim, N., Strange, C., and Erlich, H. A.,** *Proc. Natl. Acad. Sci. U.S.A.,* 82, 6970, 1985.
22. **Cohen-Haguenauer, O., Robbins, E., Massart, C., Busson, M., Deschamps, I., Hors, J., Lalouel, J.-M., Dausset, J., and Cohen, D.,** *Proc. Natl. Acad. Sci. U.S.A.,* 82, 3335, 1985.
23. **Nepom, B. S., Palmer, J., Kim, S. J., Hansen, J. A., Holbeck, S. L., and Nepom, G. T.,** *J. Exp. Med.,* 164, 345, 1986.
24. **Schreuder, G. M. T., Maeda, H., Koning, F., and D'Amaro, J.,** *Hum. Immunol.,* 16, 127, 1986.
25. **Scharf, S. S., Friedmann, A., Brautbar, C., Szafer, F., Steinman, L., Horn, G., Gyllensten, U., and Erlich, H. A.,** *Proc. Natl. Acad. Sci.,* 85, 3504, 1988.
26. **Lee, B. S. M., Bell, J. I., Rust, N. A., and McDevitt, H. O.,** *Immunogenetics,* 26, 85, 1987.
27. **Thomson, G.,** *Am. J. Hum. Genet.,* 36, 1309, 1984.
28. **Thomson, G., Robinson, W. P., Kuher, M. K., and Joe, S.,** *Genet. Epidemiol.,* in press.
29. **Acha-Orbea, H. and McDevitt, H. O.,** *Proc. Natl. Acad. Sci. U.S.A.,* 84, 2435, 1987.
30. **Todd, J. A., Bell, J. I., and McDevitt, H. O.,** *Nature (London),* 329, 599, 1987.
31. **Bugawan, T., Horn, G. T., Long, C. M., Michelson, E., Hansen, J., Ferrara, G. B., Angelini, G., and Erlich, H. A.,** *J. Immunol.,* 141, 4024, 1988.
32. **Sheehy, M., Scharf, S., Rowe, J., Neme de Gimenez, M., Meske, L., Erlich, H., and Nepom, B.,** *J. Clin. Invest.,* 83, 830, 1988.
33. **Erlich, H., Scharf, S., and Horn, G.,** in preparation.
34. **Chung, S., et al.,** *Proc. of the 3rd Asia-Oceania Histocompatibility Workshop Conf.,* Aizawa, M., Ed., Hokkaido University Press, Sapporo, Japan, 1986.

35. **Yoon, J. W.,** *Microbiol. Pathogen.,* in press.
36. **Rubinstein, P., Walker, M. E., Fedun, N., Wit, M. E., Cooper, L. Z., and Ginsberg-Fellner, F.,** *Diabetes,* 31, 1088, 1982.
37. **Pak, C. Y., Eun, H. M., McArthur, R., and Yoon, J. W.,** *Lancet,* in press.
38. **Kim, K. W., Erlich, H. A., and Yoon, J. W.,** submitted.
39. **Baer, R., Bankier, A. T., Biggin, M. D., Deininger, P. L., Farrell, P. J., Gibson, T. J., Hatfull, G., Hudson, G. S., Satchwell, S. C., Seguin, C., Tuffnell, P. S., and Barrell, B. G.,** *Nature (London),* 310, 207, 1984.
40. **Zavala, F., Cochrane, A. H., Nardin, E. H., Nussenzweig, R. S., and Nussenzweig, V.,** *J. Exp. Med.,* 157, 1947, 1983.
41. **Srinvasappa, J., Saegusa, J., Prabhakar, B. S., Gentry, M. K., Buchmeier, M. J., Wiktor, T. J., Koprowski, H., Oldstone, M. B. A., and Notkins, A. L.,** *J. Virol.,* 57, 397, 1986.
42. **Oldstone, M. B. A.,** *Cell* 50, 819, 1987.
43. **Roudier, J., Rhodes, G., Peterson, J., Vaughen, J. H., and Carson, D. A.,** *Scand. J. Immunol.,* 27, 367, 1988.
44. **Razzaque, A.,** *Dermatology,* 1, 1, 1983.
45. **Brautbar, C., Moscovitz, M., Livshits, T., Haim, S., Hacham-Zadeh, S., Cohen, H. A., Saron, R., and Nelken, D.,** *Tissue Antigens,* 16, 238, 1985.
46. **Szafer, F., Brautba, C., Tzfoni, E., Frankel, G., Sherman, L., Cohen, I., Hacham-Zadeh, S., Aberer, W., Tappeiner, G., Holubar, K., Steinman, L., and Freidmann, A.,** *Proc. Natl. Acad. Sci. U.S.A.,* 84, 6542, 1987.
47. **Gorski, J. and Mach, B.,** *Nature (London),* 322, 67, 1986.
48. **Horn, G. T., Bugawan, T. L., Long, C. M., Manos, M. M., and Erlich, H. A.,** *Hum. Immunol.,* 21, 249, 1988.
49. **Scharf, S. J., Long, C., and Erlich, H. A.,** *Hum. Immunol.,* 22, 61, 1988.
50. **Hansen, J. A., Beaty, P. G., Anasetti, C., Martin, P. J., Mickelson, E., and Thomas, E. D.,** *Br. Med. Bull.,* 43, 203, 1987.
51. **Nepom, G. T., Hansen, J., and Nepom, B.,** *J. Clin. Immunol.,* 7, 1, 1987.
52. **Sinha, A. A., et al.,** *Science,* 239, 1026, 1988.
53. **Brown, J., Jardetzk, T., Sapen, M. A., Samraous, B., Bjorkman, P., and Wiley, P. C.,** *Nature (London),* 332, 845, 1988.
54. **Scharf, S. J., Friedmann, A., Steinman, L., Brautbar, C., and Erlich, H. A.,** *Proc. Natl. Acad. Sci. U.S.A.,* 86, 6215, 1989.
55. **Begovich, A. et al.,** unpublished observations.
56. **Begovich, A., Bugawan, T. L., Nepom, B. S., Klitz, W., Nepom, G. T., and Erlich, H. A.,** *Proc. Natl. Acad. Sci. U.S.A.,* in press.

Chapter 8

FUNCTIONAL IMPORTANCE OF MHC CLASS II POLYMORPHISM IN NORMAL IMMUNE RESPONSES AND AUTOIMMUNE DISEASE

Animesh A. Sinha, Hans Acha-Orbea, John Todd, Luika Timmerman, and Hugh O. McDevitt

TABLE OF CONTENTS

I. INTRODUCTION

Our understanding of the biological function of major histocompatibility complex (MHC) class II molecules has expanded considerably since the discovery that immune response genes controlling the antibody response to synthetic branched polypeptides in mice map to the MHC,[1-3] and that genes regulating immune responsiveness in man map to the HLA-D (class II) region of the human MHC.[4,5] Class II molecules in mice and humans are heterodimeric transmembrane glycoproteins consisting of noncovalently associated α (34 kDa) and β (29 kDa) chains.[6] The extracellular portion of each domain is folded into two discrete domains, of approximately 90 amino acids, stabilized by a single disulfide bridge (except for the α1 domain). It is now known that humans express three distinct class II isotypes encoded by separate αβ gene pairs in the DP, DQ, and DR subregions of the HLA-D complex, which under normal circumstances are expressed only on the surface of antigen-presenting B lymphocytes, macrophages, dendritic cells, and activated T lymphocytes.

The most striking feature of both class I and class II MHC genes is their extensive allelic polymorphism.[7,8] Our understanding of the biological significance of this polymorphism has been clarified by recent biochemical demonstrations that MHC molecules serve as cell surface receptors for peptide fragments of protein antigens.[9,10] It is in this capacity that class II molecules regulate immune responsiveness; particular allelic variants differ in their ability to interact with a given peptide. Moreover, it is now clear that antigen binding by class II molecules on the surface of antigen-presenting cells is a prerequisite for the activation of thymus-derived (T) lymphocytes. Unlike B cells, which are able to bind native antigen in solution, T cells demonstrate dual specificity; they are at once antigen specific and MHC restricted.[11] In other words, T cells only recognize a particular peptide fragment of foreign proteins when the peptide is bound to a self class II MHC molecule. Class II molecules also regulate immune responsiveness at the level of the developing T-cell repertoire. Intrathymic interactions (still not well understood) between immature T cells and class II proteins on thymic antigen-presenting cells are fundamental in deleting T cells which react with self from the repertoire.[12,13]

Several years following the discovery that MHC class II genes regulate immune responsiveness to foreign antigens, it was observed that genetic predisposition to several human autoimmune diseases also maps to the class II region of the MHC.[14] Thus, genetic responsiveness to self antigens also appears to be determined by the MHC. However, the human class II region is considerably more complex than its murine counterpart, and elucidation of the role played by MHC genes in susceptibility to autoimmune disease has had to await a more complete characterization of the genes and proteins encoded by the many class II loci.

Recent advances in molecular biology have greatly enhanced the resolution and speed with which we can characterize individual class II alleles. In this chapter, we will describe the use of the polymerase chain reaction and nucleotide sequencing to characterize many of the DR and DQ alleles found in normal individuals. These sequences have provided fundamental information concerning the nature and extent of class II allelic polymorphism, as well as insights into mechanisms of generating such remarkable genetic variation. We have also characterized numerous class II alleles from patients with autoimmune diseases such as insulin-dependent diabetes mellitus (IDDM), myasthenia gravis (MG), and *pemphigus vulgaris* (PV). These studies address the role of class II allelic polymorphism in disease susceptibility at the molecular level. This information suggests new strategies for clinical screening, diagnosis, and therapeutic intervention in human autoimmune disease.

II. GENOMIC ORGANIZATION

The human class II region, which is located centromeric to class I loci on chromosome six, spans approximately 1.1 megabases of DNA (Figure 1A). Organization of the α and β

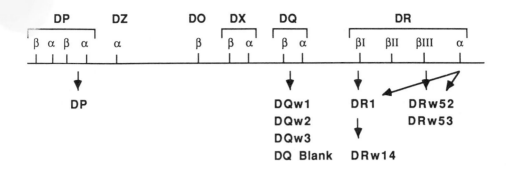

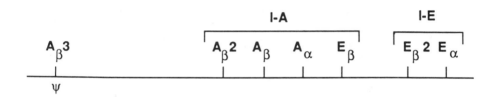

B

FIGURE 1. (A) Genomic organization of class II genes in the human. DPα2 and DPβ2 are both pseudogenes,[82,83] as is the DRβII gene (at least in DR3 and DR4 haplotypes). Transcripts of DXα and DXβ have not been detected, but their genomic sequences show no obvious defects.[84] The DZα and DOβ genes are transcribed, but their products have thus far not been detected at the cell surface.[85,86] DRβIII mRNA and protein expression is lower than that for DRβI' suggesting differential regulation of these two loci. The number of DRβ genes found on different haplotypes can vary, ranging from one in DR1 and DR8 individuals to four in DR4 individuals.[20,87] (B) Genomic organization of class II genes in the mouse. Aβ3 is a pseudogene. Eβ2 is transcribed, but whether it encodes a protein that is expressed on the cell surface is unknown (reviewed in Reference 15). The overall similarity in the class II genomic organization of mouse and man suggests that duplication and divergence of a primordial αβ gene pair most likely occurred in a common ancestor. Subsequent to mouse-man speciation, it appears that the class II region in both species continued to evolve by duplications and/or deletions of individual genes.

genes within the DP, DQ, and DR subregions was initially determined by genetic linkage studies of overlapping cosmid clones (reviewed in Reference 15). Studies of recombination in families, and deletion mutants, located DP as being centromeric to DR,[16] but the relative positions of all the subregions were only recently established definitively by pulsed-field gel electrophoresis.[17] This technique allows the separation of DNA fragments up to 1000 kb in length. By digestion of DNA with enzymes that have six or eight base-pair recognition sequences and, therefore, cut very infrequently, large DNA fragments spanning several subregions were obtained and hybridized with various locus-specific DNA probes to establish a map order. The relative order of DP and DZα could not be established from this work, but further studies with deletion mutants[16] and pulsed-field gel electrophoresis[18] indicate that the DP subregion is centromeric to DZα. This map order (DP - DZα- DOβ - DX - DQ - DRβ - DRα) has recently been confirmed, and an additional class II pseudogene (DVβ) has been found between DXα and DQβ.[19]

The organization of the murine class II MHC (H-2) loci is shown in Figure 1B. Nucleotide and amino acid sequence similarities between individual murine and human α and β loci suggest that Aβ3 and DPβ, Aβ2 and DOβ, Aβ and DQβ, Aα and DQα, EβI and DRβ, and Eα and DRα are homologs.[6] Thus, both the murine and human class II regions are similarly organized.

III. POLYMORPHISM

A. RESTRICTION FRAGMENT LENGTH POLYMORPHISM STUDIES

We initially used restriction fragment length polymorphism (RFLP) analysis to characterize allelic variants of class II genes at the DNA level, in an attempt to define a more accurate assessment of the extent of polymorphism than had been obtained by serological or cellular typing. The goal was to determine whether standard band patterns could be established for each serologically defined DR allele, and if so, whether RFLP could define new polymorphisms within standard haplotypes. The large number of bands that hybridize with a given class II probe necessitated the use of cell lines homozygous for serological DR specificities. However, serological homozygosity does not ensure genetic homozygosity; therefore, only cell lines homozygous by consanguinity were used.

Overall, these studies[20,21] have provided a catalog of standard DRβ and DQβ RFLP patterns which can be used to clarify or supplement standard typing methods. The identification of subtypes within most DR and DQ haplotypes indicates a level of polymorphism beyond that detected by standard techniques. These studies indicate that RFLP analysis could provide a rapid method of defining further polymorphism at the population level.

B. NUCLEOTIDE SEQUENCE ANALYSIS

The most direct and informative way to investigate the nature of allelic polymorphism is to directly sequence the proteins or genes. Recently, the development of *in vitro* gene amplification (or the polymerase chain reaction) technology[22] has obviated the need for the time-consuming process of constructing and screening cDNA libraries. The technique was originally developed for use on genomic DNA, but we have found that cDNA is amplified more efficiently.[23] With this modification, use of oligonucleotide primers complementary to sequences in the leader peptide region (5′) and in the second domain (3′) allows the entire first (and most polymorphic) domain to be amplified and sequenced. The technique is outlined schematically in Figure 2. Synthetic oligonucleotides which flank the region to be sequenced are annealed to a denatured cDNA substrate (made from total RNA) and extended by DNA polymerase (Klenow fragment). Multiple rounds of denaturation, annealing of primers, and extension produce large quantities of the target sequence which can then be gel-purified and directly blunt-end ligated into the M13 sequencing vector. Recently, the availability of heat-stable DNA polymerase from *Thermus aquaticus* has simplified the technique[24] and made it amenable to automation.

1. Extent of Class II Polymorphism

In both mice and humans, the extent of polymorphism differs among the different class II molecules. With respect to the number of alleles identified, murine I-A molecules appear to be slightly more polymorphic than I-E molecules. In humans, there have been many more DR than DQ molecules identified by serology. However, RFLP and nucleotide sequence analysis has uncovered additional polymorphism in DQα and DQβ loci.

Individual genes also differ in their extent of polymorphism. Aα, Aβ, and Eβ in the mouse, and their human homolog DQα, DQβ, and DRβI are all highly polymorphic. DRβ molecules contain slightly more first domain polymorphic residues than DQβ molecules (38 versus 31). DRβI and DRβIII loci within a given haplotype are as similar to each other as they are to alleles of the same loci. DRβIII shows more limited allelic variation, but this differs among haplotypes defined by the supertypic specificities DRw52 and DRw53. DR3 and DRw6 haplotypes are found in association with both DRw52a and DRw52b DRβIII alleles, which differ by 10 residues.[25] A third allele, DRw52c, has been found in a DRw6 cell line.[26] In contrast, the DRβIII deduced amino acid sequence in DRw53 haplotypes is invariant.[27,28] Overall, however, there is sequence similarity within DRw52 haplotypes (DR3,

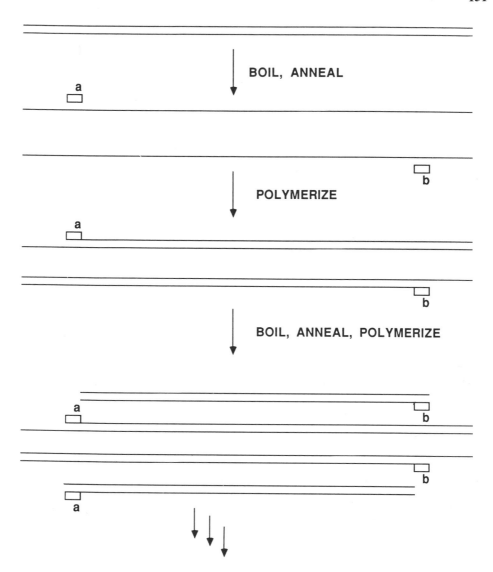

BOIL, ANNEAL

POLYMERIZE

BOIL, ANNEAL, POLYMERIZE

FIGURE 2. *In vitro* gene amplification technique. Oligonucleotide primers (a and b), which flank the region to be specifically amplified, are annealed to a denatured genomic DNA or cDNA substrate. Annealing is followed by extension with the Klenow fragment of DNA polymerase I (or, more recently, the heat-stable polymerase from *Thermus aquaticus*). This cycle of denaturation, annealing, and extension is repeated approximately thirty times to generate a large amount of the region of DNA that is contained within the amplifying primers. Precise conditions for amplification are described in References 22, 23, 24, and 78.

DR5, DRw6, and DR8) and within DRw53 haplotypes (DR4, DR7, and DR9) at both DRβIII and DRβI.[28]

RFLP and sequence similarities within the DR2 associated subtypes [DR2(Dw2), DR2(Dw12), and DR2(AZH)] suggest that this group also represents a supertypic family (although no serological specificity defining all the members of this group has been identified to date).[29] These similarities most likely reflect common evolutionary origins of the haplotypes within each supertypic group.

The most centromeric and telomeric class II genes are much less polymorphic than the centrally located genes. Only six DP alleles have been distinguished by primed lymphocyte typing.[30] However, nucleotide sequence analysis has recently discovered 2 DPα alleles and 15 DRβ alleles.[91] The Eα and DRα genes are almost invariant. Moreover, the variation among Eα alleles is limited to 4 amino acids,[31] in contrast to alleles at DQα, DQβ, and DRβI, which differ from each other (in pair-wise combinations) by 8 to 20 amino acids. This pattern of polymorphism suggests that there are perhaps structural or functional constraints which strongly select against variation in Eα, DRα, and to some extent DRβIII chains. Alternatively, there may be localized mechanisms for generating polymorphism. Both mechanisms may operate, but in support of the latter possibility, recombinational hot spots have been found at the boundaries of variable and conserved tracts of DNA within the murine MHC.[32,33] Recombinational hot spots could generate allelic variability by promoting frequent unequal crossing-over events during meiosis to facilitate the joining of the proximal end of the MHC of one haplotype to the distal end of the MHC of another haplotype. At this point, the reasons for isotypic variation in terms of extent of polymorphism remain unclear.

2. Location of Allelic Polymorphism in Class II Molecules

The location of polymorphic amino acids in class II molecules has important implications for correlating structure with function. First, polymorphic amino acids are not randomly distributed over the whole molecule. The membrane-distal (amino terminal) domains of the α and β molecules contain the majority of the polymorphic residues. For example, residues in the membrane-distal domain of Aα chains show 12% variability on average, versus only 2% for the membrane-proximal domain.[7] Similarly, comparison of the deduced amino acid sequences for human DRβI alleles from the DR3 and DR1 haplotypes shows that 17/21 polymorphic residues are found in the amino-terminal domain.[28] Moreover, the ratios of productive (leading to a change in amino acid sequences) versus silent (no resulting change in amino acid sequence) nucleotide substitutions in the amino-terminal domains are very high relative to those in the carboxy-terminal domains of different alleles (e.g., 33:4 for Aβ NH_2-terminal domain versus 9:17 for the COOH-terminal domain).[7] The ratio of productive to nonproductive nucleotide changes due to simple random point mutations is between 2 to 3. This suggests that strong pressures exist that select for variability in the first domain and against variability in the second domain. The frequency of silent substitutions (mutation rates) is identical in the first and second domains.[34]

The second, and potentially most revealing, feature of class II polymorphism in terms of structure-function predictions is the clustering of allelic variation within discrete hypervariable regions (HVR).[7] For example, 84% of the variability in Eβ loci is contained in only 17% of the total protein sequence, comprising residues 1 to 13, 27 to 39, 68 to 75, and 87 to 93. Similarly, variation among DRβI alleles is largely clustered into 3 HVR that span residues 9 to 14, 25 to 38, and 67 to 74.[28] There is additional variation, however, at positions 37 to 38, 47, and 56 to 60. This general pattern of clustered polymorphisms is also seen in DQ and DP α and β alleles, although the number and location of clusters varies somewhat (reviewed in Reference 35). However, the degree of polymorphism within class II allelic HVR does not approach that seen in immunoglobulin HVR. Usually, there are less than four different amino acids found at each variable class II residue. In DRβ, only residues 11, 13, 30, and 37 exhibit polymorphism of greater than four amino acids. There are no residues with more than four amino acids in DQβ. Moreover, entire allelic HVR are often shared among several different haplotypes (Figure 3 and Reference 28). There seems to be a limited number of HVR sequences that are shuffled between alleles to generate "patchwork" allelic polymorphism. Thus, variation within each allelic hypervariable region, as well as at each residue, is limited. The restrictions on variation may be due to the structural constraints placed on class II molecules by their dual function — to bind peptide antigens and to be recognized by T-cell receptors. Such constraints would not apply to immunoglobulins, since they act only as antigen ligands.

Several different mechanisms may be responsible for generating this pattern of class II

153

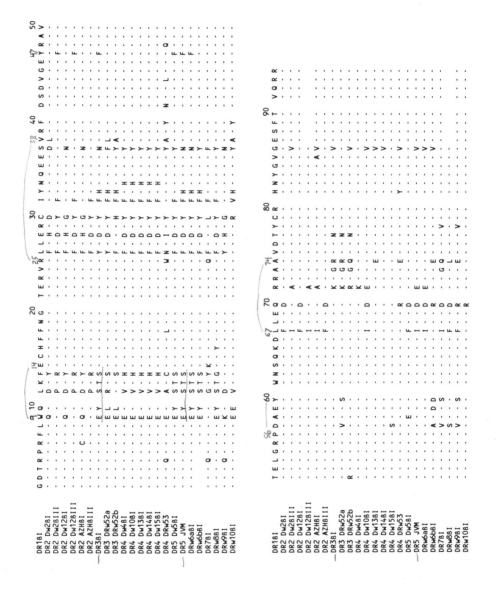

FIGURE 3. First domain amino acid sequences of DRβI and DRβIII alleles. The single letter code is used. Sources of the sequences are detailed in References 28 and 90.

allelic polymorphism. First, the higher nucleotide substitution rate in protein-coding exons versus noncoding introns[36] suggests a powerful mechanism for fixing selected random point mutations. This mechanism presumably reflects the importance of allelic polymorphism in controlling immune responsiveness to pathogens. Obviously, it would be beneficial at the population level to have a large number of alleles available as restriction elements for foreign proteins. However, the high frequency of di- and tri-nucleotide changes within a given codon and the "patchwork" nature of allelic polymorphism provide compelling circumstantial evidence that other mechanisms of generating polymorphisms exist. One likely mechanism is gene conversion, originally described in yeast. During such an event, flanking regions of interlocus homology mediate the transfer of a stretch of nucleotide sequence from a donor to a recipient gene during DNA replication.[37] Although it is difficult to provide conclusive evidence for true gene conversion in mammalian cells, several donor sequences for potential gene-conversion events have been identified. The murine $A\beta^{bm12}$ molecule shares nucleotides 68 to 75 with the $E\beta^b$ molecule, and differs from the parental $A\beta^b$ molecule by only three amino acids.[36,38,39] Similarly, examination of nucleotide sequences of several DRβ alleles supports the gene-conversion hypothesis. For example, the DR5βI allele (JVM) shares HVR1 with DR3βI and DRw6aβI, HVR2 with DR2βIII, and HVR3 with DR4(Dw10)βI (see Figure 3 and Reference 28). All members of the DRw52 supertypic family share HVR1, which may form an epitope for some DRw52 antibodies. DR2, DR4, DR5, and DRw6 share HVR2. Similarly, HVR3 of DR4(Dw10) is also found in some DR5, DRw6, and DR1 haplotypes.[28,40] HVR can also be shared between different loci. For example, the human DRw6aβIII and DR3βI alleles have identical third HVRs.[25] Furthermore, there is a TAC at position 78 of all DRβ chains except DR7 and DR9 which have GTG. Several DQβ alleles also have the nucleotides GTG at residue 78, suggesting that this locus may have served as a donor or recipient in a gene conversion-like event.

Double homologous recombination has also been implicated in creating new class II alleles by scrambling existing sequences; the DRβI and DRβIII loci in a DR2 cell line appear to have exchanged the COOH-terminal halves of their membrane-distal domains.[29] Both recombination and gene conversion likely serve to maximize variability by placing variable residues in different contexts, thus providing a sort of combinatorial diversity to class II proteins.

IV. FUNCTIONAL RELEVANCE OF ALLELIC POLYMORPHISM IN NORMAL IMMUNE RESPONSES

The realization that variability within class II alleles is clustered in discrete HVR brought immediate speculation that these regions were analogous in structure and function to immunoglobulin HVR. The antigen-binding site of immunoglobulin molecules is formed by the juxtaposition, through portein-folding, of noncontiguous residues in the heavy and light chain HVR. This model for class II has yet to be directly confirmed, but there are several supporting lines of evidence.

Direct antigen-Ia interactions have long been suspected. However, Babbit et al. were the first to show that class II molecules serve as receptors for peptide fragments of protein antigens.[9] This work provides a biochemical basis for immune response gene phenomena. Subsequent work by Buus and colleagues has confirmed that this is a general mechanism by which class II genes control the specificity of the immune response.[10] Furthermore, a peptide derived from HLA-A2 can activate HLA-A2-specific T cells under certain conditions,[41] suggesting that T-cell activation by allogeneic class I (and by implication, class II) molecules follows the same fundamental rules as activation by foreign antigen.

Other antigen-binding molecules of the immune system (immunoglobulins, and the T-cell receptor) display a vastly greater amount of diversity than do class II molecules. How the relatively few class II molecules present in an individual can bind a large variety of peptide fragments is unclear. In heterozygous individuals, *trans*-association of α and β molecules from different haplotypes could increase the diversity of class II molecules.[42] Furthermore, earlier studies suggested that each class II molecule may contain several distinct antigen-binding sites, but recent data support the existence of a single binding site per class II molecule.[43] However, an individual combining site can tolerate substantial variation in peptide sequence before binding is abolished.[44] Thus, a particular class II molecule apparently displays broad specificity and can probably accommodate a large variety of peptide fragments. Moreover, a typical foreign protein of 100 to 200 amino acids will likely be processed into several different fragments, of which at least one could be expected to interact with the binding site.

As yet, there is no direct evidence that the allelic HVRs participate in antigen-binding by class II molecules. However, it is clear that allelic HVR sequences can serve as functional epitopes for T cells.[39,45] The potential importance of allelic HVRs in antigen binding can be inferred from the immune response phenotype of the bm12 mutant. Residues 67, 70, and 71 in the third allelic HVR distinguish $A\beta^{bm12}$ from the parental $A\beta^b$ molecule and correlate with altered immune responsiveness to several antigens in the bm12 mutant (reviewed in Reference 7).

Recently, the three-dimensional structure of a class I molecule, HLA-A2, has been determined by X-ray crystallography at 3.5 Å resolution.[46] Because class I and class II are evolutionarily related and exhibit functional and sequence similarity, class II structure can be modeled from class I. In the class I molecule, a putative antigen-binding cleft at the top of the molecule is formed by a groove between the C-terminal α helixes of the α1 and α2 domains. The base of the cleft is formed by the NH_2-terminal β-pleated sheets of each domain. A prediction of the class II three-dimensional structure, made by comparing the patterns of conserved and variable amino acids of multiple class I and class II sequences, conforms closely to the class I structure and also includes a putative antigen-binding groove[47] (Figure 4). Two observations suggest an alignment of class II α1 with class I α1 and class II β1 with class I α2. First, the glycosylation site (Asn, x, ser/thr) at residue 86 (class I) or residue 82 (class II α1 domain) is completely conserved in all class I α1 and class II α1 sequences. Second, disulfide-linked cysteines at residues 101 and 164 (class I) and 15 and 78 (class II β chain) are found in all class I α2 and class II β1 sequences.

In this model, the majority of the polymorphic residues (located in the allelic HVR) are found in or near the antigen binding site. There is a pattern of conserved residues every third or fourth amino acid in the α-helical region of class II α1 (64, 68, 71, 74, and 80 — corresponding to 68, 72, 75, 78, and 84 in class I α1) and class II β1 (62, 68, 69, 72, 73, 76, 79, 80, and 83 — corresponding to 148, 153, 154, 157, 158, 161, 164, 165, and 168 in class I α2). On the predicted class II 3-D structure, these residues align on the outer face (away from the antigen-binding site) of the α-helices. Most of the polymorphic residues on the helices (class II α1: 70, 75, 76, 77 — corresponding to 74, 79, 80, 81 in class I α1; β1: 67, 70, 71, 74 — corresponding to 152, 155, 156, 159 in class I α2) are interspersed between the conserved residues and face either up or into the binding site. The β strand residues exhibit a striking periodicity; every other residue (9, 11, and 13 in S1′ and 26 and 28 in S2′ — corresponding with 95, 97, 99, 114, 116 in class I α2) is polymorphic and points into the site to form the bottom of the antigen-binding cleft. There are no residues (polymorphic or nonpolymorphic) in the membrane proximal domains that contribute to the antigen-binding site. However, it is possible that variation in areas of class II outside of α1 or β1 could affect conformation of the molecule and thus influence antigen-binding and/or T-cell recognition properties of the molecule. A mutation in α3 at position 227 (located outside the antigen binding site of class I) does affect CTL recognition.[48] Furthermore, contributions from the β2 domain have been shown to be necessary to produce the native

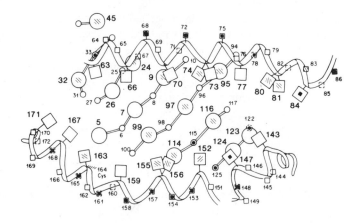

FIGURE 4.   A predicted model of class II structure. HLA-A2 sequence numbers are used. Residues conserved (solid) or polymorphic (hatches) in both
class I and class II are indicated. Residues forming (large symbols) or adjacent
to (small symbols) the putative antigen-binding site are displayed on α-helices
(squares) or β-strands (circles). The break between helices at residue 149 is
the site of insertions and deletions. Although different isotypes have somewhat
different distributions of polymorphic residues, the sequence similarities val-
idate the construction of a single overall class II model. (From Brown, H.,
Jardetsky, T., Saper, M. A., Samraoui, B., Bjorkman, P., and Wiley, D. C.,
Nature (London), 332, 845, 1988. With permission.)

configuration for an Aβ molecule.[49] Together with functional evidence that class II molecules
bind peptides and the correlation of immune responsiveness with variation in allelic HVR
sequences, this model of class II provides a structural basis for understanding the functional
relevance of allelic polymorphism in class II molecules.

The localization of polymorphic residues within the cleft supports the hypothesis that
allelic HVRs are in fact contact residues for either a foreign peptide fragment, the T-cell
receptor, or both. This contention is further supported by studies of natural and site-directed
mutants. Mice made transgenic with site-specific mutated class II constructs should permit
a critical test of structure-function correlation and provide insight into the mechanism by
which class II MHC molecules regulate immune responsiveness to foreign peptides.

V. RELEVANCE OF ALLELIC POLYMORPHISM IN SUSCEPTIBILITY TO HUMAN AUTOIMMUNE DISEASE

Genetic predisposition to over forty human diseases has been associated with specific
class II MHC alleles.[14] While class II alleles are correlated with autoimmune disease, it is
clear that susceptibility is polygenic in nature. Genetic studies in nonobese diabetic (NOD)
mice indicate that at least two recessive MHC-unlinked genes contribute to disease development.[50] In humans, there is only 10 to 15% concordance for IDDM in MHC-identical
siblings compared to 50% concordance in monozygotic twins, indicating a substantial contribution by loci not linked to the MHC.[51] In most diseases, these unlinked loci have not
been identified, but polymorphisms linked with the human insulin gene on chromosome
eleven have been correlated with IDDM.[52] The low (25 to 50%) concordance rate for several
diseases in monozygotic twins indicates that environmental factors are also likely to play a
significant role. Concordance values range from 30 to 50% for IDDM[51] and are less than
30% for rheumatoid arthritis (RA).[53]

Two hypotheses have been proposed to account for genetic associations between class
II MHC alleles and autoimmune disease. The first implicates unidentified disease suscep-

tibility genes present in linkage disequilibrium with particular class II alleles. Non-MHC genes do map within the MHC, and some are associated with MHC-linked non-autoimmune diseases (e.g., 21-hydroxylase deficiency and congenital adrenal hyperplasia). The second hypothesis attributes disease susceptibility directly to class II MHC genes. The central immunoregulatory role of class II molecules suggests they are likely to be directly involved in autoimmune processes. Furthermore, in some diseases, such as RA, association with a particular class II allele (DR4) is maintained across different ethnic groups.[54] Chances that a non-MHC susceptibility locus would have remained linked to the same class II allele over such a long evolutionary time span are slight. Strong evidence implicating class II genes as susceptibility loci also comes from animal models in which spontaneous or experimentally induced autoimmune diseases are blocked by treatment with monoclonal anti-I-A antibodies.[55,56]

Several problems remain to be understood before class II molecules can be unequivocally implicated as susceptibility loci. First, only a small proportion of individuals carrying a susceptibility allele actually develop disease. For instance, although 90 to 95% of IDDM patients are DR3 and/or DR4, these alleles are present in about half of the normal population.[14,57] Second, several different class II alleles are often found associated with an individual disease. This may reflect multiple mechanisms of developing a particular disease, disease heterogeneity, or shuffling of common short sequences (susceptibility epitopes) among otherwise distinct alleles by a gene conversion-like process. Third, strong linkage disequilibrium between particular DR and DQ alleles makes it difficult to unequivocally attribute susceptibility to either locus.

A. RESTRICTION FRAGMENT LENGTH POLYMORPHISM STUDIES

To clarify these issues, and to better understand the role of class II allelic polymorphism in susceptibility to autoimmune disease, we sought to identify subtypes of standard alleles at a molecular level which might better correlate with disease. Our first approach (and that of several other groups) was to search for DNA RFLPs that might differentiate between class II alleles of diseased and normal individuals. Disease-associated subtypes of DR2, DR3, DR4, and DRw6 were identified by RFLP typing. Of DR4 diabetics, 90% carry a BamHI 12.0 kb DQβ RFLP found in only 60% of DR4 controls.[57-59] DR2 diabetics have an EcoRI 16 kb DQβ RFLP pattern found only rarely in normal DR2 individuals,[60] indicating that only a subset of DR2 individuals are susceptible to IDDM. We found that half of DR3 myasthenia gravis patients possess a HincII DQβ RFLP variant found in <5% of normal DR3 individuals.[61] All the RFLP variants identified in these studies showed significant associations with autoimmune disease, but none were disease specific. In contrast, Szafer et al. recently identified PvuII and BamHI DQβ RFLP variants in DR4 and DRw6 Israeli *pemphigus vulgaris* patients which were virtually absent in healthy DR-matched Israeli controls.[62]

B. NUCLEOTIDE SEQUENCE ANALYSIS

The RFLP studies were valuable in identifying disease-associated subtypes of class II alleles. In some cases, the RFLP subdivisions correlated with known serological and cellular subtypes. However, this was not true for the myasthenia and PV RFLP. To delineate the extent and location of polymorphism detected by RFLP typing, we used the *in vitro* gene amplification technique to facilitate nucleotide sequence analysis of class II alleles in many different autoimmune patients and controls.

1. Insulin-Dependent Diabetes Mellitus

IDDM is caused by selective destruction of the insulin-producing beta cells in the pancreatic islets of Langerhans. The presence of cellular pancreatic infiltrates and autoantibodies to islet cells and insulin suggest an autoimmune etiology for the disease. In animal models, disease development has been shown to be T cell dependent.[63] Overt diabetes is not detected until >95% of islet-cell function is obliterated, but glucose-tolerance tests can identify declining beta cell function prior to the onset of symptoms. This interval between commencement of the destructive process and irreversible islet-cell damage may provide opportunities for prophylactic intervention if appropriate strategies can be devised.

The incidence of IDDM is 0.5 to 1.0% in Caucasians. It has been estimated that about 60% of susceptibility to IDDM is attributable to genes within the MHC.[51] DR3, DR4, and DR1 alleles are positively associated with susceptibility (reviewed in Reference 64). Other class II alleles (e.g., DR2 [Dw2] and DR2 [Dw12]) are negatively associated and appear to confer resistance to IDDM. DR3/DR4 heterozygotes have the highest relative risk for IDDM, suggesting a synergistic interaction between susceptibility loci on the DR3 and DR4 haplotypes. This could be the result of *trans*-association of α and β chains to generate a more potent susceptibility molecule. Serologic and RFLP studies implicate DQ, rather than DR, as the susceptibility locus.[57-60] The DR4-associated DQw3 serologic specificity contains two subtypes (3.1 and 3.2) identified by DQβ RFLP and the monoclonal antibody TA10.[58,59,65] Over 90% of DR4 diabetics are DQw3.2 (TA10$^-$) vs. 70% of normals. The IDDM-associated RFLP previously identified in DR4 individuals reflects this subdivision.

We sequenced DQα, DQβ, DRβI, and DRβIII genes from DR3/DR3, DR3/DR4, and DR4/DRw6 IDDM patients in search of coding region differences from alleles found in healthy individuals that may not have been detected by serological, cellular, or RFLP typing. However, in all cases, the deduced amino acid sequences were no different from those in normal controls.[23] Thus, in IDDM, disease is not associated with mutations in class II alleles. Other genetic and environmental factors are required to produce disease in individuals with class II susceptibility alleles.

Nevertheless, upon inspection of amino acid sequences from several DQβ alleles, a striking correlation between the identity of residue 57 and susceptibility or resistance to IDDM[23] was noted (Figure 5A). Susceptibility alleles (DR4,DQw3.2, DR3,DQw2, DR1,DQw1.1, and DR2,DQw1.AZH) encode either serine, alanine, or valine at this position. DQβ alleles from neutral or resistant haplotypes (DR4,DQw3.1, DR2,DQw1.2, and DR2,DQw1.12) encode aspartic acid at position 57. The correlation of IDDM susceptibility with Asp-negative DQβ chains is also seen in American blacks, who unlike white Americans, have both Asp-57 positive (DQBLANK) and Asp-57 negative (DQw2) DQβ alleles found with equal frequency on the DR3 haplotype. A recent study found that 28/28 DR3 American black IDDM patients were DQw2.[66] Of Caucasian DR2 individuals, 80 to 90% have an Asp-positive DQβ chain, perhaps explaining the negative correlation between DR2 and IDDM. Interestingly, this correlation is maintained in a murine model of IDDM; the Aβ allele in NOD mice has serine at position 57, whereas all other (diabetes resistant) strains have aspartic acid.[67]

The correlation of Asp-57 negative DQβ chains with IDDM across many different DR haplotypes and across racial differences strongly suggests that DQβ is the MHC-linked susceptibility locus. However, definitive proof will require knowledge of how certain class II alleles predispose to autoimmune disease. Proposed mechanisms of susceptibility must account for the finding that in both mice[68] and humans, the MHC-linked IDDM susceptibility locus is recessive; 90% of patients analyzed had Asp-57 negative DQβ alleles on both chromosomes.[23] One possibility is that Asp-57 positive DQβ molecules could activate suppressor T cells specific for a putative islet cell autoantigen. Alternatively (or in addition), the Asp-57 positive DQβ molecules could prevent the intrathymic maturation of autoreactive T cells. Class II-mediated clonal deletion has been noted during intrathymic ontogeny of murine T cells.[69] NOD mice made transgenic with an Asp-57 positive Aβ gene may serve as a useful model to evaluate these two possibilities.

2. Rheumatoid Arthritis

Genetic studies in multiplex families with seropositive RA suggest the presence of an MHC-linked RA susceptibility gene.[54] Of RA patients, 65 to 75% carry the DR4 serologic specificity, even across ethnic differences. However, the DR4 haplotype is heterogeneous; there are five different DRβI molecules associated with distinct cellular subtypes (Dw4, Dw10, Dw13, Dw14, and Dw15) and three different DQβ molecules (DQw3.1, DQw3.2 and DQw-blank).[70,71] Two arguments favor DRβI rather than DQβ as the susceptibility locus in RA. First, there appears to be selection for particular DRβI subtypes in patients; Dw4 and Dw14 are increased, but Dw10 is not. Second, although Dw14 haplotypes contain only DQw3.2β, DQw3.1β and DQw3.2β are found at similar frequencies in Dw4 normals and RA patients.[54] Furthermore, DQw3.1β and DQw3.2β alleles are also present on some non-DR4 haplotypes. DQw3.1β is commonly associated with DR5 and DQw3.2β with DR7, haplotypes that are not associated with RA. Thus, there does not appear to be selection for a particular DQβ allele in RA.

Among DR4⁻ RA patients, DR1 and DRw10 haplotypes are increased.[64] DR1 RA patients share the serological determinant MC1 with 90% of DR4 RA patients but only 70% of DR4 controls.[72] This epitope may be contained in HVR3 of DRβI chains; residues 70 and/or 71 are shared among the DR4(Dw4), DR4(Dw14), DR1, and DRw10 susceptibility alleles. Remarkably, DR4(Dw10), which is not associated with RA,[73] differs from DR4(Dw4) by only four amino acids and from DR4(Dw14) by only two amino acids in HVR3 of DRβI (Figure 5B). A positively charged residue (lysine or arginine) at position 71 in the susceptibility alleles is replaced by a negatively charged glutamic acid in DR4(Dw10). Residue 71 (corresponding to position 156 in class I α2) points into the antigen binding site in the predicted class II structure (Figure 4). Thus, as for IDDM, comparison of polymorphic residues present in susceptibility alleles with nonsusceptibility alleles provides strong evidence that allelic differences among MHC molecules are critical for genetic predisposition to RA.

3. Pemphigus Vulgaris

Pemphigus vulgaris is a potentially fatal skin disease in which autoantibodies cause loss of cellular adhesion and severe blistering.[74] The disease is most prevalent in people of Jewish descent. PV is associated with DR4,DQw3 in Ashkenazi Jews and with both DR4,DQw3 and DRw6,DQw1 in non-Ashkenazi Jews and in Europeans.[62] Nearly all Israeli DR4 PV patients type as Dw10 by cellular methods.[75] The DR4(Dw10) allele shares HVR3 of the DRβI locus with a subset of DRw6 haplotypes (DRw6a)[25,70] (Figure 5C); it has been suggested that this HVR3 sequence (the Dw10 epitope) may be a susceptibility epitope for PV.[76]

To determine whether the Dw10 epitope could explain the association of PV with both DR4 and DRw6 haplotypes, we hybridized genomic DNA (that had been amplified *in vitro* from several DR4⁻/DRw6⁺ PV patients with an oligonucleotide probe complementary to the region spanning codons 69 to 74 of the DR4(Dw10) DRβI gene. Following a high stringency wash such that the oligonucleotide probe remained bound only to sequences with zero base-pair mismatches, we found that 19/19 DR4⁺/DRw6⁻ and 0/7 DR4⁻/DRw6⁺ PV patients were Dw10⁺.[77] These results confirm that DR4 PV patients are Dw10, but clearly indicate that the Dw10 sequence is not necessary for susceptibility to PV in DRw6 individuals.

Recently, disease-specific RFLP have been identified in both DR4 and DRw6 Israeli PV patients using a DQβ, but not a DRβ, probe.[62] We therefore sequenced the DQβ alleles from a DR4/DRw6 Israeli PV patient who carried both disease-specific RFLPs in an effort to identify possible coding-region polymorphisms. We obtained one first domain nucleotide sequence that was identical to a common DQβ allele (DQw3.2) associated with DR4(Dw10). However, the DRw6-associated DQβ allele we identified was different from all other DQw1β alleles previously sequenced, including two found on the DRw6 haplotype (DQw1.18 and

A.

IDDM - DQB

POLYMORPHIC RESIDUES

DR HAPLOTYPE	DQ ALLELE	IDDM ASSOCIATION	3	9	13	14	26	28	30	37	38	45	46	47	52	53	55	56	57	66	67	70	71	74	75	77	84	85	86	87	89	90
DR4	w3.2	POSITIVE	S	Y	G	M	L	T	Y	Y	A	G	V	Y	P	L	P	P	A	E	V	R	T	E	L	T	Q	L	E	L	T	T
DR3	w2	POSITIVE	-	-	.	.	S	S	I	V	.	-	E	F	L	.	-	L	-	D	I	-	K	A	V	R	-	.	.	.	.	.
DR1	w1.1	POSITIVE	-	-	.	.	L	G	.	H	.	V	.	-	.	-	Q	R	-	.	G	.	G	A	S	V	R	E	V	A	Y	G	I
DR2	w1.AZH	POSITIVE	-	-	.	.	L	G	.	H	.	V	.	-	.	-	Q	R	S	.	G	.	G	A	S	V	R	E	V	A	Y	G	I
DRw6	w1.19	POSITIVE	-	-	.	.	.	.	.	H	.	.	.	-	.	-	Q	R	V	.	.	.	.	.	.	.	E	V	G	Y	G	I
DR4	w3.1	NEUTRAL/NEGATIVE	-	A	.	Y	.	.	.	.	E	.	.	-	.	-	.	.	D	.	.	.	.	.	.	.	.	.	.	.	.	.
DR5	w3.1	NEUTRAL/NEGATIVE	-	A	.	Y	.	.	.	.	E	.	.	-	.	-	.	.	D	.	.	.	.	.	.	.	.	.	.	.	.	.
DRw9	w3.3	NEUTRAL	-	.	.	.	.	.	.	.	.	.	.	-	.	-	.	.	D	.	.	.	.	.	.	.	.	.	.	.	.	.
DRw8	BLANK	NEUTRAL	-	F	.	G	.	.	.	.	.	.	.	-	.	-	R	L	D	D	I	E	S	V	.	.	.	.	.	.	.	.
DRw6	w1.18	NEUTRAL/NEGATIVE	-	F	.	.	.	H	.	.	.	.	.	-	.	-	Q	R	D	.	G	.	.	.	.	.	E	V	A	F	G	I
DRw6	w1.9	NEUTRAL/NEGATIVE	-	.	.	.	L	G	.	H	.	.	.	-	.	-	Q	R	D	.	G	A	S	V	R	E	V	A	F	G	I
DR2	w1.12	NEUTRAL/NEGATIVE	P	L	A	.	Y	.	D	V	.	.	.	-	.	-	Q	R	D	D	I	.	.	.	.	.	E	V	A	F	G	I
DR2	w1.2	NEGATIVE	-	F	.	.	.	.	.	.	.	.	.	-	.	-	Q	R	D	.	G	.	.	.	.	.	E	V	A	F	G	I
MOUSE																																
NOD	ABNOO	POSITIVE	-	H	.	E	.	.	.	L	.	E	.	E	.	-	R	H	S	Y	*	.	.	.	.	.	E	E	T	E	P	.
B10	Abb	NEGATIVE	-	.	.	E	Y	.	.	V	.	E	H	E	.	-	.	R	.	D	E	I	.	.	.	.	E	G	P	E	H	.

B. Rheumatoid arthritis - DR$_{\beta}$I

	65					70					75	
	Lys	Asp	Leu	Leu	Glu	Gln	Lys	Arg	Ala	Ala	Val	
DR4-Dw4	Lys	Asp	Leu	Leu	Glu	Gln	Lys	Arg	Ala	Ala	Val	
-Dw14	---	---	---	---	---	---	Arg	---	---	---	---	
-Dw15	---	---	---	---	---	---	Arg	---	---	---	---	
DR1-Dw1	---	---	---	---	---	---	Arg	---	---	---	---	Susceptibility
DRw10	---	---	---	---	---	Arg	Arg	---	---	---	---	
DRw53βIII	---	---	---	---	---	Arg	Arg	---	Glu	---	---	
DR4-Dw10	---	---	Ile	---	---	Asp	Glu	---	---	---	---	No positive association

C. Pemphigus vulgaris - DR$_{\beta}$I

	65					70					75	
	Lys	Asp	Ile	Leu	Glu	Asp	Glu	Arg	Ala	Ala	Val	
DR4-Dw10	Lys	Asp	Ile	Leu	Glu	Asp	Glu	Arg	Ala	Ala	Val	
DR6a	---	---	---	---	---	---	---	---	---	---	---	
DR6b	---	---	Leu	---	---	Arg	Arg	---	---	Glu	---	

FIGURE 5. (A) Analysis of polymorphic residues in HLA DQβ between susceptible and nonsusceptible DR haplotypes for IDDM. Amino acid sequences from the susceptible (NOD) and nonsusceptible mouse strains are included. (B) Analysis of polymorphic residues in HLA DRβI between susceptible and nonsusceptible DR haplotypes for RA. (C) Comparison of DRβI third allelic hypervariable region from the DR4-Dw10 subtype and DRw6a and DRw6b alleles. Sources of the sequences are detailed in Reference 64.

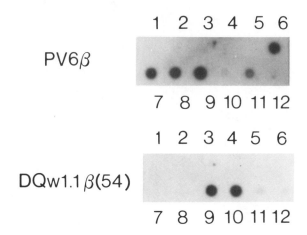

FIGURE 6.   Dot-blot hybridization of amplified genomic DNA from PV patients and controls with PV6β and DQw1.1β[54] oligonucleotide probes. PV6β (GGCGGCCTGACGCCGAG) is complementary to the DRβ sequence derived from a DRw6 PV patient (DRw6,DQw1.9). DQw1.1β[54] (GGCGGCCTGTTGCCGAG) is complementary to the DQβ sequence from a DR1 HTC.[78] Samples 1 to 8: PV patients DR4,DQw3/DR5,DQw3; DR5,DQw3/DRw6,DQw1; DR4,DQw3/DR5,DQw3; DR5,DQw3/DRw6,DQw1; DR4,DOw3/DRw6,DQw1; DR5,DQw2/DRw6,DQw1; DR5,DQw3/DRw6,DQw1; and DR5,DQw3/DR7,DQw2. Samples 9 and 10: DR1,DQw1 HTC: MVL and BVR. Samples 11 and 12: DRw6,DQw1 HTC: WVD and APD.

DQw1.19).[78] This sequence was subsequently found in a non-Israeli DRw6(Dw9) HTC; we have therefore designated the PV-associated allele DRw6,DQw1.9. DQw1.1β and DQw1.AZHβ (two alleles not associated with PV) differ from DQw1.9 only at position 57 of the DQβ molecule (valine and serine versus aspartic acid, respectively).

To facilitate the screening of a large number of PV patients and controls, we designed an oligonucleotide probe specific for DQw1.9 (Figure 6). By dot-blot analysis we found that 13/13 DRw6 PV patients have this allele, versus 1/13 matched controls.[78] The extreme sensitivity of this technique is demonstrated by the fact that DNA from DRw6 PV patients does not hybridize with an oligonucleotide probe specific for the DQw1.1β allele that differs by only two nucleotides from DQw1.9 in this region.

These studies provide further evidence that nonconservative changes at DQβ-57 have dramatic effects on disease susceptibility. However, it must be stressed that disease cannot be attributed exclusively to the presence of aspartic acid at position 57. The DRw6,DQw1.18 allele shares Asp-57 with DQw1.9, but differs outside this region and is not associated with PV.

4. Other Diseases

We have sequenced class II genes from patients with several other autoimmune diseases (reviewed in Reference 64). The DRβI, DQα, and DQβ genes from a DR3[+] myasthenia gravis patient were identical to those on the normal haplotype, despite the presence of the disease-associated HincII DQβ RFLP. Similarly, DRβ, DQα, and DQβ genes sequenced from patients with celiac disease (DR3/DR7), multiple sclerosis (DR2/DR3), and systemic lupus erythematosus (DR2/DR3 and DR2/DR7 patients) were all normal. A comprehensive list of class II genes sequenced in this laboratory from autoimmune patients is given in Table 1.

TABLE 1
HLA Class II Disease Association

Disease	Class II antigen	Relative risk	Patient DR type	Genes sequenced
Coeliac disease	DR3	8—12	DR3,7	DR3βI DR7βI
	DR7			DQw2(α,β)
	DOw2			
Insulin-dependent	DR3	4—6	DR3,3	DR3βI DRw52a
diabetes mellitus	DR4	4—6	DR3,4(2)	DR4βI Dw10
	DR3,4	20	DR4,w6(2)	DR4βI Dw14
	DR2	0.25		DRw53
				DQw3.2 (α,β)
				DQw2 (α,β)
				DQw3.1 (β)
				DQw1.19 (β)
Rheumatoid arthritis	DR4	4—6	DR2,7	DQw1.2
	DR1			DQw2
Myasthenia gravis	DR3	2.5	DR1,3	DR1βI DR3βI
	DR7		DR2,7	DRw52a DR2βI-AZ
			DR5,9	DR7βI DRw53
				DQw1.1 (α,β)
				DQw2 (DR3,α,β)
				DQw2 (DR7,β)
				DQw3.1 (DR5,α,β)
				DQw3.3 (DR9,α,β)
				DQw1.AZH (β)
				DX (α,β)
Multiple sclerosis	DR2	4	DR2,3(2)	DR2βI Dw2, DR2βI
				DR3βI, DRw52a
				DQw1.2 (α,β)
				DQw2 (α,β)
Pemphigus vulgaris	DR4	24	DR4,w6	DR4βI Dw10
	DRw6	1.5		DQw1.9
				DQw3.2
Systemic lupus	DR2	3	DR2,3(2)	DQw1.2(β)
erythematosus	DR3	3	DR2,7	DQw2(DR3,β)
				DQw2(DR7,β)
				DQw1.AZH(β)

Note: Relative risk and genes sequenced are indicated. Sources of sequences are detailed in References 23, 64, 77, 78, 88, and 89.

5. Implications

We conclude from our studies that autoimmune diseases are not caused by mutant class II alleles found exclusively in diseased patients. Nevertheless, particular polymorphic regions or individual residues within allelic HVR show very strong correlations with susceptibility to IDDM, RA, and PV. Allelic HVR control interactions between class II molecules, foreign antigens, and the T-cell receptor. It is therefore likely that alterations in allelic HVR may also influence immune responses to self antigens, or to environmental antigens which trigger autoimmune disease.

For IDDM and PV, amino acid 57 of the DQβ molecule (in HVR2) was identified as the most important MHC-linked determinant of disease susceptibility. In a predicted three-dimensional structure of class II molecules, residue 57 (corresponding to position 143 in class I α2) forms a salt bridge with a conserved Arg at position 79 in the DQ α1 domain (position 84 in class I α1 domain)[47] (Figure 4). Therefore, nonconservative changes at position 57 (Asp [not susceptible] → Ser, Ala, Val in IDDM, or Val, Ser, Ala [not susceptible] → Asp in DRw6 PV patients) would cause major structural changes in the DQ molecule. Such changes could affect antigen binding and/or T-cell recognition, but studies with murine class II genes implicate HVR2 residues as critical for antigen binding.[79] Conversely, single amino acid changes within HVR3 have been shown to abolish responsiveness of several T-cell clones that use a particular class II restriction element, regardless of their antigen specificity.[80]

For most diseases, the autoantigens are unknown. However, once peptides that trigger human autoimmune responses are identified, the precise mechanisms which govern interactions with class II susceptibility alleles can be delineated. The availability of T-cell clones from autoimmune patients will further facilitate the *in vitro* characterization of the trimolecular complex (TcR-Ag-Ia) involved in the inductive (afferent) phase of immune responses to self antigens. Ultimately, it may be possible to design peptides that bind a class II susceptibility allele with much higher affinity than the autoantigen and thus competitively inhibit the induction of autoreactive T cells.

The observation that diseases are often associated with more than one class II allele indicates that at the amino acid level, several sequences may be permissive for, or compatible with, activation of the autoimmune cascade. Consistent with this, there is more than one responder haplotype for most antigens under Ir gene control in mice. Furthermore, recent biochemical data which show the broad specificity of Ia-Ag interactions support the contention that several different allelic sequences may be able to functionally interact with a putative peptide involved in triggering a particular autoimmune disease.

Our findings showing that MHC alleles responsible for disease susceptibility are found in normal individuals further suggest that these alleles participate in a permissive sense and reinforce the importance of other genetic and environmental factors. Loci controlling class II expression, T cell, and immunoglobulin receptor genes are obvious candidates for further investigation as susceptibility loci because of their relevance in immune regulation. The recent availability of a large number of molecular probes specific for variable number tandem repeat sequences scattered throughout the genome should facilitate the search for non-MHC markers of disease susceptibility.[81]

VI. CONCLUSIONS

One major factor in understanding class II function has been the systematic and detailed characterization of class II polymorphism. Applications of recent advances in molecular biology to the study of the MHC have greatly enhanced our appreciation of the extent and precise nature of class II polymorphism. This information has elucidated the mechanisms by which class II molecules determine the specificity of the normal immune response. With the characterization of class II alleles associated with susceptibility to particular human autoimmune diseases, we are now in a position to ask questions regarding class II function in immune responses directed against self antigens. Furthermore, the identification of clear-cut class II susceptibility alleles for some diseases, including IDDM and PV, is of clinical relevance. It should be possible in some cases to design molecular probes for early clinical screening to identify individuals at risk of developing autoimmune disease. The identification of candidate class II disease-susceptibility alleles and further elucidation of the mechanisms by which these alleles predispose to disease states should bring us closer to designing effective strategies of disease prevention and therapy.

ACKNOWLEDGMENTS

We thank Z. Fronek, J. Holloman, C. Jacob, B. S. Lee, T. Lopez, M. McDermott, L. Steinman and D. Wraith for critical review of the manuscript. We also thank K. Moody for final preparation of the manuscript. A. A. Sinha is a Centennial Fellow of the Medical Research Council of Canada and a recipient of an Independent Research Allowance from the Alberta Heritage Foundation for Medical Research. H. Acha-Orbea acknowledges support from the Juvenile Diabetes Foundation. This work was supported by a grant from the National Institutes of Health, AI-07757.

REFERENCES

1. **McDevitt, H. O. and Sela, M.,** Genetic control of the antibody response. I. Demonstration of determinant-specific differences in response to synthetic polypeptide antigens in two strains of inbred mice, *J. Exp. Med.,* 122, 517, 1965.
2. **McDevitt, H. O. and Chinitz, A.,** Genetic control of the antibody response: relationship between immune response and histocompatibility (H-2) type, *Science,* 163, 1207, 1969.
3. **McDevitt, H. O., Deak, B. D., Schreffler, D. C., Klein, J., Stimpfling, J. H., Snell, G. D.,** Genetic control of the immune response. Mapping of the Ir-1 locus, *J. Exp. Med.,* 135, 1259, 1972.
4. **Sasazuki, T., Kaneoka, H., Nishimura, Y. et al.,** An HLA-linked immune suppression gene in man, *J. Exp. Med.,* 152, 2975, 1980.
5. **Sasazuki, T., Nishimura, Y., Muto, M. and Obert, N.,** HLA linked genes controlling immune response and disease susceptibility, *Immunol. Rev.,* 70, 51, 1983.
6. **Kaufman, J. F., Auffray, L., Korman, A. J., Shackelford, D. A., and Stominger, J.,** The class II molecules of the human and murine major histocompatibility complex, *Cell,* 36, 1, 1984.
7. **Mengle-Gaw, L. and McDevitt, H. O.,** Genetics and expression of mouse Ia antigens, *Annu. Rev. Immunol.,* 3, 367, 1985.
8. **Steinmetz, M. and Hood, L.,** Genes of the major histocompatibility complex, *Science,* 222, 727, 1983.
9. **Babbitt, B. P., Allen, P. M., Matsueda, G., Haber, E., and Unanue, E.,** Binding of immunogenic peptides to Ia histocompatibility molecules, *Nature (London),* 317, 359, 1985.
10. **Buus, S., Sette, A., Colon, S. M., Miles, C., and Grey, H. M.,** The relation between major histocompatibility complex (MHC) restriction and the capacity of Ia to bind immunogenic peptides, *Science,* 235, 1353, 1987.
11. **Schwartz, R. H.,** T-lymphocyte recognition of antigen in association with gene products of the major histocompatibility complex, *Annu. Rev. Immunol.,* 3, 237, 1985.
12. **Marrack, P. and Kappler, J.,** The T cell receptor, *Science,* 238, 1073, 1987.
13. **Lo, D., Yacov, L., and Sprent, J.,** Induction of MHC-restricted specificity and tolerance in the thymus, *Immunol. Res.,* 5, 221, 1986.
14. **Svejgaard, A., Platz, P., and Ryder, L. P.,** HLA and disease 1982 — a survey, *Immunol. Rev.,* 70, 193, 1983.
15. **Travers, P. and McDevitt, H. O.,** Molecular genetics of class II (Ia) antigens in *The Antigens,* Vol. 7, Sela, M., Ed., Academic Press, San Diego, CA, 1987, 147.
16. **Erlich, H., Leu, J. S., Peterson, J. L., Bugawan, T., and DeMars, R.,** *Hum. Immunol.,* 16, 205, 1986.
17. **Hardy, D. A., Bell, J. I., Long, E. O., Linsten, T., and McDevitt, H. O.,** Pulsed-field gel electrophoresis mapping of the class II region of the human major histocompatibility complex, *Nature (London),* 323, 453, 1987.
18. **Inoko, H., Tsuji, K., Groves, V., and Trowsdale, J.,** Mapping of HLA region genes by pulsed-field gel electrophoresis and cosmid walking, Proc. 10th Int. Histocompatibility Conf. and 13th Annu. Meet. Histocompatibility and Immunogenetics, Abst. 160, New York, 1987, A11.
19. **Ando, A., Inoko, H., Kawai, J., Awataguchi, S., Nakatsuji, T., Tsuji, K., and Trowsdale, J.,** Nucleotide sequence polymorphism of a new genomic clone of HLA-DV light chain, Proc. 10th Int. Histocompatibility Conf. and 13th Annu. Meet. Histocompatibility and Immunogenetics, Abst. 213, New York, 1987, A20.
20. **Bell, J. I., Denney, D., MacMurray, A., Foster, L., Watling, D., and McDevitt, H. O.** Molecular mapping of class II polymorphism in the human major histocompatibility complex. I. DRβI, *J. Immunol.,* 139, 562, 1987.
21. **MacMurray, A. J., Bell, J. I., Denney, D., Watling, D., Foster, L., and McDevitt, H. O.,** Molecular mapping class II polymorphisms in the human major histocompatibility complex. II. DQβ, *J. Immunol.,* 139, 574, 1987.

22. **Saiki, R. K., Scharf, S., Faloona, F., Mullis, K. B., Horn, G. T., Erlich, H. A., and Arnheim, N.,** Enzymatic amplification of β-globin genomic sequences and restriction site analysis for diagnosis of sickle cell anemia, *Science,* 230, 1350, 1985.

23. **Todd, J. A., Bell, J. I., and McDevitt, H. O.,** HLA-DQβ contributes to susceptibility and resistance to insulin-dependent diabetes mellitus, *Nature (London),* 329, 599, 1987.

24. **Saiki, R. K., Gelfand, D. H., Stoffel, B., Scharf, S. J., Higuchi, R., Horn, G. T., Mullis, K. B., and Erlich, H. A.,** Primer-directed enzymatic amplification of DNA with a therostable DNA polymerase, *Science,* 239, 487, 1988.

25. **Gorski, J. and Mach, B.,** Polymorphism of human Ia antigens: Gene conversion between two DRβ loci results in a new HLA-D/DR specificity, *Nature (London),* 322, 67, 1986.

26. **Knudsen, P. J. and Strominger, J. L.,** Analysis of the DRβ chains from two DRw6 cell lines (WT46 and WT52): Recombination in vivo may have generated new haplotypes, *Immunogenetics,* 25, 209, 1987.

27. **Gregersen, P. K., Moriuchi, T., Karr, R. W., Obata, F., Moriuchi, J., Maccari, J., Goldberg, D., Winchester, R. J., and Silver, J.,** Polymorphism of HLA-DRβ chains in DR4, -7, and -9 haplotypes: implications for the mechanism of allelic variation, *Proc. Natl. Acad. Sci. U.S.A.,* 83, 9149, 1986.

28. **Bell, J. I., Denney, D., Foster, L., Bell, T., Todd, J. A., and McDevitt, H. O.,** Allelic variation in the DR subregion of the human major histocompatibility complex, *Proc. Natl. Acad. Sci. U.S.A.,* 84, 6234, 1987.

29. **Lee, B. S. M., Rust, N. A., McMichael, A., and McDevitt, H. O.,** HLA-DR2 subtypes from an additional supertypic family of DRβ alleles, *Proc. Natl. Acad. Sci. U.S.A.,* 84, 4591, 1987.

30. **Kappes, D. and Strominger, J. L.,** Human class II major histocompatibility complex genes and proteins, *Annu. Rev. Biochem.,* in press.

31. **Ayane, M., Mengle-Gaw, L., McDevitt, H. O., Benoist, C., and Mathis, D.,** Eαu and Eβu chain association: where lies the anomaly?, *J. Immunol.,* 137, 948, 1986.

32. **Steinmetz, M., Stephan, D., and Fischer-Lindahl, L.,** Gene organization and recombinational hot spots in the murine major histocompatibility complex, *Cell,* 44, 895, 1986.

33. **Kabori, J. A., Stauss, E., Minard, K., and Hood, L.,** Molecular analysis of the hot spot of recombination in the murine major histocompatibility complex, *Science,* 234, 173, 1986.

34. **Gustafsson, K., Winman, L., Emmoth, E., Larhammar, D., Bohme, J., Hyldig-Hielson, J. J., Ronne, H., and Peterson, P. A.,** Mutations and selections in the generation of class II histocompatibility antigen polymorphism, *EMBO J.,* 3, 1655, 1984.

35. **Bell, J. I., Todd, J. A., and McDevitt, H. O.,** Molecular structure of human class II alleles, in *Immunology of HLA,* Dupont, B., Ed., Springer-Verlag, New York, 1988.

36. **Widera, G. and Flavell, R. A.,** Nucleotide sequence of the murine I-Eβb immune response gene: evidence for gene conversion events in class II genes of the major histocompatibility complex, *EMBO J.,* 3, 1221, 1984.

37. **Radding, C. M.,** Genetic recombination: strand transfer and mismatch repair, *Annu. Rev. Biochem.,* 47, 847, 1978.

38. **McIntyre, K. R. and Seidman, J. G.,** Nucleotide sequence of mutant I-Aβbm12 gene is evidence for genetic exchange between mouse immune response genes, *Nature (London),* 308, 557, 1984.

39. **Mengle-Gaw, L., Conner, S., McDevitt, H. O., and Fathman, C. G.,** Gene conversion between murine class II MHC loci: functional and molecular evidence from the bm12 mutant, *J. Exp. Med.,* 160, 1184, 1984.

40. **Erlich, H.,** personal communication.

41. **Clayberger, C., Parham, P., Rothbard, J., Ludwig, D. S., Schoolnick, G. K., and Krensky, A. M.,** HLA-A2 peptides can regulate cytolysis by human allogeneic T lymphocytes, *Nature (London),* 330, 763, 1987.

42. **Charron, D. J., Lotteau, V., and Turmel, P.,** Hybrid HLA-DC antigens provide molecular evidence for gene *trans*-complementation, *Nature (London),* 312, 157, 1984.

43. **Guillet, J.-G., Lai, M.-Z., Briner, T. J., Smith, J. A., and Gefter, M. L.,** Interaction of peptide antigens and class II major histocompatibility complex antigens, *Nature (London),* 324, 260, 1986.

44. **Sette, A., Buus, S., Colon, S., Smith, J. A., Miles, C., and Grey, H. M.,** Structural characteristics of an antigen required for its interaction with Ia and recognition by T cells, *Nature (London),* 328, 395, 1987.

45. **Cairns, J. S., Cursinger, J. M., Dahl, C. A., Freeman, S., Alter, B. J., and Bach, F. H.,** Sequence polymorphism of HLA DRβI alleles relating to T cell-recognized determinants, *Nature (London),* 317, 166, 1985.

46. **Bjorkman, P. J., Saper, M. A., Samraoui, B., Bennet, W. S., Strominger, J. L., and Wiley, D. C.,** Structure of the human class I histocompatibility antigen, HLA-A3, *Nature (London),* 329, 506, 1987.

47. **Brown, H., Jardetzky, T., Saper, M. A., Samraoui, B., Bjorkman, P., and Wiley, D. C.,** A hypothetical model of the foreign antigen binding site of class II histocompatibility antigens, *Nature (London),* 332, 845, 1988.

48. **Potter, T. A., Bluestone, J. A., and Rajan, T. V.,** A single amino acid substitution in the α3 domain of an H-2 class I moelcule abrogates reactivity with CTL, *J. Exp. Med.,* 166, 956, 1987.

49. **Germain, R. N., Bentley, D. M., and Quill, H.,** Influence of allelic polymorphism on the assembly and surface expression of class II MHC (Ia) molecules, *Cell,* 43, 233, 1985.

50. **Wicker, L. S., Miller, B. J., Coker, L. Z., McNally, S. E., Scott, S., Mullen, Y., and Appel, M. C.,** Genetic control of diabetes and insulinitis in the nonobese diabetic (NOD) mouse, *J. Exp. Med.,* 165, 1639, 1987.

51. **Rotter, J. I. and Landau, E. M.,** Measuring the genetic contribution of a single locus to a multilocus disease, *Clin. Genet.,* 26, 529, 1984.

52. **Eisenbarth, G. S.,** Type I diabetes mellitus: a chronic autoimmune disease, *N. Engl. J. Med.,* 314(21), 1360, 1986.

53. **Lawrence, J. S.,** Rheumatoid arthritis — nature or nurture?, *Annu. Rheum. Dis.,* 24, 357, 1970.

54. **Nepom, G. T., Hansen, J. A., and Nepom, B. S.,** The molecular basis for HLA class II association with rheumatoid arthritis, *J. Clin. Immunol.,* 7, 1, 1987.

55. **Rosenbaum, J. T., Adelman, N. E., and McDevitt, H. O.,** *In vivo* effects of antibodies to immune response gene products. I. Haplotype specific suppression of humoral immune responses with monoclonal anti-I-A, *J. Exp. Med.,* 154, 1694, 1981.

56. **Steinman, L., Rosenbaum, J. T., Sriram, S., and McDevitt, H. O.,** *In vivo* effects of antibodies to immune response gene products. II. Prevention of experimental allergic encephalitis, *Proc. Natl. Acad. Sci. U.S.A.,* 78, 7111, 1981.

57. **Owerbach, D., Lemmark, A., Platz, P., Ryder, L. P., Rask, L., Peterson, P. A., and Ludvigsson, J.,** HLA-D region β chain DNA endonuclease fragments differ between HLA-DR identical healthy and insulin-dependent individuals, *Nature (London),* 303, 815, 1983.

58. **Cohen-Haguennaves, D., Robbins, E., Massart, C., Musson, M., Deschamps, I., Hors, J., Lalouch, J. M., Dausset, J., and Cohen, D.,** A systematic study of HLA class II β DNA restriction fragments in insulin-dependent diabetes mellitus, *Proc. Natl. Acad. Sci. U.S.A.,* 82, 3335, 1985.

59. **Nepom, B. S., Palmer, J., Kim, S. J., Hansen, J. A., Holbeck, S. L., and Nepom, G. T.,** Specific genomic markers for the HLA-DQ subregion discriminate between DR+ insulin-dependent diabetes mellitus and DR4+ seropositive juvenile rheumatoid arthritis, *J. Exp. Med.,* 164, 345, 1986.

60. **Cohen, N., Brautbar, C., Font, M. P., Dausset, J., and Cohen, D.,** HLA-DR2-associated Dw subtypes correlate with RFLP culsters: most DR2 IDDM patients belong to one of these clusters, *Immunogenetics,* 23, 84, 1986.

61. **Bell, J., Rassenti, L., Smoot, S., Smith, K., Newby, C., Hohlfeld, R., Toyka, K., McDevitt, H. O., and Steinman, L.,** HLA-DQ beta chain polymorphism linked to myasthenia gravis, *Lancet,* 1058, May 10, 1986.

62. **Szafer, F., Brautbar, C., Tzfoni E., Frankel, G., Sherman, L., Cohen, I., Hacham-Zadeh, S., Aberer, W., Tappliner, G., Holubar, K., Steinman, L., and Friedmann, A.,** Detection of disease-specific restriciton fragment length polymorphisms in *pemphigus vulgaris* linked to the DQw1 and DQw3 alleles of the HLA-D region, *Proc. Natl. Acad. Sci. U.S.A.,* 84, 6542, 1987.

63. **Rossini, A., Mordes, J. P., and Like, A. A.,** Immunology of insulin-dependent diabetes mellitus, *Annu. Rev. Immunol.,* 3, 289, 1985.

64. **Todd, J. A., Acha-Orbea, H., Bell, J. I., Chao, N., Fronek, Z., Jacob, C. O., McDermott, M., Sinha, A. A., Timmerman, L., Steinman, L., and McDevitt, H. O.,** A molecular basis for MHC class II-associated autoimmunity, *Science,* 240, 1003, 1988.

65. **Schreuder, G. M. Th., Tilanus, M. G. L., Bontrop, R. E., Bruining, G. J., Giphardt, M. J., van Rood, J. J., and de Vries, R. R. P.,** HLA-DQ polymorphism associated with resistance to type I diabetes detected with monoclonal antibodies, isoelectric point differences, and restriction fragment length polymorphism, *J. Exp. Med.,* 164, 938, 1986.

66. **Hurley, C. H., Gregersen, P. K., Steiner, N., Bell, J., Hartzman, R., Nepom, G., Silver, J., and Johnson, A. H.,** Polymorphism of the HLA-D region in American blacks: a DR3 haplotype generated by recombination, *J. Immunol.,* 140, 885, 1988.

67. **Acha-Orbea, H. and McDevitt, H. O.,** The first external domain of the nonobese diabetic mouse class II I-Aβ chain in unique, *Proc. Natl. Acad. Sci. U.S.A.,* 84, 3425, 1987.

68. **Hattori, M., Buse, J. B., Jackson, R. A., Glimcher, L., Dorf, M. E., Minami, M., Makima, S., Moriwaki, K., Kuzuya, H., Imura, H., Strauss, W. M., Seidman, J. G., and Eisenbarth, G. S.,** The NOD mouse: recessive diabetogenic gene in the major histocompatibility complex, *Science,* 231, 733, 1986.

69. **Kappler, J. W., Roehm, N., and Marrack, P.,** T cell tolerance by clonal elimination in the thymus, *Cell,* 49, 273, 1987.

70. **Gregersen, R. K., Shen, M., Song, Q.-L., Merryman, P., Degar, S., Seki, T., Maccari, J., Goldberg, D., Murphy, H., Schwenger, J., Wang, C. Y., Winchester, R. J., Nepom, G. T., and Silver, J.,** Molecular diversity of HLA-DR4 haplotypes, *Proc. Natl. Acad. Sci. U.S.A.,* 83, 2642, 1986.

71. **Kim, S. J., Holbeck, S. L., Nisperos, B., Hansen, J. A., Maeda, H., and Nepom, G. T.,** Identification of a polymorphic variant associated with HLA-DQw3 and characterized by specific restriction sites within the DQβ chain gene, *Proc. Natl. Acad. Sci. U.S.A.,* 82, 8139, 1985.

72. **Duquesnoy, R. J., Marrari, M., Hackbarth, S., and Zusi, A.,** Serological and cellular definition of a new HLA-DR associated determinant, MCl, and its association with rheumatoid arthritis, *Hum. Immunol.,* 10, 165, 1984.

73. **Zoschke, D. and Segall, M.,** Dw subtypes of DR4 in rheumatoid arthritis: evidence for a preferential association with Dw4, *Hum. Immunol.,* 15, 118, 1986.

74. **Singer, K. H., Hashimoto, K., Jensen, P. J., Morioka, S., and Lazarus, G. S.,** Pathogenesis of autoimmunity in pemphigus, *Annu. Rev. Immunol.,* 3, 87, 1985.

75. **Amar, A., Rubinstein, N., Hacham-Zadah, S., Cohen, O., Cohen, T., and Brautbar, C.,** Is predisposition to *pemphigus vulgaris* in Jewish patients mediated by HLA-Dw10 and DR4?, *Tissue Antigens,* 23, 17, 1984.

76. **Gregersen, P. K., Silver, J., and Winchester, R. J.,** The shared epitope hypothesis: an approach to understanding the molecular genetics of susceptibility to rheumatoid arthritis, *Arthritis Rheum.,* 30, 1205, 1987.

77. **Sinha, A. A., Brautbar, C., Szafer, F., Tzofini, E., Friedmann, A., Steinman, L., and McDevitt, H. O.,** Nucleotide sequence and oligonucleotide dot-blot analysis of HLA-DRβ alleles associated with *pemphigus vulgaris,* in preparation.

78. **Sinha, A. A., Brautbar, C., Szafer, F., Tzfoni, E., Friedmann, A., Todd, J., and McDevitt, H. O.,** A newly characterized HLA DQβ allele associated with pemphigus vulgaris, *Science,* 234, 1026.

79. **Ronchese, F., Schwartz, R. H., and Germain, R. N.,** Functionally distinct subsites on a class II major histocompatibility complex molecule, *Nature (London),* 329, 254, 1987.

80. **Ronchese, F., Brown, M. A., and Germain, R. N.,** Structure-function analysis of the Aβbm12 mutation using site-directed antigenesis and DNA-mediated gene transfer, *J. Immunol.,* 139, 629, 1987.

81. **Lander, E. S. and Botstein, D.,** Mapping complex genetic traits in humans: new methods using a complete RFLP linkage map, *Cold Spring Harbor Symp. Quant. Biol.,* 51, 49, 1986.

82. **Servenius, B., Gustafsson, K., Widmark, E., Emmoth, E., Andersson, G., Larhammar, D., Rask, L., and Peterson, P. A.,** Molecular map of the human HLA-SB (HLA-DP) region and sequence of an SBα (DPα) pseudogene, *EMBO J.,* 3, 3209, 1984.

83. **Kappes, D. J., Arnot, D., Okada, K., and Strominger, J. L.,** Structure and polymorphism of the HLA class II SB light chain genes, *EMBO J.,* 3, 1985, 1984.

84. **Larhammar, D., Servenius, B., Rask, L., and Peterson, P. A.,** Characterization of an HLA-DRβ pseudogene, *Proc. Natl. Acad. Sci. U.S.A.,* 82, 1475, 1985.

85. **Auffray, C., Lillie, J. W., Arnot, D., Grossberger, D., Kappes, D., and Strominger, J. L.,** Isotypic and allotypic variation of human class II histocompatibility antigen α chain genes, *Nature (London),* 308, 327, 1984.

86. **Trowsdale, J., Lee, J. S., Kelly, A., Carson, S., Austin, P., and Travers, P.,** A linkage map of two HLA-SB alpha and beta genes: an intron in one of the SB beta genes contains a processed pseudogene, *Cell,* p.38, 1984.

87. **Tonnelle, C., Demars, R., and Long, E. O.,** A gene encoding an MHC class II beta chain distinct from DR, DQ, and DR beta chains is expressed in a human B cell line, Proc. American Societies for Experimental Biology, 69th annual meeting, abstracts of papers, *Fed. Proc.,* 44, 3, 1985.

88. **Fronek, Z., Timmerman, L., and McDevitt, H. O.,** unpublished data.

89. **Jacob, C. and McDevitt, H. O.,** unpublished data.

90. **Bell, J. I., Todd, J. A., and McDevitt, H. O.,** The molecular basis of HLA disease association, *Adv. Hum. Genet.,* in press.

91. **Erlich, H.,** personal observation.

Chapter 9

PROBING FOR DISEASE SUSCEPTIBILITY

Barbara S. Nepom and Gerald T. Nepom

TABLE OF CONTENTS

I. INTRODUCTION

Allelic variability among human leucocyte antigen (HLA) class II genes engenders the tremendous structural and functional polymorphism characteristic of the major histocompatibility complex (MHC). In addition, interlocus variation both in structure and expression contributes to a complex and diverse array of class II phenotypes encoded on each human haplotype. Key to the ability to analyze phenotypic polymorphisms associated with immune response and with disease is the ability to identify and characterize individual class II genes within a given haplotype. Although the first associations of HLA specificities with autoimmune diseases were noted over a decade ago, the recent explosion of specific DNA sequence information in the HLA region has only recently allowed a precise investigation of the HLA genes responsible for these associations. As molecular genetic techniques have permitted the demonstration of important amino acid diversity even within what had heretofore been defined serologically as a single specificity, the importance of defining individual alleles critical for disease susceptibility has emerged.

Our laboratory has approached the issue of HLA and disease associations by asking the following questions: What is the degree of diversity found within haplotypes known to be disease associated? Which of these defined alleles is most closely correlated with susceptibility to specific diseases? What methods can be developed to most accurately and precisely identify these genes of interest in disease populations? What are the specific nucleotide (and amino acid) variations which are characteristic of disease-associated alleles? And what is the contribution of such variations to aberrant immune recognition?

By using as a model the HLA-DR4 specificity, which is known to be associated with a number of prevalent autoimmune diseases such as rheumatoid arthritis (RA) and insulin-dependent diabetes mellitus (IDDM), we have learned much about the diversity of the HLA class II genetic region and its relationship to disease susceptibility. The demonstration that, in certain cases, the difference between alleles may be only a few nucleotide changes has necessitated the development of exquisitely precise methods for distinguishing among susceptibility alleles. We have, therefore, utilized synthetic oligonucleotide probes, short stretches of nucleotides complementary to unique sequences on genes of interest, which are capable of discriminating single-base changes. We will describe the theoretical and practical advantages of this methodology, its application to model autoimmune diseases such as RA and IDDM, and the resulting implications for the understanding of the major histocompatibility complex in man.

II. HLA-SPECIFIC OLIGONUCLEOTIDE PROBES

Genetic mechanisms of diversification within the HLA class II region apparently have involved a series of gene duplication and recombination events. As a result, nucleotide sequence sharing among different genes is rampant and cross-hybridization with class II gene probes is a frequent observation. Although this poses a technical problem for some methods of analysis, such as restriction fragment length polymorphism (RFLP), it proves to be a technical advantage for methods of oligonucleotide probe analysis, as described in this section. We have utilized three different strategies to exploit short nucleotide sequences as precise gene markers: (1) the use of locus specific probes which distinguish among DR, DQ, DX, DO, and DP loci, yet which hybridize to all alleles of a particular locus; (2) the use of allele-specific probes to identify individual allelic variants present in genomic DNA; and (3) the use of "shared sequence" probes to localize distinct series of nucleotides which are present on otherwise unrelated genes, and which encode "mobile epitope" structures on disparate class II polypeptides.

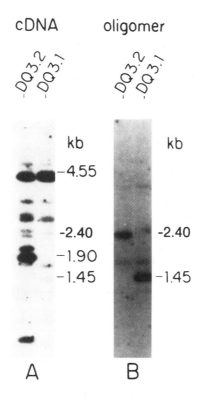

FIGURE 1. Comparison of hybridization with cDNA and oligonucleotide probes for analysis of genomic DNA. A nick-translated full-length DQβ cDNA probe (Panel A) and a ^{32}P end-labeled oligonucleotide probe to a DQβ consensus sequence in the second exon (Panel B) were used as hybridization probes on Taq I digested genomic DNA. DNA from homozygous typing cells carrying the DQ3.2β gene were compared with digests containing the DQ3.1β gene in each panel. Hybridizing bands at 2.4 kb and 1.45 kb represent fragments containing the second DQβ exon from DQ3.2 and DQ3.1, respectively (Panel B). These bands are only faintly visualized using the cDNA probe (Panel A). Many other bands are visualized with the cDNA probe, with the dark bands at 4.5 and 1.9 kb corresponding to third exon fragments, migrating on electrophoresis very close to invariant DXβ fragments. The DQβ oligonucleotide used in this study, 5'-GCCCTTAAACTGGTACACGAAATCCT-3', readily distinguishes among different DQβ alleles using a variety of restriction enzymes as previously described.[1,2] Results shown were obtained using an oligonucleotide probe hybridization technique for analysis of restriction digested genomic DNA, as described in the text.

A. LOCUS-SPECIFIC OLIGONUCLEOTIDE PROBES

We have constructed a series of locus-specific oligonucleotides, between 20 and 46 nucleotides long, which are homologous to DNA sequences that differ among different loci, but are conserved among all alleles at that locus.[1,2] In this manner, locus-specific oligonucleotide probes have been derived for DRβ, DQβ, DQα, DXβ, DXα, DOβ, DPβ, and DPα. These probes identify one gene per haplotype, visualized as a single DNA fragment hybridizing to each probe, which provides a much more simplified and unambiguous identification of restriction fragments in the analysis of genomic DNA, as compared to conventional cDNA probes.

Figure 1 demonstrates that when DNA genomic digests are loaded onto an agarose gel, electrophoresed, and then hybridized with a conventional cDNA probe, multiple gene fragments are visualized, both because the probe is nearly full length and because of homologous sequences at cross-reactive loci. For example, DQβ cDNA probes also hybridize to the closely homologous DXβ gene. On the other hand, a DQβ-specific oligonucleotide probe hybridizes to a sequence within the β2 exon which is unique to the DQβ gene. The use of this probe identifies a single restriction fragment representing the DQβ gene on each haplotype.[1]

Such locus-specific probes are of obvious value for gene identification, gene mapping, and linkage studies. For example, the DXβ oligonucleotide probe has been used to assign unambiguously an unknown RFLP fragment to the DX locus,[3] and the DOβ and the DXα oligonucleotide probes have been used to document an unusual intraclass II recombination in the DX/DQ region.[2] In addition, these locus-specific probes have been very valuable for identifying and assigning specific polymorphisms associated with class II variation. For allelic polymorphisms at a given locus, the oligonucleotide probes can recognize fragments of different size depending on the allele identified. In this way, for example, we can readily distinquish between two variants within the DQw3 specificity, DQ3.1 and DQ3.2 (see Figure 1). Following Taq I digestion, a DQβ oligonucleotide probe hybridizes to a 1.45-kb fragment on DQ3.1 genes and a 2.4-kb fragment on DQ3.2 genes;[1] no DX-associated bands nor any other homologous class II bands cross-hybridize. In an extension of this type of analysis of genomic polymorphisms, we have reported a rapid and specific "DQ gene typing" methodology utilizing a single restriction enzyme and a single DNA probe.[2] The same locus-specific DQβ oligonucleotide probe described above also distinguishes DQ1 and DQ2 genes in Taq I digested DNA, which have hybridizing fragments at 1.8 kb and 1.65 kb, respectively.

In addition to facilitating genomic identification of known polymorphisms, locus-specific probes have been very helpful in the rapid identification of new polymorphisms useful for linkage studies. Thus, two alleles at DXα and two alleles at DOβ were identified and used to extend linkage analysis of class II haplotypes centromeric of DR and DQ.[2]

B. ALLELE-SPECIFIC OLIGONUCLEOTIDE PROBES

This type of technology has been extended to develop *allele*-specific oligonucleotide probes which are necessary for the analysis of very subtle nucleotide differences between highly homologous HLA alleles. We have reported the use of 20 or 21 base probes for hybridization to unamplified genomic DNA for the detection of single-base mismatches. For example, in a study of patients with classical adult RA (described below),[4] it became necessary to distinguish between two different DR4-associated DRβ genes, called Dw4 and Dw14. These differ by only three nucleotides in the β1 exon.[5] We therefore constructed oligonucleotides for use as probes which were mismatched at either one or two nucleotides for each sequence. Using these probes we were able to distinguish easily between the two alleles, and demonstrated that both Dw4 and Dw14 are highly prevalent among these patients. This study validated the notion of using allele-specific oligonucleotides as probes for specific HLA genes, and illustrated the use of this technology in analysis of disease susceptibility.

C. "SHARED-SEQUENCE" OLIGONUCLEOTIDE PROBES

In addition to the construction of oligonucleotide probes which can be locus-specific or allele-specific, some probes can be useful in detecting other subtle structural polymorphisms. One example is the analysis of related genes at different loci which share sequences. For instance, we can distinguish between different DR2-positive haplotypes by oligonucleotide probe analysis of DRβ alleles associated with the DR2 specificity. At least three DR subtypes within DR2 have been characterized by two-dimensional gels,[6] mixed lymphocyte culture,[7] and nucleotide sequence differences.[8,9] Interestingly, two DRβ genes are expressed in each DR2 haplotype, but they are present in different combinations, which define the subtype. That is, some haplotypes contain the same DRβI gene, but possess differing DRβIII genes,

TABLE 1
DR2 Gene Analysis Using a Single
Oligonucleotide Probe, DRβ-AS83

	3.2-kb band (βI)	3.8-kb band (βIII)
Dw2	+	−
Dw12	+	+
AZH	−	+

Note: The presence of genes associated with DRβI and DRβIII can be ascertained by hybridization with a 3.2 and/or a 3.8-kb fragment of EcoRI-digested DNA with this synthetic probe, indicating the Dw subtype of each DR2-positive cell (see text).

and vice versa. An oligonucleotide probe, called DRβAS83, is directed against a polymorphic sequence on DRβ genes centered on codon 86. This 20 nucleotide sequence occurs in some of the DR2 DRβI genes and in some of the DR2 DRβIII genes, and sometimes in both on a single haplotype. Conveniently, the three DR2 subtypes, referred to as Dw2, Dw12, and AZH, can be distinguished by virtue of characteristic hybridization patterns with this single oligonucleotide probe. DRβI genes which carry the AS83 sequence occur on a Taq I restriction fragment of 3.8 kb; DRβIII genes which carry the sequence occur on a 3.2-kb band. Thus, with a Taq I digest of genomic DNA, Dw2-like cells are positive for a 3.2-kb band but negative for a 3.8-kb band; Dw12-like cells are positive for both bands, and AZH-like cells are positive for the 3.8 but not the 3.2-kb band. (See Table 1.)

The second example of "shared sequence" oligonucleotide probes illustrates a different point. Codons 68 to 74 of the HLA Dw14 DRβ gene sequence distinguish the Dw14 gene from other DRβ genes on the other DR4 subtypes.[5] Analysis of a large number of haplotypes with an oligonucleotide probe to this region demonstrated that this nucleotide sequence was also present in several different, apparently unrelated, DRβ genes.[10] In addition to the Dw14 gene, this oligonucleotide probe hybridized to most Dw1 (DR1) and Dw16 (DRw14 [w6]) DRβ genes. Thus, this probe identifies a shared sequence present on different alleles of the same DRβ locus present on haplotypes with diverse serologic specificities. There are two important conclusions which derive from this observation. First, the shared nucleotides identified by this oligonucleotide probe corresponded precisely to the ability of the same haplotypes to stimulate a single alloreactive proliferative T-cell clone raised against a Dw14 positive stimulator cell. Comparative analysis of T-cell clone stimulation with oligonucleotide hybridization indicated that both were directed against the same DRβ determinants. In other words, the oligonucleotide probe appeared to identify the presence of a sequence encoding a shared epitope present on serologically distinct DR alleles that is recognized as a specific allodeterminant by a single T-cell clone. This second conclusion, discussed in more detail below, is that this shared sequence epitope is a good candidate for a disease susceptibility epitope,[11] which may account for genetic contributions to rheumatoid arthritis by diverse HLA haplotypes.

D. METHODOLOGIC ASPECTS OF THE USE OF OLIGONUCLEOTIDE PROBES

Our methods for analyzing genomic DNA with oligonucleotide probes have been reported elsewhere,[4,10] but certain points are worth emphasizing. Briefly, genomic DNA is extracted from cells without amplification, purified, and digested with restriction endonucleases in a standard fashion. After electrophoresis on a 1% agarose gel in TAE buffer, the gels are denatured, neutralized, and dried on Whatman 3MM paper. After soaking briefly in water to remove the backing paper, they are prehybridized in 6X NET and then hybridized with the appropriate ^{32}P end-labeled oligonucleotide probe in 6X NET, under conditions previously determined to allow hybridization only with a 100% nucleotide match. The gels are then washed, first with 5X SSC and 0.5% SDS, then with 3.2 mol/l tetramethylammonium chloride (TMACL) containing 0.5% SDS. Gels are wrapped in plastic and exposed to Kodak XAR film with intensifying screens.

Two points within this method are worth reiterating, as they are departures from other similar methods using oligonucleotide probes.[12-14] First, the gels themselves are dried and then probed, with no transfer step to the nitrocellulose filter. This not only simplifies the procedure considerably, but tends to decrease background dramatically. These gels can be rehybridized several times. Second, the final washes using the quaternary salt TMACL allow stringency conditions to be based solely on the length of the probe, as the GC base content is irrelevant. These modifications have simplified the procedure and made it more accurate. Application of this method to DNA slot-blotting is described below; RNA analysis has also been performed using these methods with equal success.

III. INSULIN-DEPENDENT DIABETES MELLITUS (IDDM)

A. OLIGONUCLEOTIDE PROBE ANALYSIS CONFIRMS THE DQβ3.2 GENE AS CRITICAL TO IDDM SUSCEPTIBILITY

Oligonucleotide probe hybridization, as described above, has been useful for genomic analysis of the susceptibility to certain autoimmune diseases. IDDM provides a good example of a common and serious disease associated with HLA-linked genes, but whose underlying immunogenetic mechanism is not understood. A decade ago, it was shown that HLA-B8 and B15 were associated with susceptibility to IDDM.[15,16] Since then, the HLA-B associations have been shown to be secondary to the linked class II specificities DR3 and DR4.[17,18] We have previously analyzed the DR4 specificity, and found that it encompasses at least five different DRβ alleles, all typing positive for DR4.[19,20] Analysis of the most common DQβ specificity linked to DR4, DQw3, reveals at least two different DQβ alleles, DQ3.1 and DQ3.2, occurring on DR4 positive haplotypes.[21] These DR4 positive DRβ chains and DQw3 positive DQβ chains can occur on the same haplotype in a number of combinations, leading to at least seven different DR4 positive haplotypes in the normal population, as outlined in Figure 2.[20,22]

Because of complex patterns of linkage and polymorphism among highly homologous genes within the HLA region, it has been difficult until now to develop gene-specific markers for the study of disease predisposition. With the more extensive understanding of the DR4 haplotype, however, we can now precisely identify the presence of alleles of interest among patient groups. In an early study, we evaluated a small group of DR4-positive IDDM patients for the presence of the DQβ3.2 allele, which was identified by a particular RFLP pattern, namely, a 12.0-kb band when genomic DNA was digested with the restriction enzyme BamHI.[23] The alternate allele, DQβ3.1, showed a pair of bands at 6.9 and 3.7-kb instead. Because a DQβ cDNA probe was utilized, a number of other bands representing homologous DX genes and DQβ genes from the non-DR4 haplotype were also revealed. In this earlier study, 95% of the DR4 positive IDDM patients possessed the DQ3.2 allele, pinpointing DQβ3.2 as the critical gene accounting for the DR4 association with IDDM.

Further analysis of the DQ3.2 gene using locus specific DQβ oligonucleotide probes identified a large number of restriction fragment polymorphisms which distinguish the DQ3.2 gene from the DQ3.1 allele.[1] Any of these enzymes, in combination with DQβ-specific probes, provides a useful method for identification of the DQ3.2 genes in patient populations. An example of this kind of analysis is shown in Figure 3, in which a number of diabetic patients were analyzed using the DQβ locus-specific oligonucleotide probe. As described

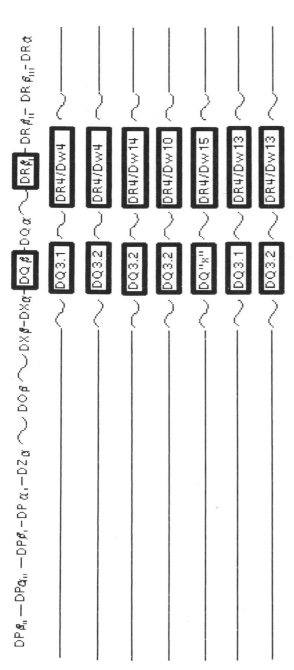

FIGURE 2. Schematic representation of seven distinct DR4+ haplotypes, highlighting allelic variation at the DQβ and the DRβI loci. Five different allelic variants of DR4+ DRβI genes are shown, labeled according to their corresponding HLA-D specificity, Dw4, Dw10, Dw13, Dw14, and Dw15. Three different allelic variants of linked DQβ genes are shown, the DQ3.1, 3.2, and DQ''X'' genes. The DQ''X'' allele has also been termed DQ''blank'', DQ''WA'' or, recently, ''DQw4''. Nucleotide sequences for each of these DRβ and DQβ genes has previously been reported.[5,32] Analysis of multiple DR4 homozygous typing cells and allele-specific oligonucleotide analysis of large numbers of DR4+ individuals indicates that the linkage between different DRβ alleles and different DQβ alleles among the DR4 family of haplotypes is not random. The linkage patterns illustrated by these seven haplotypes are the dominant linkages observed in our studies. Of particular interest are the first two haplotypes shown, which are identical for DRβI genes but carry different DQβ genes. As discussed in the text, only the second haplotype, carrying the DQ3.2 gene, is associated with Type I diabetes, although both haplotypes, which have identical DRβI genes, are associated with classic rheumatoid arthritis. A number of observations derive from these nonrandom linkage patterns: for instance, the finding that the DQ3.2β gene is linked to the Dw4, Dw10, and Dw14 gene on different haplotypes is consistent with the observation that all three of these haplotypes are observed in our IDDM patient population, since the DQ3.2 gene is the primary IDDM susceptibility gene associated with DR4. Many of the linkage patterns, as illustrated for DQβ and DRβI genes, extend further within the HLA region. Each of the seven haplotypes shown shares identical DQα genes and identical DRβIII and DRα genes. The first haplotype (Dw4, DQ3.1) is often linked to HLA-Bw44 and, on the centromeric side, is usually linked to a DXα ''lower'' allele, characterized by a 2.1 kb band following Taq I digestion.[2] The Dw4, DQ3.2 haplotype is often linked to HLA-Bw62, and can carry either the DXα ''upper'' or ''lower'' alleles. The Dw14, DQ3.2 haplotype is usually linked to HLA-Bw60, and the Dw10, DQ3.2 haplotype is usually linked to HLA-Bw38. The Dw15, DQ''X'' haplotype is usually found in Oriental populations linked to HLA-Bw54.

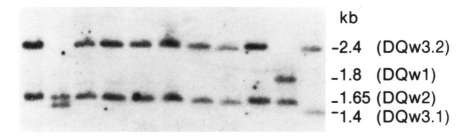

FIGURE 3.   Oligonucleotide hybridization analysis of DQβ genes in IDDM. Genomic DNA from 12 individuals was digested with Taq I and electrophoresed in an agarose gel. A DQβ locus-specific oligonucleotide probe was end-labeled with ^{32}P and used for hybridization analysis as previously described.[1] All individuals were heterozygous, with control individuals in the two lanes on the right illustrating the discrimination between the DQ3.1 and DQ3.2 alleles in one individual, and the DQ1 and DQw2 alleles in another. Ten IDDM patients heterozygous for HLA-DR3/DR4 were analyzed; all were positive for the DQw2 gene fragment at 1.65 kb, characteristic of the DR3 haplotype. Nine of the ten carried the 2.4 kb DQw3.2 gene fragment, and one the 1.4 kb Dw3.1 gene fragment.

above, this probe, in combination with Taq I digestion, provides rapid and simple "DQ gene typing".

To simplify this methodology for analysis of large populations, we have more recently developed allele-specific oligonucleotide probes to identify the DQβ3.2 and DQβ3.1 genes.[24] Because these probes are allele-specific, and do not hybridize to any other human genes, they can be used to identify the presence or absence of specific DQβ alleles using genomic DNA prepared from peripheral blood lymphocytes (or B-cell lines) which have been applied directly to nitrocellulose paper without digestion or amplification. Using these locus- and allele-specific probes, we recently analyzed 31 DR4-positive IDDM patients representing 23 unrelated sibships. All 31 carried the DQβ3.2 gene. Three patients were also positive for DQβ3.1; these three patients were homozygous for DR4, and heterozygous for DQβ3.2 and DQβ3.1. Thus, this analysis confirms, in a simplified and precise manner, the presence of the DQβ3.2 gene among DR4-positive IDDM patients.

As illustrated in Figure 2, the DQβ3.2 gene is present on several different DR4-positive haplotypes. In order to evaluate exactly which haplotypes are carried in the IDDM population, we then analyzed the DRβ alleles linked to DQβ3.2 in each IDDM patient. For this analysis, we used allele-specific oligonucleotide probes distinguishing among different DR4 positive DRβ genes, called Dw4, Dw14, and Dw10, as described in Reference 4 and below. Of the 23 unrelated DR4-positive haplotypes in the IDDM population studied, 17 (74%) carried a Dw4 gene, four carried a Dw14 gene, and 2 carried other DRβ genes. Thus, in this study, all patients carried a DQβ3.2 allele, while several different DR4-positive DRβ alleles were represented.

B. RESULTS FROM FAMILY ANALYSIS

Family analysis can also be used to identify the gene(s) segregating with disease. We recently studied parents and offspring from five families containing more than one affected diabetic sibling, using our oligonucleotide probe methodology.[25] In all five families, a DR4-positive haplotype segregated with IDDM. Thirteen affected siblings from the five families were analyzed, and all caried DQβ3.2 genes. In three families DQ3.2 was linked to a Dw14 gene and in the other two families it was linked to a Dw4 gene, again demonstrating that DQ3.2, rather than a linked DRβ allele, is the primary HLA genetic element contributing to disease susceptibility.

In addition, a recent international workshop investigated the basis of the HLA class II-related susceptibility to IDDM utilizing DR and DQ RFLP markers.[26] Of 87 families evaluated, 60 families showed segregation of DR4[+] haplotypes in all diabetic family members.

Of these 60, 58 or 97% of these DR4$^+$ haplotypes, carried the DQβ3.2 allele, again confirming that the DQβ3.2 gene appears to be sufficient to account for the association of IDDM with DR4. These 58 DR4/DQβ3.2-positive families accounted for 133 out of 163 (82%) of the total DR4-positive diabetics among the 87 families studied. Two DR4/DQβ3.1 haplotypes were identified in these families; however, the diabetic siblings were heterozygous for DR4/DQβ3.1 and either DR3 or DR1, both known to be susceptibility alleles for IDDM. Therefore, when a diabetic individual carried the DR4/DQβ3.1 haplotype, the other haplotype segregated with disease and was apparently responsible for the HLA susceptibility to IDDM.

The DQβ3.2 gene itself, and its product, have been thoroughly studied. The serologic profile of the DQ3.2 product (anti-DQw3$^+$, TA10$^-$) distinguishes it from the DQ3.1 (TA10$^+$) allele, and provides for a simple phenotyping method using monoclonal antibodies.[21,27-29] In addition to its characteristic RFLP pattern, as described above, two dimensional polyacrylamide gel electrophoresis also has been used to identify a characteristic DQ3.2β molecule.[27,28,30] The DQ3.2β polypeptide is more basic than DQ3.1 on electrofocusing gels, a pI shift shown by sequence analysis to be due to charged amino acid substitutions at codons 45 and 57.[31,32]

A large number of different DQ3.2 genes have now been cloned and sequenced, both from diabetic individuals and from unaffected haplotypes.[31-34] All reported DQ3.2 sequences have been identical.

C. GENE COMPLEMENTATION IN IDDM

HLA-DR3 haplotypes represent the second most prevalent associated HLA phenotype in IDDM after DR4, and DR3/4 heterozygotes demonstrate the highest risk for the disease.[35-36] The reason for increased genetic susceptibility in heterozygotes is unknown. Possibly two independent class II-associated genes contribute synergistic risk, or alternatively, one haplotype may contribute a class II-associated risk and the other haplotype some nonspecific disease accelerating factor. Another hypothesis suggests that since the HLA class II molecule is a dimer, each haplotype contributes one of the two class II polypeptide chains. This latter model, that of transcomplementary class II genes, implies that relevant structural polymorphisms occur on at least one alpha chain as well as on at least one beta chain involved in such a specific dimer. Because the contribution of the DQ3.2β gene seems primary on DR4$^+$ haplotypes, we have investigated the types of structural polymorphisms associated with the DQα gene present on DR3 haplotypes. By immunoprecipitation followed by two dimensional gel electrophoresis, we analyzed the class II molecules present on heterozygous DR3/4 cell lines.[37] Immunoprecipitation with a monoclonal antibody specific for the DQ3.2β chain demonstrated two different α chains associated with it, implying two types of cell surface dimers. One was the expected DQ3.2β chain associated with its *cis* acting α element, also encoded on the DR4 haplotype. In addition, however, a high level of protein expression was seen for the mixed dimer consisting of DQ3.2β with the DQα chain normally associated with DQw2 present on the DR3 haplotype. We identified the presence of such hybrid molecules in lymphoblastoid cell lines from diabetic individuals, from their unaffected HLA identical siblings, and from normal DR3/4 individuals. In addition, we have recently expressed a cloned DQ3.2β gene in DQw2 homozygous cell lines and have again observed high levels of surface expression of the DQ3.2β in association with the DQw2 associated alpha gene.[38]

Analysis of DQα genes encoding potential transcomplementary polypeptides has also identified other structural polymorphisms of potential importance. Using conventional RFLP analysis as well as DQα specific oligonucleotide probes, we described three major families of DQα genes with distinct structural characteristics.[39] Additional variation within each of these families creates a large number of individual DQα alleles. The DQα gene which encodes the transcomplementary polypeptide associated with DQ3.2β in the DR3/4 heterozygotes is a member of the DQα family called DQα5$^+$. DQα5 is a serologic specificity defined by a specific monoclonal antibody directed against DQα polymorphisms, which correlates completely with a DQα allele present on haplotypes carrying the DR3, 5, and 8 specificities.[39] This allele does not segregate with the known serologically defined DQ types, which are primarily based on β chain determinants. The other DQα gene families, which are DQα5$^-$, are even more heterogeneous with respect to DR and DQ phenotyping. In addition to clarifying the structural basis for transcomplementary class II molecules, such structural variation among DQα genes implies that an extensive complement of allelic variants at this locus exists which may be of importance to immune function and disease association mechanisms, but which is phenotypically silent with respect to standard HLA typing.

IV. RHEUMATOID ARTHRITIS

A. PARTICULAR DR4 POSITIVE DRβI ALLELES ARE ASSOCIATED WITH ADULT RA AND SEROPOSITIVE JRA

Rheumatoid arthritis, another disease whose association with HLA-DR4 has been known for a long time,[40] provides a contrasting example of the uses of oligonucleotide probes for evaluation of immunogenetic disease susceptibility. Early work by ourselves and others, as mentioned above, defined at least five different "subtypes" of the DR4 specificity.[19,20,41] Initially identified by mixed lymphocyte cultures and two-dimensional gel electrophoresis analysis of the class II protein products, the diversity of DR4 positive alleles at the DRβ locus invited comparisons of these subtypes with susceptibility to rheumatoid arthritis. We initially investigated a group of seropositive juvenile rheumatoid arthritis patients, clinically similar to adult RA, and demonstrated a dramatic increase of both Dw4 and the less common subtype Dw14 by a combination of MLC and 2D gel analysis.[42] Not only are these methods cumbersome, difficult to interpret, and difficult to apply to a large population, but even the subsequent application of Southern blotting techniques were not helpful in this case; while most DQβ alleles can readily be identified by specific RFLP, DR4 positive DRβI alleles could not be distinguished by standard Southern blotting, as most restriction sites are conserved.[43] Evaluation of the question of DR4 susceptibility was made dramatically simpler and more precise by the use of allele-specific oligonucleotide probes.

DNA sequencing of representative members of different DR4 subtypes by Gregersen et al. revealed that many DR4 positive alleles are extremely homologous;[5] Dw4 and Dw14, for example, differ by only three nucleotide changes, leading to two amino acid differences. We took advantage of these mutations by designing allele-specific oligonucleotide probes to distinguish these two alleles: the Dw4 probe includes two nucleotide differences between Dw4 and Dw14, and the Dw14 probe encompasses a single nucleotide change. We applied these probes first to analysis of genomic DNA from a small group of DR4 positive RA patients.[4] Among seven who were homozygous for DR4, a striking five were, in fact, heterozygous for both Dw4 and Dw14, a finding similar to the seropositive JRA patients but unexpected in the adults.

On the basis of this preliminary information, we next evaluated a larger group of 45 Caucasian patients who had been selected on clinical grounds, that of having classical rheumatoid factor positive erosive disease. Among these patients, 38 were positive for DR4, with 31 positive for Dw4 and 12 positive for Dw14.[44]

In contrast, when these same patients were evaluated for their linked DQβ alleles, the distribution of DQ3.1 and DQ3.2 was essentially identical to that of normals, indicating that unlike IDDM, the HLA gene of importance for susceptibility to RA appeared to be the DRβ gene, not the DQβ gene.

While the Dw14 oligonucleotide probe was designed primarily to distinguish the Dw14

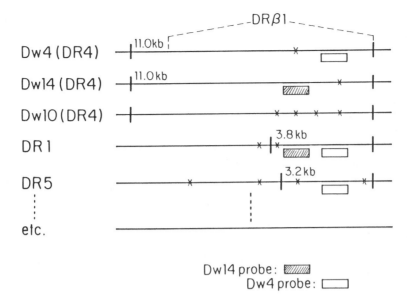

FIGURE 4. Schematic representation of recurrent nucleotide sequence motifs discriminating the Dw4 and Dw14 DRβI genes. Nucleotide variation (designated by X's) in just
two codons distinguishes the Dw4 from the Dw14 coding sequence. Variation at codon
71 is specifically detected with the oligonucleotide probe labeled "Dw14 probe," in the
figure; sequence variation at codon 86 is determined with a "Dw4 probe,"as illustrated.
Neither of these oligomers hybridizes to the Dw10 DRβI sequence, a haplotype not
associated with rheumatoid arthritis, because of several nucleotide mismatches. The
sequence identified by the Dw4 probe recurs in multiple other DRβI genes, as illustrated,
although not in Dw14 or Dw10. Interestingly, the Dw14 probe sequence is not found in
most other DRβI genes, but does occur in DR1-associated and DRw16-associated DRβI
genes. Restriction fragment length polymorphisms distinguish among different DRβI
genes as illustrated. Thus, the Dw14 probe identifies the Dw14 gene on an EcoR1 11.0
kb fragment, and the DR1 DRβI gene on a 3.8 kb fragment.[10] This Dw14 nucleotide
sequence appears to be sufficient to generate a Dw14 epitope recognized by alloreactive
T cells.[10] Thus, the short gene segment centered on codon 71 in the Dw14 DRβI gene
is an excellent candidate for a "mobile epitope" in which a recurring nucleotide sequence
motif generates identical or cross-reactive epitopes on otherwise unrelated gene products.

from the Dw4 allele, it was noted to hybridize also with DNA from patients who possessed
the DR1 specificity (see Figure 4).[10] This probe is an example of a "shared sequence"
oligonucleotide identifying nucleotides present in more than one allele, as mentioned above.
Although the Dw14 and DR1 genes could be readily distinguished because of the hybridization of this probe with different-sized fragments of EcoRI digested DNA, the occurrence
of this identical stretch of DNA sequence on DR1 positive DRβ genes is intriguing, because
DR1 is also known to be associated with rheumatoid arthritis.[45] In fact, 11 patients among
our study group were positive for the DR1 allele. When taken together, the Dw4 and Dw14
probes, identifying DR4/Dw4, DR4/Dw14, and DR1 patients, accounted for 93% (40/43)
of our patient population.[44]

An obvious question is whether this sequence shared between Dw14 and DR1 DRβ
alleles represents an expressed epitope which could potentially contribute directly to the
expression of disease. We have demonstrated that the presence of this nucleotide sequence
correlates with the ability of these haplotypes to stimulate an alloreactive proliferative T-
cell clone raised against a Dw14⁺ stimulator.[10] DRβ alleles that differ by as little as one
nucleotide neither hybridize to the Dw14 oligonucleotide probe nor stimulate this T-cell
clone. This comparison is particularly intriguing since it may indicate that disparate alleles

such as Dw14 and DR1 are functionally equivalent by virtue of a shared sequence encoding a shared epitope. As noted above, almost all the non-DR4 patients in our study can be accounted for by the presence of this shared sequence, suggesting that a single genetic pathway, although not associated with any single DRβ allele, may account for almost all the HLA associated genetic susceptibility to RA.

The key to this analysis is the use of oligonucleotide probes for rapid and exquisite discrimination among highly homologous alleles. Specific genes have been identified which are most closely associated to disease susceptibility among an array of haplotypes positive for a particular specificity. In the case of IDDM, the primary gene of interest appears to be a specific allele of the DQβ locus, DQβ3.2. In contrast, in RA the genes of interest are alleles of the DRβ locus, namely Dw4 and Dw14, and to a lesser degree DR1.

V. IMPLICATIONS FOR UNDERSTANDING THE MHC

A. STRUCTURAL ASPECTS

The pursuit of a better understanding of the HLA genetic contribution to disease susceptibility as outlined above has led concurrently to conceptual progress in our understanding of the major histocompatibility region. One of the concepts highlighted by the above IDDM and RA studies is the relative contribution of alleles at different loci to disease risk. As we have shown above, the primary allele important in susceptibility to IDDM appears to be DQ3.2, implicating a specific variant at the DQβ locus as the key polymorphism putatively involved in immunopathogenic mechanisms of disease. In contrast, studies of rheumatoid arthritis have instead implicated genes at the DRβ1 locus as being the primary contributors to HLA susceptibility in that disease. The critical locus implicated in susceptibility to most of the other HLA-associated diseases which we have not discussed here are as yet unknown. It is quite possible, and in fact preliminary information has suggested, that alleles of other loci, such as those in the DP region, may be involved in other diseases.[46] In fact, certain autoimmune diseases may be demonstrated to have primary associations with, for instance, DQα alleles, which have heretofore not been well characterized or tested for, but which could conceivably be important mechanistically. It is evident that, in spite of their common phenotypic association with the HLA-DR4 specificity, IDDM and RA do not necessarily derive from similar mechanisms of susceptibility and not even from homologous alleles.

The "shared sequence" DRβ oligonucleotide probe which hybridizes not only to a portion of the Dw14 gene but also hybridizes to an identical DNA sequence stretch on DRβ chians of DR1 genes (see Figure 4) exemplifies the idea of "mobile epitopes", where functionally recognizable protein epitopes are encoded by genes that otherwise are widely divergent. It remains to be seen whether this particular oligonucleotide sequence will be shown to be important functionally for disease susceptibility, or whether it merely is a convenient marker for haplotypes at risk for rheumatoid arthritis. Nevertheless, it validates the concept that such small pieces of DNA can in fact move between alleles and potentially could be critical for function.

B. EVOLUTIONARY ASPECTS

Although the above studies imply that specific genes within particular disease-associated haplotypes are responsible for susceptibility, identification of the individual genes of interest is not always straightforward because of the tight linkage found within the HLA region. In addition to the methods described above, another way to approach this problem is to evaluate haplotypes which have undergone recombination, thus allowing a separate analysis of genes segregating with disease. The pursuit of this type of information also adds considerably to our understanding of the evolution of the MHC in general. One example we have recently described centers on the comparison of the nucleotide sequences of DQβ3.1 and DQβ3.2

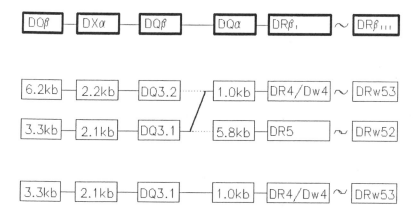

FIGURE 5. Schematic representation of a predicted ancestral recombination event which generated the DR4/Dw4, DQ3.1-positive haplotype. The DQβ and DRβI genes are illustrated as for Figure 2. Also shown are characteristic RFLP fragment sizes discriminating a distinct set of alleles at DQα,[39] and at DXα and DOβ which flank the DQβ gene.[2] Two common haplotypes are illustrated, in which DR4/Dw4 is linked to DQ3.2, and another in which DR5 is linked to DQ3.1. These are the most frequent linkage combinations seen in these haplotypes. A single recombination event between DQα and DQβ would be sufficient to generate the bottom haplotype illustrated, in which DR4/Dw4 is linked to a DQ3.1 gene. The RFLP fragments corresponding to DQα, DXα, and DOβ typically associated with these haplotypes are consistent with this model.[32] As discussed in the text, the bottom haplotype (Dw4, DQ3.1) is not associated with IDDM, whereas the top haplotype (Dw4, DQ3.2) is the one most highly associated in IDDM. One of the implications of the recombinational model illustrated in this figure is that susceptibility to IDDM was lost coincident with the recombination event between DQα and DQβ. This inferred recombination effectively maps susceptibility to IDDM centromeric of DQα and strongly implicates the DQβ gene itself as the key permissive element responsible for DR4 associated IDDM genetic susceptibility.

alleles.[32] We have reported genomic DNA sequencing of the DQ3.1β alleles present on DR4 and DR8 haplotypes. These data, summarized below, suggest a likely recombinational event between DQβ and DQα in an ancestral DR4+ haplotype (see Figure 5), which effectively maps IDDM susceptibility centromeric of DQα: the DQ3.1 nucleotide sequence on non-IDDM-associated DR4 and DR8 haplotypes differs from the DQ3.2 allele linked to diabetes-prone DR4 haplotypes by four amino acid differences in the β1 exon and two amino acid differences in the β2 exon. In addition, there are numerous similarities in the introns of the DQβ3.1 genes from both DR4 and DR8 haplotypes which distinguish them from the DR4/DQβ3.2 sequence, notably a 90 to 100-bp duplicated sequence inserted into the DQβ3.1 intron. Also, the coding region differences between the two genes are dispersed throughout the β1 and β2 exons. These data argue that simple mutation does not account for the differences between DQ3.1 and DQ3.2 on DR4 haplotypes; rather, a likely interpretation is that a recombination between a DR4/DQβ3.2 and a DR8 (or DR5)/DQβ3.1 haplotype generated the DR4/DQβ3.1 haplotype. Restriction mapping and the fact that the DR4/DQβ3.1 haplotype studied differs at DQα from the DR8/DQβ3.1 haplotype, yet shares both DQβ and DXα RFLPs, localizes the putative crossover to the region between the DQα and DQβ genes, a span of about 10 kb, as diagrammed in Figure 5. Furthermore, preliminary data suggest that other haplotypes might also have arisen by similar mechanisms within this same region.[47] So, while gene conversion has been one mechanism implicated recently for shuffling sequences between HLA genes,[48] our data suggest this is unlikely to be the case here. Furthermore, this model signifies that when this putative recombinational event of substituting the DQβ3.1 for the 3.2 allele occurred, simultaneously the DR4+ haplotype lost its associated susceptibility to IDDM, effectively mapping the DR4 associated susceptibility locus centromeric of DQα, and strongly implicating the DQβ3.2 allele itself as the HLA gene most closely associated with IDDM.

C. FUNCTIONAL APPLICATIONS

The molecular and structural approaches to identifying the HLA genes critical to disease susceptibility will lead to direct assays for the function of the newly identified alleles. A wide variety of mechanisms have been proposed for how these alleles lead to disease manifestation. One view hypothesizes that epitopes on individual HLA class II molecules which are critical for the ability of a cell to present antigens (either exogeneous antigen or autoantigen) selectively associate with a determinant on the antigen and trigger pathogenic T-cell responses. An alternative theory holds that these individual alleles are important at the level of thymic selection of potentially responsive T-cell clones. Thus, expression of the disease-associated allele in early T-cell development may be critical, and not necessarily involved in antigen presentation at the autoimmune target itself. This alternative allows the possibility that even HLA genes not expressed in the adult could play a role in early thymic selection. A third model is that of molecular mimicry, where the three-dimensional conformation of the class II molecule mimics an exogeneous antigen such as a viral protein, or triggers an aberrant response rendering the individual incapable of responding to it. It is quite possible that in fact different mechanisms may exist in different autoimmune diseases. In any event, it seems clear than an understanding of the HLA contribution to autoimmune diseases awaits progress in resolving these functional issues. The types of molecular studies discussed here provide the first step toward this goal by identifying the individual alleles responsible for particular HLA associations with individual diseases.

REFERENCES

1. **Holbeck, S. L. and Nepom, G. T.,** Exon-specific oligonucleotide probes localize HLA DQ beta allelic polymorphisms, *Immunogenetics,* 24, 251, 1986.
2. **Amar, A., Holbeck, S. L., and Nepom, G. T.,** Specific allelic variation among linked HLA class II genes, *Transplantation,* 44, 831, 1987.
3. **Hurley, C., Gregersen, P., Steiner, N., Silver, J., Bell, J., Hartzman, R. S., Nepom, G. T., and Johnson, A.,** Polymorphisms of the HLA-D region in American blacks: a DR3 haplotype generated by recombination, *J. Immunol.,* 140, 885, 1988.
4. **Nepom, G. T., Seyfried, C. E., Holbeck, S. L., Wilske, K. R., and Nepom, B. S.,** Identification of HLA-Dw14 genes in DR4+ rheumatoid arthritis, *Lancet,* 2(8514), 1002, 1986.
5. **Gregersen, P. K., Shen, M., Song, Q-L., Merryman, P., Degar, S., Seki, T., et al.,** Molecular diversity of HLA-DR4 haplotypes, *Proc. Natl. Acad. Sci. U.S.A.,* 83, 2642, 1986.
6. **Nepom, G. T., Nepom, B., Wilson, M., Mickelson, E., Antonelli, P., and Hansen, J.,** Multiple Ia-like molecules characterize HLA-DR2-associated haplotypes which differ in HLA-D, *Hum. Immunol.,* 10, 143, 1984.
7. **Richiardi, P., Belvedere, M., Borelli, I., DeMarchi, M., Curtoni, E. S.,** Split of HLA-Drw2 into subtypic specificities closely correlated to two HLA-D products, *Immunogenetics,* 7, 57, 1978.
8. **Wu, S., Yabe, T., Madden, M., Saunders, T. L., and Bach, F. H.,** cDNA cloning and sequencing reveals that electrophoretically constant DR β₂ molecules, as well as the variable DR β₁ molecules, from HLA-DR2 subtypes have different amino acid sequences including a hypervariable region for a functionally important epitope, *J. Immunol.,* 138, 2953, 1987.
9. **Lee, B. S. M., Rust, N. A., McMichael, A. J., and McDevitt, H. O.,** HLA-DR2 subtypes form an additional supertypic family of DRβ alleles, *Proc. Natl. Acad. Sci. U.S.A.,* 84, 4591, 1987.
10. **Seyfried, C. E., Mickelson, E., Hansen, J. A., and Nepom, G. T.,** A specific nucleotide sequence defines a functional T cell recognition epitope shared by diverse HLA-DR specificities, *Hum. Immunol.,* 21, 289, 1988.
11. **Gregersen, P. K., Silver, J., and Winchester, R. J.,** The shared epitope hypothesis: an approach to understanding the molecular genetics of susceptibility to rheumatoid arthritis, *Arthritis Rheum.,* 30, 1205, 1987.

12. **LeGall, I., Millasseau, P., Dausset, J., and Cohen, D.,** Two DR beta allelic series defined by exon II-specific synthetic oligonucleotide genomic hybridization: a method of HLA typing?, *Proc. Natl. Acad. Sci. U.S.A.,* 83, 7836, 1986.

13. **Angelini, G., dePreval, C., Gorski, J., and Mach, B.,** High-resolution analysis of the human HLA-DR polymorphism by hybridization with sequence-specific oligonucleotide probes, *Proc. Natl. Acad. Sci. U.S.A.,* 83, 4489, 1986.

14. **Saiki, R. K., Bugawan, T. L., Horn, G. T., Mullis, K. B., Erlich, H. A.,** Analysis of enzymatically amplified beta-globulin and HLA-DQ alpha DNA with allele-specific oligonucleotide probes, *Nature (London),* 324, 163, 1986.

15. **Nerup, J., Platz, P., Anderson, O. O., Christy, M., Lyngsoe, J., Poulson, J. E., Ryder, L. P., Staub-Neilsen, L., Thomsen, M., and Svejgaard, A.,** HLA-antigens and diabetes mellitus, *Lancet,* 2, 864, 1974.

16. **Cudworth, A. G. and Woodrow, J. C.,** HL-A system and diabetes mellitus, *Diabetes,* 24, 345, 1975.

17. **Thomsen, M., Platz, P., Anderson, O. O., Christy, M., Lyngsoe, J., Nerup, J., Rasmussen, K., Ryder, L. P., Staub-Nielsen, L., and Svejgaard, A.,** MLC typing in juvenile diabetes mellitus and idiopathic Addison's disease, *Transplant Rev.,* 22, 125, 1975.

18. **Svejgaard, A. and Ryder, L. P.,** Associations between HLA and disease: notes on methodology and a report from the HLA and Disease Registry, in *HLA and Disease,* Dausset, J. and Svejgaard, A., Eds., Munksgaard, Copenhagen, 1977, 46.

19. **Nepom, B. S., Nepom, G. T., Mikelson, E., Antonelli, P., and Hansen, J. A.,** Electrophoretic analysis of human HLA-DR antigens from HLA-DR4 homozygous cell lines: correlation between beta-chain diversity and HLA-D, *Proc. Natl. Acad. Sci. U.S.A.,* 80, 6962, 1983.

20. **Nepom, G. T., Nepom, B. S., Antonelli, P., Mickelson, E., Silver, J., Goyert, S. M., and Hansen, J. A.,** The HLA-DR4 family of haplotypes consists of a series of distinct DR and DS molecules, *J. Exp. Med.,* 159, 394, 1983.

21. **Kim, S-J., Holbeck, S. L., Nisperos, B., Hansen, J. A., Maeda, H., and Nepom, G. T.,** Identification of a polymorphic variant associated with HLA-DQw3 and characterized by specific restriction sites within the DQ beta-chain gene, *Proc. Natl. Acad. Sci. U.S.A.,* 82, 8139, 1985.

22. **Nepom, G. T., Palmer, J., and Nepom, B.,** Specific HLA class II variants associated with IDDM, in *Immunology of Diabetes Mellitus,* Jaworski, M., et al., Eds., Elsevier, Amsterdam, 1986, 9.

23. **Nepom, B. S., Palmer, J., Kim, S-J., Hansen, J. A., Holbeck, S. L., and Nepom, G. T.,** Specific genomic markers for the HLA-DQ subregion discriminate between DR4+ insulin-dependent diabetes mellitus and DR4+ seropositive juvenile rheumatoid arthritis, *J. Exp. Med.,* 164, 345, 1986.

24. **Nepom, G. T., Seyfried, C., Holbeck, S., Byers, P., Wilske, K., Palmer, J., Robinson, D. M., and Nepom, B.,** HLA-DR4-associated disease: oligonucleotide probes identify specific class II susceptibility genes in Type I diabetes and rheumatoid arthritis, in *Immunology of HLA, Vol. 2, Immunogenetics and Histocompatibility,* Springer-Verlag, New York, 1988, 404.

25. **Robinson, D. M., Holbeck, S., Seyfried, C., Byers, P., Palmer, J., and Nepom, G. T.,** HLA class II typing using oligonucleotide probes, *Genet. Epidemiol.,* 6, 27, 1989.

26. **Robinson, D. M., Holbeck, S., Palmer, J., and Nepom, G. T.,** HLA DQβ3.2 identifies subtypes of DR4+ haplotypes permissive for insulin-dependent diabetes mellitus, *Genet. Epidemiol.,* 6, 149, 1989.

27. **Maeda, H., Hirata, R., Thompson, A., and Mukai, R.,** Molecular characterization of three HLA class II molecules on DR4 and DRw9 haplotypes: serological and structural relationships at the polypeptide level, *Hum. Immunol.,* 15, 1, 1986.

28. **Schreuder, G. M., Tilanus, M. G., Bontrop, R. E., Bruining, G. J., Giphart, M. J., van Rood, J. J., and de Vries, R. R.,** HLA-DQ polymorphism associated with resistance to type I diabetes detected with monoclonal antibodies, isoelectric point differences, and restriction fragment length polymorphism, *J. Exp. Med.,* 164, 938, 1986.

29. **Radka, S. F., Scott, R. G., and Stewart, S. J.,** Molecular complexity of HLA-DQw3: the TA10 determinant is located on a subset of DQw3 β chains, *Hum. Immunol.,* 18, 287, 1987.

30. **Nepom, G. T., Seyfried, C. A., and Nepom, B. S.,** Immunogenetics of disease susceptibility: new perspectives in HLA, *Pathol. Immunopathol. Res.,* 5, 37, 1986.

31. **Michelsen, B. and Lernmark, A.,** Molecular cloning of a polymorphic DNA endonucleus fragment associates insulin dependent diabetes mellitus with HLA DQ, *J. Clin. Invest.,* 79, 1144, 1987.

32. **Holbeck, S. L. and Nepom, G. T.,** Molecular analysis of DQβ3.1 genes, *Hum. Immunol.,* 21, 183, 1988.

33. **Larhammer, D., Hyldig-Nielsen, J., Servenius, B., Anderson, G., Rask, L., and Peterson, P.,** Exon intron organization and complete nucleotide sequence of a human major histocompatibility antigen DC beta gene, *Proc. Natl. Acad. Sci. U.S.A.,* 80, 7313, 1983.

34. **Todd, J. A., Bell, J. I., and McDevitt, H. O.,** HLA-DQβ gene contributes to susceptibility and resistance to insulin-dependent diabetes mellitus, *Nature (London),* 329, 599, 1987.

35. **Bertrams, J. and Bauer, M.,** Insulin-dependent diabetes mellitus, in *Histocompatibility Testing,* Albert, E., Bauer, M., and Mayer, W., Eds., Springer-Verlag, Berlin, 1984, 348.

36. **Tiwari, J. and Terasaki, P.,** *HLA and Disease Associations,* Springer-Verlag, New York, 1985.
37. **Nepom, B. S., Schwarz, D., Palmer, J. P., and Nepom, G. T.,** Transcomplementation of HLA genes in IDDM. HLA-DQ alpha- and beta-chains produce hybrid molecules in DR3/4 heterozygotes, *Diabetes,* 36, 114, 1987.
38. **Kwok, W., Thurtle, P. S., and Nepom, G. T.,** Transfer and expression of an IDDM susceptibility gene into lymphoblastoid cell lines by retroviral vectors, in *Immunology of HLA,* Vol. 2, *Immunogenetics and Histocompatibility,* Springer-Verlag, New York, 1988, 406.
39. **Amar, A., Radka, S. F., Holbeck, S. L., Kim, S.-J., Nepom, B. S., Nelson, K., and Nepom, G. T.,** Characterization of specific HLA-DQ alpha allospecificities by genomic, biochemical, and serologic analysis, *J. Immunol.,* 138, 3986, 1987.
40. **Stastny, P.,** Association of the B-cell alloantigen DRw4 with rheumatoid arthritis, *N. Engl. J. Med.,* 298, 869, 1978.
41. **Reinsmoen, N. and Bach, F.,** Five HLA-D clusters associated with HLA DR4, *Hum. Immunol.,* 4, 249, 1982.
42. **Nepom, B. S., Nepom, G. T., Mickelson, E., Schaller, J. G., Antonelli, P., and Hansen, J. A.,** Specific HLA-DR4-associated histocompatibility molecules characterize patients with seropositive juvenile rheumatoid arthritis, *J. Clin. Invest.,* 74, 287, 1984.
43. **Holbeck, S. L., Kim, S-J., Silver, J., Hansen, J. A., and Nepom, G. T.,** HLA-DR4-associated haplotypes are genotypically diverse within HLA, *J. Immunol.,* 135, 637, 1985.
44. **Nepom, G. T., Byers, P., Seyfried, C., Healey, L. A., Wilske, K. R., Stage, D., and Nepom, B. S.,** HLA genes associated with rheumatoid arthritis: identification of susceptibility alleles using specific oligonucleotide probes, *Arthritis Rheum.,* 32(1), 15, 1989.
45. **Winchester, R. J.,** The HLA system and susceptibility to diseases: an interpretation, *Clin. Aspects Autoimmunity,* 1, 9, 1986.
46. **Howell, M. D., Smith, J. R., Austin, R. K., Kelleher, D., Nepom, G. T., and Kagnoff, M. F.,** An extended HLA-D region haplotype associated with celiac disease, *Proc. Natl. Acad. Sci. U.S.A.,* 85, 222, 1988.
47. **Nepom, G. T.,** Structural variation among MHC class II genes which predispose to autoimmunity, *Immunol. Res.,* 8, 16, 1989.
48. **Gorski, J. and Mach, B.,** Polymorphism of human Ia antigens: gene conversion between two DR beta loci results in a new HLA-D/DR specificity, *Nature (London),* 322, 67, 1986.

Chapter 10

GENE CONVERSION AND THE GENERATION OF POLYMORPHISM IN CLASS I GENES OF THE MOUSE MAJOR HISTOCOMPATIBILITY COMPLEX

Jean-Pierre Abastado and Philippe Kourilsky

TABLE OF CONTENTS

I. INTRODUCTION

Gene conversion was initially defined[1] as a nonreciprocal transfer of genetic information.[2-9] Gene conversion has been particularly well studied in yeast and several fungi, such as *Ascobolus*,[4] where tetrad and octad formation after meiosis allows an easier observation of phenotypes in haploid spores and, thereby, an easy measurement of nonreciprocal recombination events. However, mitotic as well as meiotic gene conversion takes place, as well documented, for example, in yeast, although mitotic conversion events are usually thought to be less frequent than meiotic ones.[10] Gene conversion has been demonstrated in multicellular organisms, particularly *drosophila* as well as mammalian cells in culture transformed with tagged genes.[11-15] Sequence comparisons of a variety of genes have suggested gene conversion within multigene families in eukaryotes (reviewed in Reference 8, see also References 16 to 23 for a nonexhaustive list of recent references) although the nonreciprocal character of the presumptive conversion event was not established. It is important to note that in all systems so far studied gene conversion is a frequent genetic event. Frequencies of up to 10^{-3} have been estimated for meiotic events between ribosomal RNA genes in *drosophila*[24] and 10^{-2} in yeast.[9] Figures of 10^{-5} are common in cultured animal cells transformed with genes in tandem.[11-15] Gene conversion has been postulated to play an important role in genetic variations over a wide time range, as emphasized, for example, by concerted evolution theories.[25-27] On mechanistic grounds, conversion has initially been interpreted as being the result of correction events operating on mismatched base pairs found on hybrid DNA built with different alleles.[2-6] Then it was suggested that gene conversion could result from DNA neosynthesis using the donor gene as template. These general explanations have not been proven and other models are equally possible (see Reference 8 and below).

Gene conversion was initially assigned a major role in homogenizing gene sequences within a multigene family. However, its potential to diversify gene sequences was recognized by several authors[29,30] and popularized by Baltimore[31] who mentioned its possible involvement in the diversification of MHC genes. Much of the initial hypotheses and evidence concerning gene conversion in the MHC involved class I genes (see below) before class II genes. It must be emphasized that the evidence available so far, albeit convincing, is mostly indirect, and that the role of point mutations has not been precisely assessed yet. At one extreme, some authors have suggested that gene conversion plays no role in the diversification of class I MHC genes.[32]

The purpose of this article is to critically analyze the evidence supporting the occurrence of gene conversion events in the mouse class I H-2 genes and to evaluate its possible role in generating polymorphic variants of class I molecules. The case of class II genes will not be discussed, but it is hoped that the present discussion on class I genes will cast light on questions involving class II genes as well.

II. DEFINITIONS

Whereas the histocompatibility antigens expressed by different cells within a given individual are identical, they usually differ when different individuals are compared, as reflected by allograft rejection. We shall first draw an important distinction between the isotypic and the allelic diversity.

By *isotypic* diversity, we mean that the same individual (even homozygous) expresses a spectrum of class I molecules, for example K^d, D^d, L^d, $Qa-1^d$, $Qa-2^d$, etc. By *allelic* diversity, we mean that several alleles (i.e., alternative forms of the same gene) are present in the natural populations (wild or commensal mice): K^d, K^b, K^k, K^s, K^f, etc. It must be noted that a rigorous definition of alleles in the mouse class I H-2 gene family is not always possible because the size of the family varies between haplotypes. Thus, alleles of the K gene can be unambiguously defined because the K region, at the centromeric end of the H-2 complex on chromosome 17, contains only two genes in the three mice strains (BALB/c, B10, and C3H) where the region has been cloned (reviewed in Reference 33). The two genes K1 and K are sufficiently different to allow the assignment of the K molecule to the K gene. (However, it is not certain that the situation is identical in all wild mice). In contrast, the (D,L) region is quite variable in the various mouse strains: BALB/c (H-2^d) mice have 4 D genes, plus 1 L gene.[34] B10 (H-2^b) mice and C3H (H-2^k) mice have a single D gene and no L gene. Under those circumstances, it is difficult to assign alleles with certainty. For example, the D^b sequence is closer to the L^d than to the D^d sequence.[35] One should be similarly cautious with genes of the Qa and TLa regions, and the difficulties of nomenclature illustrate the point.

It is important to emphasize that the notion of polymorphism relates to population genetics. Following Klein,[36] a gene is said to be polymorphic when the frequency of the dominant allele is less than 0.99. According to this definition, globin genes are not polymorphic despite the fact that more than 300 variant hemoglobins have been described, because, in France and the U.S. at least, more than 99% of the population harbors the same hemoglobin. Globin genes thus display a high allelic diversity but no polymorphism.

III. EXTENT OF ISOTYPIC DIVERSITY, ALLELIC DIVERSITY AND POLYMORPHISM

A. ISOTYPIC DIVERSITY

The first serological studies indicated the existence of at least two "allelic series" (reviewed in Reference 34). Refinement led to the identification of several surface antigens: K, D, L, Qa-1, and TL. When cDNA probes became available,[37-39] Southern blotting experiments on mouse DNA immediately revealed more complexity at the genetic level than anticipated from the number of serologically defined products.[38,40] Cosmid cloning has led to the isolation of the almost complete MHC from DNA of two different mice.[41,42] So far, 37 class I genes have been found in BALB/c[43] and only 25 in B10. By Southern blotting experiments, it seems that certain subspecies of mice, such as *Mus cervicolor popaens* and *Mus platythrix*, may have up to 60 to 80 class I genes.[44,45] One should note that the definition of the H-2 class I multigene family is merely operational: it is based on the cross-hybridization with a probe encompassing the most conserved gene region, that encoding the 3rd extra-cellular domain of the class I transplantation antigens. Under relaxed hybridization conditions, other members might appear. In fact, two recent reports describe class I (or class I-like) genes with somewhat higher divergence in sequence.[46,47] In summary, the isotypic diversity at the gene level is high, and varies between mouse strains, presumably as the result of unequal crossing-overs, duplications, and deletions which cause contraction and expansion of the family. These findings and the fact that, contrary to HLA class I genes,[48] the nucleotide sequences have so far revealed few or no pseudogenes, have prompted further investigations of proteins encoded by these genes, some of them not being serologically defined yet. Such studies have demonstrated that the product of the Q10^d gene is made by the liver and secreted into the serum,[48] and that the Qa-2 antigen is encoded by Q7^d and Q7^b.[50,51] Also, the product of the newly discovered "37" gene[43] has been identified.[116] Nevertheless, much work is still needed to identify the possible products of all mouse class I genes.

B. ALLELIC DIVERSITY

Much of the initial work on the serological specificities which indicate the existence of a variety of alleles was carried out on a number of mice from which inbred and congenic strains were derived. It so happened that this collection of mice was derived from a small number of founder animals.[52] This gave a new impetus to the analysis of wild mice caught in various parts of the world (U.S., Chile, Scotland, Denmark, France, Spain, Poland, Italy, U.S.S.R., F.R.G., Egypt, Israel, China, Japan, etc.) as done by several groups, including Klein's.[36,53-55] These mice were then serologically typed with anti-H-2 reagents and/or analyzed for restriction fragment polymorphisms with a panel of probes. More than 100 distinct H-2 haplotypes are now known, and more than 100 alleles have been found for either K or D. Further analytical refinements might well subdivide certain allelic classes. These counts do not include the so-called ''alien'' antigens found on the surface of certain tumor cells (e.g., Reference 56 and references therein), the definition of which has remained somewhat elusive. Indeed, some of them could reflect the existence of additional alleles created by mutation.

It is thus clear that there is broad allelic diversity of the K and D genes. It is equally clear that the other class I genes mapping in the Qa-TL region show much less allelic diversity. Serological studies have so far revealed four Qa-1 variants, three Qa-2 and five TL ones (reviewed in Reference 57). The Qa-TLa reactivities have been less thoroughly analyzed than the K, D, and L ones, and the interpretations are more difficult, due to ambiguities in the identity of the genes involved. Thus, it is not firmly established that some of the observed serological reactivities are not due to the activation of an otherwise unexpressed gene in the cohort of Qa and TLa genes. At the DNA level, however, it is well documented that the Q10, Q7, and 37 genes (and, perhaps, T3) display little or no diversity. The Q10 gene has been sequenced in two haplotypes (H-2^d and H-2^b) with only a few nucleotide differences.[58] The 37 gene, as appreciated from restriction fragment polymorphism displays just a few variants in the many mice analyzed.[43] It is indeed impossible to assess for minor variations but, by and large, it may be taken as granted that allelic diversity in genes of the Qa-TLa region is poor.[59]

C. EXTENT OF POLYMORPHISM

Studies with wild mice also allow an appreciation of the extent of polymorphism in the following sense: the most frequent of the D alleles is D^d which is found in 12% of animals. Similarly, K^d is found in 14% of animals. Other alleles are found with frequencies ranging from 11% (K^k) to less than 1%.[36] These figures decrease as new searches bring about new haplotypes.

IV. EVIDENCE OF GENE CONVERSION

A. INITIAL SEQUENCE COMPARISONS

As soon as nucleotide sequences of class I cDNAs or genes were obtained, their comparisons revealed several interesting features. Lalanne et al.[60] noted that one cDNA which was sequenced looked as if it were a patchwork of three different sequences. Furthermore, the homology between two sequences was much higher in the 3' untranslated region than in the coding region of the cDNAs. It was thus suggested that the 3' untranslated region had been homogenized by a conversion event. Steinmetz et al.[38] made a similar observation. Gachelin et al.[61] and Brégégère[62] and others[63] formalized the hypothesis that the less polymorphic genes could serve as a reservoir for the transfer of blocks of sequences into the more polymorphic ones. The potential for diversification was emphasized by Kourilsky[64] (the combinatorial potential in a random process being at least in the order of n (n-1)/2 where n is the number of nucleotide divergences between the two sequences).

B. ANALYSIS OF THE BM MUTANTS

These observations and thoughts were, at best, suggestive. A convincing piece of evidence was brought about by a molecular analysis of the K^{bm} mutants (reviewed in References

65, 66). Mutant mice, detected by skin graft incompatibility with mice of the parental haplotypes, were found to have alterations in their class I molecules. Mutants in the K^b gene were more frequent (about 2×10^{-4} per gamete) and this rate was not significantly affected by mutagens. In other genes, rates of about 4×10^{-5} were observed. The K^{bm} mutants have been thoroughly studied and many of them were shown to display more than one amino acid substitution.[65] Sequence analysis of cloned mutant genes (initially of the K^{bm1} mutant) showed multiple nucleotide substitutions (7 in K^{bm1} as compared to K^b). Synthetic oligonucleotides matching the mutant sequence detected potential donor genes in the class I H-2 multigene family. This was confirmed by sequencing the potential donor genes. For example, the Q10 gene displays a stretch of 50 residues identical to the K^{bm1} mutant region and located in the homologous position. Q10 thus appears as the potential donor of the bml mutation by some kind of conversion or double recombination event.[67] Similarly, Q4 might have given two nucleotide substitutions (in 15) to K^{bm6} and K^{bm9}; Q10 might have given 4 (in 7) to K^{bm23}, D^b 3 (in 11) to K^{bm11}, etc. (reviewed in Reference 65). The fact that, for most nonpunctual K^b mutants, one can find one or several donor genes is very suggestive of the transfer of blocks of sequences. A compilation of data concerning the mutant mice has indicated that many of the mutation have probably occurred in females[55,68] at the premeiotic stage.[69] It is noteworthy that one mutation in class II genes, the bm12 mutation, probably involves the transfer of a block of nucleotides.[70,71] However, not all mutations can qualify as conversion-like events: certain bm mutations could be point mutations;[72] in addition, the dml and dm2 mutations involve recombinations and deletions in the D^d and L^d genes.[73,74]

C. SEQUENCE COMPARISONS

As the body of amino acid sequences (mainly in HLA molecules, see Reference 75) and nucleotide sequences grew, the recurrence of substitutions in the series of available variants became more and more striking. We carried out a statistical analysis to determine whether variations in the sequences of class I H-2 genes (available in 1985) could be due exclusively to random point mutations. The results were clearly negative, leading to the definition of ''concerted'' substitutions which, in probability, were unlikely to have occurred by chance.[76] These concerted substitutions, therefore, were likely to be the imprint of genetic exchanges, the molecular nature of which, albeit compatible with conversion, was, of course, not specified by this type of analysis. Two points, however, deserve mention. First, an analysis of silent vs. nonsilent, nonconcerted substitutions failed to support the existence of selective constraints, except in the third domain coding region. Second, there was a striking correlation between the location of concerted substitutions and that of the rare dinucleotide CpG. CpGs have been previously noticed to be concentrated predominantly in the first half of the gene, in a region encompassing exons 2 and 3, which encode the more polymorphic first and second domains.[77] So did the concerted substitutions, which has led us to propose that CpGs might be involved in promoting conversion events (References 8, 76 and below).

We have looked at whether presumptive conversion events in other genes (murine and human class II MHC genes, human alpha-globulin genes, immunoglobulin genes) correlate with the presence of a CpG in the neighboring region. The correlation seems to exist, although it has not yet been validated by a statistical analysis. More generally, from studies in prokaryotic systems (reviewed in Reference 7), the concept emerges that certain sequences are more ''penetrant'' than others in gene conversion processes.

V. MOLECULAR MECHANISMS OF GENE CONVERSION

The above analyses strongly suggest the existence of genetic exchanges between numbers of the class I multigene family. The nonreciprocal character of these exchanges has not been demonstrated so that the term ''gene conversion'' may be somewhat careless (but no example of reciprocal exchange has been found so far). Klein, perhaps one of the first to invoke this mechanism, argues that other mechanisms, including looping out of imperfect hairpins and mismatched repair in the systems, could be responsible for the observed mutations in this particularly GC rich part of H-2 genes.[32] It was argued before that repair in DNA hairpins could generate self-instability by a kind of "flip-flop" correction of alternative strands.[7] Caution is recommended, although Klein's conclusion that "the observed sharing of oligonucleotide stretches between genes far apart from one another is most likely not the result of gene conversion; it is probably the consequence of evolution by multiple mutations", appears unlikely to be correct. On the one hand, it would require a special, so far undocumented mutational system for the emergence of the K^{bm} mutations. On the other hand, it would also require that H-2 alleles are organized in some kind of an evolutionary tree. This is, however, difficult to assume particularly for the K gene, isolated as it is at one end of the H-2 complex, without postulating multiple unequal crossing-overs. Indeed, a double unequal crossing-over is formally equivalent to a conversion event. The above arguments emphasize the need to fill the concept of "conversion" with molecular mechanisms, otherwise it would remain an empty notion.[8,32,64]

A. POSSIBLE MECHANISMS

We will underline at least three types of mechanisms which can lead to the nonreciprocal transfer of a block of nulceotides between two genes.[8]

1. A double cross-over between mispaired genes would be equivalent to a conversion event. The size of the transfered block is potentially large, since there is no *a priori* reason that both cross-overs occur close to each other.
2. Mismatch repair of heteroduplex DNA formed by the inappropriate pairing of the strands of related genes would also yield conversion. This is the most common accepted interpretation of gene conversion events in fungi (Figure 1A).
3. "Unequal double-stranded repair", as we named it, results from the repair of a double-strand break by the sister chromatid.[6] By the action of nucleases, the break may enlarge into a gap, both strands of which can be repaired from the homologous DNA sequence of the other chromatid. Conceivably, repair could occasionally use a wrong template, that is a gene of related sequence present on the same or another chromosome, rather than the homologous gene.[8,76] The result, again, would be the transfer of a block of nucleotides from one gene into another, the block presumably being rather small, having a size similar to that of the gap created by nucleases.

In all three models, single-stranded or double-stranded breaks would play a role in initiating cross-overs or abortive cross-overs resulting in the transfer of DNA strands. We have proposed that CpGs in class I genes could be fragile DNA sites, the breakage involving deamination of the methylated CpG. This would provide a possible explanation for the correlation between CpGs and the position of "concerted substitutions".[8,76]

B. TESTS IN MODEL SYSTEMS

We wished to verify certain of the assumptions underlying the above models. This could not be achieved directly on H-2 genes and was studied in model systems which relate to a more general question: that of genetic exchanges between homologous sequences displaying a "significant amount" of heterology. This field of genetics[64] has been poorly documented to date, because most analyses have been carried out in prokaryotic systems, with a relatively small number of genetic markers, mostly point mutations. By contrast, aligned H-2 sequences display 10-20% divergent nucleotides (not counting gaps) in the most variable regions.

We first asked whether and how heteroduplexes made with H-2 sequences and displaying

UNEQUAL PAIRING

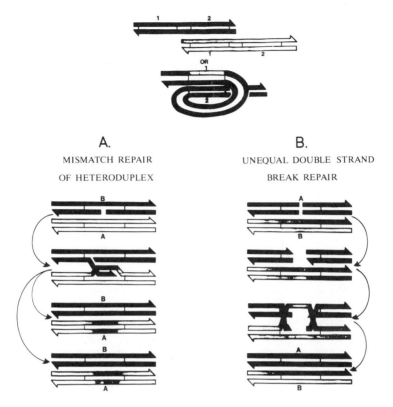

A.

MISMATCH REPAIR
OF HETERODUPLEX

B.

UNEQUAL DOUBLE STRAND
BREAK REPAIR

FIGURE 1.   Proposed molecular mechanisms for gene conversion in multigene families. Genes 1 and 2 are homologous members of a multigene family. They may eventually undergo unequal pairing (top) either within the same or between two different chromatids. Single strand (A, left) as well as double strand (B, right) breaks may lead to gene conversion. However, the two models differ with respect to the localization of the break relative to the donor sequence: whereas in A, the broken chromatid is the donor, in B the broken chromatid is the acceptor. See text for other explanations.

numerous mismatches would be corrected.[78,79] Heteroduplexes were prepared *in vitro* and transformed into *Escherichia coli* or *Cos-1* monkey cells in culture. In the latter case, DNA was extracted after two days and transformed into *E. coli* with appropriate controls in order to score and clone the corrected molecules. *E. coli* clones were probed by *in situ* hybridization with a panel of labeled oligonucleotides capable of specifically detecting one or the other parental sequence. Results showed that, in the limit of resolution of such experiments, correction of heteroduplexes, both in *E. coli* and monkey cells, was frequent, random, and sometimes generated patchworks;[79] the apparent result was a transfer of one or several short blocks of nucleotides. The most complex event which we observed was eight strand switches within 600 bp presumably resulting from four corrected patches. These experiments strongly suggested that, if heteroduplexes form on a sufficient length of DNA, heteroduplex correction can indeed be a mode of generating of diversity.[79] It must be emphasized, however, that, although the formation of hybrid DNA is generally and logically postulated in recombination processes, the occurrence of such heteroduplex structures remains to be established in all processes of recombination.

We next tried to measure the length of the DNA tract which is exchanged during a recombination event promoted by a double-stranded break, known, in yeast and other organisms to strongly stimulate recombination.[80,81] To address this question, we modified an experimental system initially devised by Weber and Weissman.[82] The so-called "snails" are linearized plasmids with (partially) homologous H-2 sequences in tandem. Such plasmids are genetically dead unless they circularize, which most often happens by intramolecular recombination. When the break is located in one of the H-2 sequences, one can measure the number of molecules which have recombined at a certain distance from the break. When doing such experiments, both in *E. coli* and *X. laevis* oocytes, we found[83] that the recombination event usually takes place rather close to the break (within a few dozen base pairs). Others have shown that a stretch of 8 to 20 homologous base pairs is enough to promote recombination in *X. laevis* oocytes.[84] In addition, we observed that a recombination signal could be propagated as far as several kilobase pairs away, in spite of the numerous heterologies.[83] Finally, we are in agreement with previous results obtained in bacteriophage system,[85] which showed that large heterology creates recombination hot spots.[83] We observed few patchworks associated with these recombination events, as would be expected if heteroduplexes were formed in the process of recombination, but too few recombinants were analyzed to draw a significant conclusion.

C. RELEVANCE TO VARIATIONS IN H-2 CLASS I GENES

The relevance of such experiments to actual genetic variations in H-2 genes is hypothetical. The available data do not prove the involvement of the postulated mechanisms, nor choose between them. One mutant human alpha-globin gene displays a patchwork which could be due to heteroduplex correction.[86] On the other hand, many other genetic changes interpreted as conversion events correspond to the exchange of a single block of nucleotides. When the block is relatively large, it can result from a double unequal recombination. In mutants of the K[bm] series, where the exchanged block is small, unequal double-strand break repair is a plausible but unproven explanation. In addition, one should not *a priori* dismiss hypotheses making use of an RNA intermediate in the formation of a DNA-RNA heteroduplex. (Certain sequence analyses,[87] not confirmed by us,[76] had given the impression that conversion could be confined to exons). Finally, it should be kept in mind that polymorphic regions in the MHC are apparently bordered by hot spots of recombination (reviewed in Reference 88). If, as shown by Abastado et al.,[83] a recombination signal can, albeit rarely, propagate far away from a break point, conversion events due to heteroduplex correction could happen far away from sites of recombination. In summary, several plausible mechanisms can explain conversion events. None of them is proven, and several could operate in different situations.

VI. HOMOGENIZATION VS. DIVERSIFICATION OF NUCLEOTIDE SEQUENCES BY CONVERSION

It may seem paradoxical that gene conversion was initially invoked to account for sequence homogenization in multigene families, and then for diversification. Several authors have studied the theoretical long-term effects of gene conversion in population genetics, and concluded that it might be an important parameter in certain models of evolution[25-27,63] as an homogenizing mechanism. Others have called attention to its potential for diversification[28,30-32,64] and we emphasized that it is not a blind mutation process, in that it saves, at the level of the primary sequence, what is identical, and had been previously selected.[64] In the class I H-2 gene family, we[61] and Brégégère[62] have suggested that gene variation in the most polymorphic genes (K, D, L) could use the less polymorphic ones (in the Qa-T1a region) as a stable reservoir of sequences. By computer simulations, Brégégère actually reached the conclusion that within certain limits of mutation and conversion rates, the polymorphism of the K, D, L genes could not be maintained in the absence of a more stable reservoir of sequences.[62]

We wish to emphasize here the importance of a so far somewhat neglected parameter, namely the length of the converted tract.[30,83,89] Intuitively, it is clear that, if converted tracts are long with regard to gene length (e.g., longer than the converted gene), conversion will rapidly lead to homogenization. In contrast, if the converted tracts are short, diversification may occur. In a simplified fashion, and for a gene family of two members, this can be formalized as follows: if AD is the allelic diversity, and IDo is the isotypic diversity at time zero, and L is the average length of converted tracts, one obtains:

$$AD = IDo + 1 - L$$

$$ID = IDo - L$$

Conversion thus appears to transform isotypic diversity into allelic diversity: significant allelic diversity can be generated with a moderate rate of point mutations which renews the isotypic diversity, provided that L is small. Otherwise, isotypic homogenization takes place. It is interesting, in this respect, that the blocks of nucleotides presumably transferred in the generation of the K^b mutants are in the range of about 50 bp or less. Similarly, the length of DNA exchanged in the double-stranded break-instructed intramolecular recombination which we measured in the snail system[83,84] was usually short: less than 200 bp in 85% of the events. These two observations are compatible with the notion that the exchange of short blocks of sequence can account for at least some of the allelic diversification of class I H-2 genes.

VII. WHY ARE SOME GENES POLYMORPHIC AND OTHERS NOT?

While directional conversion as emphasized by Brégégère[62] requires a pool of nonpolymorphic genes, there is so far no experimental evidence in favor of such a process. In fact, directional conversion may take place, being ''allowed'' by the existence of nonpolymorphic genes, but the model does not explain why the latter exists. Furthermore, there is evidence that nonpolymorphic genes (namely Q7 and Q8) have undergone conversion events[90] and it is suspected that polymorphic genes also exchange sequences (D^b being the likely donor in the K^{bm11} mutation). This raises the question of whether the directionality of conversion events is real, or only apparent.

Leaving apart, for the time being, the question of selection, we shall extend here our dicussion to the length of converted tracts. We feel that it is likely that conversion events can take place in all H-2 genes, either between alleles or between nonallelic genes. It may be, however, as suggested by certain studies[91] that the length of the converted tract is a function of the initial homology between the two partners: if homology is high, conversion will occur over a longer distance and will tend to suppress the effects of genetic drift. If homology is low, conversion will involve a few dozen base pairs and generate diversification. Thus, for example, a Qa gene may be converted by K^b, over a short distance, but this conversion event may often be suppressed by a second conversion event involving its true allele. Other sources of directionality, discussed in Kourilsky[8] involve the state of methylation of CpGs in the different genes as well as the distance from the centromers which could create a gradient of probabilities along the chromosome.[92,93]

Whatever mechanism is postulated, it is difficult to reach a satisfactory understanding without involving selection. Several groups have looked for functional constraints in the polymorphic genes by scoring the silent *versus* nonsilent nucleotide substitutions.[76,94,95] Several of these analyses do indicate functional constraints on the third external domain, and suggest that some selective pressure may exist in the first and second domain. The evidence, however, is not compelling. Theoretical (e.g., Reference 96) as well as experimental population genetics studies, particularly in man, have suggested that MHC antigens are subject to selection (see, for example, References 97, 98). The crystal structure of HLA-A2 and the probable structure of H-2 class I molecules including the K[b] mutants[99,100] provide strong arguments in favor of selection. Apparently, a major function of the major transplantation antigens is to present peptides derived from processed proteins to the receptors displayed by T cells. Many of the polymorphic residues, and many of the amino acids altered in the K[b] mutants, most probably map in the peptide binding site.[100] Experiments in which peptides are presented by chimeric H-2 molecules made between K[d] and D[d] to specific CTL clones also direct the specificity of presentation to residues located in the presumptive site.[101,117] The correlation thus established between the specific biological activity and the polymorphic residues obviously reinforces the plausibility of selective processes associated with polymorphism.

These selective processes fall into two major classes. First, the classical view has spread that polymorphism is related to a selective pressure exerted by pathogens, because resistance to the latter is haplotype dependant. Second, more recently, it has been discovered that fetal growth is activated by histocompatibility differences[102-105] providing a possible driving force for allelic diversification and polymorphism.

We have recently proposed that class I MHC molecules present self peptides as well as foreign ones.[106,107] This "peptidic self" model could be of some relevance here, because it broadens the function of MHC molecules to self determinants and enlarges the concept of immune surveillance. Thus, "internal" diseases (not due to infection) could contribute to selection over MHC molecules as well. Further, we raised the hypothesis that Qa and TLa antigens may have some specialized presenting function (being, perhaps, involved in the presentation of the so-called idiopeptides; Reference 107). It thus seems possible that the poorly polymorphic genes of the Qa,TLa region, although denied any function by certain authors,[108] do have an essential function and are subject to strong selective pressure. The fact that in certain mice, many of these genes can be deleted without any apparent loss of viability may mean that their essential functions are exerted by alternative genes or group of genes. In summary, there appears to be numerous possible targets for selection on many of the class I genes.

VIII. LONG-RANGE VS. SHORT-RANGE VARIATIONS

It must be stressed that polymorphism in the H-2 region is not only seen in genes: the polymorphism of restriction site clearly involves larger blocks of DNA, including several genes as well as intergenic sequences.[33,34,41,42,59] One may thus have to distinguish two levels of polymorphism, possibly reflecting distinct diversification mechanisms: on a smaller scale, gene conversion and point mutations might be preeminent. On the larger scale, variations result in differences in gene numbers (as observed in BALB/c and B10 mice), and almost certainly involve unequal crossing-overs (see, for example, Reference 109). Other poorly understood or documented mechanisms, may also intervene (see below). So far, there is no obvious link between both types of variations, except if one relates conversion to recombination through long range effects such as heteroduplex zipping over large distances (as suggested in Reference 83, and by the existence of hot spots of recombination seemingly bordering several polymorphic H-2 regions: Reference 88). In this respect, it would be important to determine more precisely the length of heteroduplex DNA formed *in vivo* and the distance to which heteroduplex DNA propagates from the site where it nucleated. It may be relevant that studies in yeast have occasionally shown co-conversion over rather large distances (several kbs to several dozens kbs)[92,93] suggesting heteroduplex formation and/or zipping over the equivalent distance. Finally, such a link between conversion and recombination should not be taken as implying a causal relationship: a conversion event might cause recombination as well as the opposite, or both conversion and recombination could be consequences of some other genetic event.

The distinction between long-range vs. short-range variations in the H-2 complex is reminiscent of recent advances in the analysis of repair in the prokaryotic systems, where a distinction is now made between "very short-patch" mismatch repair, and "long-patch" mismatch repair (reviewed in Reference 6). In essence, the former appears to be a highly specialized system that conserves certain bases or short sequences (for example, methylated cyctosines in the GCA or TGG sequences in *E. coli*, or, possibly, the ATTAAT sequence in *S. pneumoniae*).[110,111] Their preferred conservation justifies the notion of "more penetrant" sequences in conversion events as stated above. Long-patch mismatch repair is usually referred to as the methyl-directed repair in *E. coli*, because the requirement for nonmethylated GATC sequences explains why the newly synthetized, unmethylated strand made during DNA replication is the one corrected. This system involves the excision of long tracts of single-stranded DNA carrying a replication error, followed by resynthesis on the naked template strand. In *S. pneumoniae*, where there is no methylation of the GATC sequences, free DNA ends apparently play a major role in identifying the sequence to be excised (reviewed in Reference 112), and could also do so in *E. coli*.[7] The relationships between long-patch and very short-patch repair and recombination are quite different: long-patch repair opposes recombination in the sense that the removal of the incoming strand destroys the first necessary intermediate (i.e., hybrid DNA) in recombination.[7,112] In contrast, very short patch repair involves hyperrecombination effects which happen, in *E. coli* and lambda phage, to interfere with genetic mapping. As a whole, it would thus appear, as proposed by Radman,[7] that long-patch mismatch repair would tend to be conservative, while short-patch mismatch repair would contribute to diversification, both by their mode of repair and action on recombination.

Such studies on prokaryotic systems, albeit now reaching molecular definition of conversion processes, may not take in account all features of eukaryotic DNA, such as the existence of repeated DNA. In this regard, it is striking that recent studies in *Neurospora crassa* have brought to light a previously unrecognized genetic process, that premeiotically rearranges duplicated DNA. This highly efficient process scrambles host sequences in a cross involving a partner with duplicated DNA, leaving unaltered the unique sequences.[113] This finding should, indeed, lead to emphasize that many mechanisms of genetic variations probably remain unknown at this time.

IX. CONCLUSIONS

The above discussion shows how little is actually known on the diversification of H-2 class I genes by gene conversion. First, the very existence of conversion events is not rigorously established and their mechanisms are totally unknown. Conversion does appear, however, as the most likely and simplest—but nonexclusive—explanation for a number of observations. Second, assuming that conversion events have taken place, the rates at which they introduce variations in H-2 genes is not determined either. The K^b gene, where most mutants were found, appears exceptional in that, for an unknown reason, it has given rise to mutants at a much higher rate (2×10^{-4}) than D^b, K^d, or D^d (about 4×10^{-5}).[65,66] Accordingly, there is no compelling reason to postulate that the MHC genes vary particularly rapidly. Polymorphism has probably preceeded speciation in the mouse.[114] It is quite possible that the evolution of class I genes has been "slow" and highly selected, rather than an uninterrupted and rapid shuffling of sequences under loose selective conditions.

This caution in interpreting the few existing data for mouse H-2 class I gene should indeed prevail in extrapolating the interpretations to class II genes of the mouse, or class I and class II genes of humans. For example, HLA class I genes[48,115] include many pseudogenes and fewer active genes than mouse H-2 genes. Class II genes of man and mouse do not show a clear discrimination between two subsets of polymorphic and nonpolymorphic genes. Finally, the presently known and/or hypothetical functions of the various genes are not equivalent, nor are the selective pressures which might be exerted over their products (it is not clear that selective pressure over class I and class II MHC antigens are equivalent). Much more work is needed to reach a reliable understanding of MHC genes evolution. In the meanwhile, caution should be exercised and dogmatic views prohibited.

ACKNOWLEDGMENTS

We are extremely grateful to many colleagues, particularly Dr. M. Radman, G. Gachelin, and M. Kieran for stimulating and helpful discussions or corrections. We are also indebted to Mrs. V. Caput for editing the manuscript. This work was supported by grants from Institut National de la Santé et de la Recherche Médicale, Institut Pasteur, Centre National de la Recherche Scientifique and the Ligue Nationale Française contre le Cancer.

REFERENCES

1. **Zickler, H.,** Genetische Untersuchungen an einem heterohalischen Askomyzeten (Bombardia Lunata nov. spec.), *Planta,* 27, 573, 1934.
2. **Fogel, S., Mortimer, R. C., and Lunsnak, K.,** Mechanism of meiotic gene conversion, or "wanderings on a foreign strand", in *The Molecular Biology of the Yeast Saccharomyces,* Strathern, Y. N., Jones, E. W., and Broach, J. R., Eds., Cold Spring Harbor Laboratory, New York, 1982, 289.
3. **Radding, C.,** Homologous pairing and strand exchange in genetic recombination, *Annu. Rev. Genet.,* 16, 405, 1982.
4. **Rossignol, J. L., Nicholas, A., Hamza, H., and Langin, T.,** Origins of gene conversion and reciprocal exchange in *Ascobolus, Cold Spring Harbor Laboratory Symp. Quant. Biol.,* 49, 13, 1984.
5. **Claverys, J. P. and Lacks, S. A.,** Heteroduplex deoxyribonucleic acid base mismatch repair in bacteria, *Microbiol. Rev.,* 50, 133, 1986.
6. **Radman, M. and Wagner, R.,** Mismatch repair in *E. coli, Annu. Rev. Genet.,* 20, 523, 1986.
7. **Radman, M.,** Mismatch repair and genetic recombination, in *Genetic Recombination,* Kucherlapati, I. R. and Smith, G. R., Eds., American Society for Microbiology, Washington, D.C., 1989.
8. **Kourilsky, P.,** Molecular mechanisms for gene conversion in higher cells, *Trends Genet.,* 2, 60, 1986.
9. **Symington, L. S. and Petes, T. D.,** Meiotic recombination within the centromere of a yeast chromosome, *Cell,* 52, 237, 1988.
10. **Klein, H. L.,** Lack of association between intrachromosomal gene conversion and reciprocal exchange, *Nature (London),* 310, 748, 1984.
11. **Song, K.-Y., Schwartz, F., Maeda, N., Smithies, O., and Kucherlapati, R.,** Accurate modification of a chromosomal plasmid by homologous recombination in human cells, *Proc. Natl. Acad. Sci. U.S.A.,* 84, 6820, 1987.
12. **Waldman, A. S. and Liskay, R. M.,** Differential effects of base-pair mismatch on intrachromosomal versus extrachromosomal recombination in mouse cells, *Proc. Natl. Acad. Sci. U.S.A.,* 84, 5340, 1987.
13. **Brenner, D. A., Smigocki, A. C., and Camerini-Otero, R. D.,** Double-strand gap repair results in homologous recombination in mouse L cells, *Proc. Natl. Acad. Sci. U.S.A.,* 83, 1762, 1986.
14. **Ayares, D., Chekuri, L., Song, K.-Y., and Kurcherlapati, R.,** Sequence homology requirements for intermolecular recombination in mammalian cells, *Proc. Natl. Acad. Sci. U.S.A.,* 83, 5199, 1986.
15. **Thomas, K. R. and Capecchi, M. R.,** Site-directed mutagenesis by gene targeting in mouse embryo-derived stem cells, *Cell,* 51, 503, 1987.
16. **Atchison, M. and Achesnik, M.,** Gene conversion in a cytochrome P450 gene family, *Proc. Natl. Acad. Sci. U.S.A.,* 83, 2300, 1986.
17. **Akimenko, M. A., Mariané, B., and Rougeon, F.,** Evolution of the immunoglobulin K light chain locus in the rabbit: evidence for differential gene conversion events, *Proc. Natl. Acad. Sci. U.S.A.,* 83, 5180, 1986.

18. **Gorski, J. and Mach, B.,** Polymorphism of human Ia antigens: gene conversion between two DR loci results in a new HLA-D/DR specificity, *Nature (London),* 322, 67, 1986.

19. **Lefranc, M. P., Forster, A., and Rabbitts, T. H.,** Genetic polymorphism and exon changes of the constant regions of the human T-cell rearranging gene gamma, *Proc. Natl. Acad. Sci. U.S.A.,* 83, 9596, 1986.

20. **Young, J. A. T., Wilkinson, D., Bodmer, W. F., and Trowsdale, J.,** Sequence and evolution of HLA-DR7 and DRW53 associated bêta chain genes, *Proc. Natl. Acad. Sci. U.S.A.,* 84, 4929, 1987.

21. **Bell, J. I., Denney, D., Foster, L., Belt, T., Todd, J. A., and McDevitt, H. O.,** Allelic variation in the DR subregion of the human major histocompatibility complex, *Proc. Natl. Acad. Sci. U.S.A.,* 84, 6234, 1987.

22. **Harada, F., Kimura, A., Iwanaga, T., Shimozawa, K., Yata, J., and Sasazuki, T.,** Gene conversion-like events cause steroid 21-hydroxylase deficiency in congenital adrenal hyperplasia, *Proc. Natl. Acad. Sci. U.S.A.,* 84, 8091, 1987.

23. **Reynaud, C. A., Anquez, V., Grimal, H., and Weill, J.-C.,** A hyperconversion mechanism generates the chicken light chain preimmune repertoire, *Cell,* 48, 379, 1987.

24. **Coen, E. S. and Dover, G. A.,** Unequal exchanges and the coevolution of X and Y rDNA arrays in Drosophila melanogaster, *Cell,* 33, 849, 1983.

25. **Dover, G.,** Molecular drive: a cohesive mode of species evolution, *Nature (London),* 299, 111, 1982.

26. **Otha, T.,** Linkage disequilibrium due to random genetic drift in finite subdivided populations, *Proc. Natl. Acad. Sci. U.S.A.,* 79, 1940, 1982.

27. **Dover, G.,** The molecular drive in multigene families, *Trends Genet.,* 2, 159, 1986.

28. **Gefter and Fox quoted by Seidman, J. G., Leder, A., Nau, M., Norman, B., and Leder, P.,** Antibody diversity: the structure of cloned immunoglobulin genes suggests a mechanism for generating new sequences, *Science,* 202, 11, 1978.

29. **Stahl, F.,** Genetic recombination: thinking about it in phage and fungi, W. H. Freeman, San Francisco, 1979.

30. **Bourguignon-Van Horen, F., Brotcorn, A., Caillet-Fauquet, P., Diver, W. P., Dohet, C., Doubleday, O. P., Lecomte, P., Maenhaut-Michel, G., and Radman, M.,** Conservation and diversification of genes by mismatch correction and SOS induction, *Biochimie,* 64, 559, 1982.

31. **Baltimore, D.,** Gene conversion: some implications for immunoglobulin genes, *Cell,* 24, 592, 1981.

32. **Klein, J.,** Gene conversion in MHC genes, *Transplantation,* 38, 327, 1984.

33. **Flavell, R. A., Allen, H., Burkley, L. C., Sherman, D. M., Waneck, G. L., and Widera, G.,** Molecular biology of the H-2 histocompatibility complex, *Science,* 233, 437, 1986.

34. **Stephan, D., Sun, H., Fischer Lindahl, K., Meyer, E., Hämmerling, G., Hood, L., and Steinmetz, M.,** Organization and evolution of D region class I genes in the mouse major histocompatibility complex, *J. Exp. Med.,* 163, 1227, 1986.

35. **Reyes, A. A., Schöld, M., and Wallace, R. B.,** The complete amino acid sequence of the murine transplantation antigen H-2D^b as deduced by molecular cloning, *Immunogenetics,* 16, 1, 1982.

36. **Klein, J.,** Natural History of the Major Histocompatibility Complex, John Wiley and Sons, New York, 1986.

37. **Kvist, S., Brégégère, F., Rask, L., Cami, B., Garoff, H., Daniel, F., Wiman, K., Larhammar, D., Abastado, J.-P., Gachelin, G., Peterson, P. A., Dobberstein, B., and Kourilsky, P.,** Structure of the C-terminal half of two H-2 antigens deduced from their cloned mRNA sequences, *Nature (London),* 292, 78, 1981.

38. **Steimnetz, M., Frelinger, J. G., Fisher, D., Hunkapiller, T., Pereira, D., Weissman, S. M., Uehara, H., Nathenson, S. G., and Hood, L.,** Three cDNA clones encoding mouse transplantation antigens: homology to immunoglobulin genes, *Cell,* 24, 125, 1981.

39. **Reyes, A. A., Johnson, M. J., Schöld, M., Ito, H., Ike, Y., Morin, C., Itakura, K., and Wallace, R. B.,** Identification of an H-2K^b-related molecular by molecular cloning, *Immunogenetics,* 14, 383, 1981.

40. **Cami, B., Brégégère, F., Abastado, J.-P., and Kourilsky, P.,** Multiple sequences related to classical histocompatibility antigens in the mouse genome, *Nature (London),* 291, 673, 1981.

41. **Steimetz, M., Winoto, A., Minard, K., and Hood, L.,** Clusters of genes encoding mouse transplantation antigens, *Cell,* 28, 489, 1982.

42. **Weiss, E. H., Golden, L., Fahrner, K., Mellor, A. L., Devlin, J. J., Bullman, H., Tiddens, H., Bud, H., and Flavell, R. A.,** Organization and evolution of the class I gene family in the major histocompatibility complex of the C57B1/10 mouse, *Nature (London),* 310, 650, 1984.

43. **Transy, C., Nash, R. S., David-Watine, B., Cochet, M., Hunt, S. W., III, Hood, L. E., and Kourilsky, P.,** A low polymorphic mouse H-2 class I gene from the *Tla* complex is expressed in a broad variety of cell types, *J. Exp. Med.,* 166, 341, 1987.

44. **Delarbre, C., Morita, T., Kourilsky, P., and Gachelin, G.,** Evolution within the multigene family coding for the class I histocompatibility antigens: the case of the mouse t-haplotypes, *Ann. Inst. Pasteur/Immunol.,* 136C, 51, 1985.

45. **Rogers, M. J., Germain, R. N., Hare, J., Long, E., and Singer, D. S.,** Comparison of MHC genes among distantly related members of the genus mus, *J. Immunol.,* 134, 630, 1985.

46. **Singer, D. S., Hare, J., Golding, H., Flaherty, L., and Rudikoff, S.,** Characterization of a new sub-family of class I genes in the H-2 complex of the mouse, *Immunogenetics,* 28(1), 13, 1988.

47. **Fischer Lindahl, K., Loveland, B. E., and Richards, C. S.,** The end of H-2, in *Major Histocompatibility Genes and Their Roles in Immune Function,* C. S. David, Ed., Plenum Press, New York, 1988.

48. **Geraghty, D. E., Koller, B. H., and Orr, H. T.,** A human major histocompatibility complex class I gene that encodes a protein with a shortened cytoplasmic segment, *Immunology,* 84, 9145, 1987.

49. **Lew, A. M., Maloy, W. L., and Coligan, J. E.,** Characteristics of the expression of the murine soluble class I molecule (Q10), *J. Immunol.,* 136, 254, 1986.

50. **Stroynowski, I., Soloski, M., Low, M. G., and Hood, L.,** A single gene encodes soluble and membrane-bound forms of the major histocompatibility Qa-2 antigen: anchoring of the product by a phospholipid tail, *Cell,* 50, 759, 1987.

51. **Sherman, D. H., Waneck, G. L., and Flavell, R. A.,** Qa-2 antigen encoded by Q7[b] is biochemically indistinguishable from Qa-2 expressed on the surface of C57 BL/10 mouse spleen cells, *J. Immunol.,* 140, 138, 1988.

52. **Moriwaki, K.,** Genetic significance of laboratory mice in biomedical research, in *Animal Models: Assessing the Scope of Their Use in Biomedical Research,* Alan R. Liss, 1987, 53.

53. **Nadeau, J. H., Wekeland, E. K., Götze, D., and Klein, J.,** The population genetics of the H-2 poly-morphism in European and North African populations of the house mouse *(Mus musculus L),* *Genet. Res. Comb.,* 37, 17, 1981.

54. **Sturm, S., Figueroa, F., and Klein, J.,** The relationship between t and H-2 complexes in wild mice. I. The H-2 haplotypes of 20 t-bearing strains, *Genet. Res. Comb.,* 40, 73, 1982.

55. **Shiroishi, T., Sagai, T., and Moriwaki, K.,** Sexual preference of meiotic recombination within the H-2 complex, *Immunogenetics,* 25, 258, 1987.

56. **Strauss, H. J., Van Waes, C., Fink, M. A., Starr, B., and Schreiber, H.,** Identification of a unique tumor antigen as rejection antigen by molecular cloning and gene transfer, *J. Exp. Med.,* 164, 1516, 1986.

57. **Tewarson, S., Zaleska-Rutczynska, Z., Figueroa, F., and Klein, J.,** Polymorphism of *Qa* and *Tla* loci of the house mouse, *Tissue Antigens,* 22, 204, 1983.

58. **Mellor, A. C., Weiss, E. H., Kress, M., Jay, G., and Flavell, R. A.,** A nonpolymorphic class I gene in the murine histocompatibility complex, *Cell,* 36, 139, 1984.

59. **Winoto, A., Steinmetz, M., and Hood, L.,** Genetic mapping in the major histocompatibility complex by restriction enzyme site polymorphism: most mouse class I genes map to the Tla complex, *Proc. Natl. Acad. Sci. U.S.A.,* 80, 3425, 1983.

60. **Lalanne, J. L., Delarbre, C., Brégégère, F., Abastado, J.-P., Gachelin, G., and Kourilsky, P.,** Comparison of nucleotide sequences of mRNAs belonging to the mouse H-2 multigene family, *Nucleic Acids Res.,* 10, 1039, 1982.

61. **Gachelin, G., Dumas, B., Abastado, J.-P., Cami, B., Papamatheakis, J., and Kourilsky, P.,** Mouse genes coding for the major class I transplantation antigens: a mosaic structure might be related to the antigenic polymorphism, *Ann. Inst. Pasteur/Immunol.,* 133C, 3, 1982.

62. **Brégégère, F.,** A directorial process of gene conversion is expected to yield dynamic polymorphism associated with stability of alternative alleles in class I histocompatibility antigens gene family, *Biochimie,* 65, 229, 1983.

63. **Hayashida, H. and Miyata, T.,** Unusual evolutionary conservation and frequent DNA segment exchange in class I genes of the major histocompatibility complex, *Proc. Natl. Acad. Sci. U.S.A.,* 80, 2671, 1983.

64. **Kourilsky, P.,** Genetic exchanges between partially homologous nucleotide sequences: possible implications for multigene families, *Biochimie,* 65, 85, 1983.

65. **Nathenson, S. G., Geliebter, J., Pfaffenbach, G. M., and Zeff, R. A.,** Murine major histocompatibility complex class I mutants: molecular analysis and structure-function implications, *Annu. Rev. Immunol.,* 4, 471, 1986.

66. **Egorov, I. K. and Egorov, O. S.,** Detection of new MHC mutations in mice by skin grafting, tumor transplantation and monoclonal antibodies: a comparison, *Genetics,* 118(2), 287, 1988.

67. **Weiss, E. H., Mellor, A., Golden, L., Fahrner, K., Simpson, E., Hurst, J., and Flavell, R. A.,** The structure of a mutant H-2 gene suggests that the generation of polymorphism in H-2 genes may occur by gene conversion-like events, *Nature (London),* 301, 671, 1983.

68. **Loh, D. Y. and Baltimore, D.,** Sexual preference of apparent gene conversion events in MHC genes in mice, *Nature (London),* 309, 639, 1984.

69. **Geliebter, T., Zeff, R. A., Melvold, R. W., and Nathenson, S. G.,** Mitotic recombination in germ cells generated two major histocompatibility complex mutant genes shown to be identical by RNA sequence analysis: K[bm9] and K[bm6], *Proc. Natl. Acad. Sci. U.S.A.,* 83, 3371, 1986.

70. **McIntyre, K. R. and Seidman, J. G.,** Nucleotide sequence of mutant I-A bm12 gene is evidence for genetic exchange between mouse immune response genes, *Nature (London),* 308, 551, 1984.

71. **Widera, G. and Flavell, R. A.,** The nucleotide sequence of the murine I-E-beta[b] immune response gene: evidence for gene conversion events of class II genes of the major histocompatibility complex, *EMBO J.,* 3, 1221, 1984.

72. **Karmann, G., Beyrenther, K. T., Cramer, M., Holtkamp, B., Proska, S., and Rajewski, K.,** A structural somatic variant of the K[k] antigen is generated by point mutation, *Immunogenetics,* 22, 35, 1985.

73. **Sun, Y. H., Goodenow, R. S., and Hood, L.,** Molecular basis of the dml mutation in the major histocompatibility complex of the mouse — a D/L hybrid gene, *J. Exp. Med.,* 162, 1588, 1985.

74. **Rubocki, R. J., Hansen, T. H., and Lee, D. R.,** Molecular studies of murine mutant BALB/c-H-2[dm2] define a deletion of several class I genes including entire H-2L[d] gene, *Proc. Natl. Acad. Sci. U.S.A.,* 83, 9606, 1986.

75. **Lopez de Castro, J. A., Strominger, J. L., Strong, D. M., and Orr, M.,** Structure of cross-reactive human histocompatibility antigens HLA-A28 and HLA-A2: possible implications for the generation of HLA polymorphism, *Proc. Natl. Acad. Sci. U.S.A.,* 79, 3813, 1982.

76. **Jaulin, C., Perrin, A., Abastado, J.-P., Dumas, B., Papamatheakis, J., and Kourilsky, P.,** Polymorphism in mouse and human class I H-2 and HLA genes is not the result of random independent point mutations, *Immunogenetics,* 22, 453, 1985.

77. **Tykocinsky, M. L. and Max, E. E.,** CG dinucleotide clusters in MHC genes and in 5′ demethylated genes, *Nucleic Acids Res.,* 12, 4385, 1984.

78. **Cami, B., Chambon, P., and Kouriksky, P.,** Correction of complex heteroduplexes made of mouse H-2 gene sequences in *E. coli* K12, *Proc. Natl. Acad. Sci. U.S.A.,* 81, 503, 1984.

79. **Abastado, J.-P., Cami, B., Dinh, T. H., Igolen, J., and Kourilsky, P.,** Processing of complex heteroduplexes in *E. coli* and *cos*-1 monkey cells, *Proc. Natl. Acad. Sci. U.S.A.,* 81, 5792, 1984.

80. **Szokzak, J. W., Orr-Weaver, T. L., Rothstein, R. J., and Stahl, F. W.,** The double strand-break repair model for conversion and crossing-over, *Cell,* 33, 25, 1983.

81. **Stahl, F. W.,** Role of double-strand breaks in generalized genetic recombination, *Prog. Nucleic Acid Res. Mol. Biol.,* 33, 169, 1986.

82. **Weber, H. and Weissman, C.,** Formation of genes coding for hybrid proteins by recombination between related, cloned genes in *E. coli, Nucleic Acids Res.,* 11, 5661, 1983.

83. **Abastado, J.-P., Darche, S., Godeau, F., Cami, B., and Kourilsky, P.,** Intramolecular recombination between partially homologous sequences in *E. coli* and *Xenopus laevis* oocytes, *Proc. Natl. Acad. Sci. U.S.A.,* 84, 6496, 1987.

84. **Grzesiuk, E. and Carroll, D.,** Recombination of DNAs in *Xenopus* oocytes based on short term homologous overlaps, *Nucleic Acids Res.,* 15, 971, 1987.

85. **Lieb, M., Tsai, M. M., and Deonier, R. C.,** Crosses between insertion and point mutations in lambda gene cI: stimulation of neighboring recombination by heterology, *Genetics,* 108, 277, 1984.

86. **Stoeckert, C. J., Jr., Collins, F. S., and Weissman, S. M.,** Human fetal globin DNA sequences suggest novel conversion event, *Nucleic Acids Res.,* 12, 4469, 1984.

87. **Weiss, E., Golden, L., Zakut, R., Mellor, A., Fahrner, K., Kvist, S., and Flavell, R. A.,** The DNA sequence of the H-2K[b] gene: evidence for gene conversion as a mechanism for the generation of polymorphism in histocompatibility antigens, *EMBO J.,* 2, 453, 1983.

88. **Steinmetz, M., Uematsa, Y., and Fischer Lindahl, K.,** Hotspot of homologous recombination in mammalian genomes, *Trends Genet.,* 3, 7, 1987.

89. **Abastado, J.-P.,** Origine et fonction du polymorphisme des antigènes de transplantation, Doctoral thesis, Paris VI, 1987.

90. **Devlin, J. J., Weiss, E. H., Paulson, M., and Flavell, R. A.,** Duplicated gene pairs and alleles of class I genes in the Qa2 region of the murine major histocompatibility complex: a comparison, *EMBO J.,* 4, 3203, 1985.

91. **Nicolas, A. and Rossignol, J. L.,** Gene conversion: point mutation heterozygoties lower heteroduplex formation, *EMBO J.,* 2, 2265, 1983.

92. **Borts, R. H. and Haber, J. E.,** Meiotic recombination in yeast: alteration by multiple heterozygosities, *Science,* 237, 1459, 1987.

93. **Symington, L. S. and Petes, T. D.,** Expansions and contractions of the genetic map relative to the physical map of yeast chromosome III, *Mol. Cel. Biol.,* 8, 595, 1988.

94. **N'guyen, C., Sodoyer, R., Truay, J., Stracham, T., and Jordan, J. R.,** The HLA-AW24 gene: sequence, surroundings and comparison with HLA-A2 and HLA-A3 genes, *Immunogenetics,* 21, 479, 1985.

95. **Gustafsson, K., Wiman, K., Emmoth, E., Larhammar, D., Böhme, J., Hyldig-Nielsen, J. J., Rone, M., Peterson, P. A., and Rask, L.,** Mutations and selection in the generation of class II histocompatibility antigen polymorphism, *EMBO J.,* 3, 1655, 1984.

96. **Walsh, J. B.,** Interaction of selection and biased gene conversion in a multigene family, *Proc. Natl. Acad. Sci. U.S.A.,* 82, 153, 1985.

97. **Dausset, J. and Pla, M., Eds.,** HLA Complexe Majeur d'Histocompatibilité, Flammarion Médecine-Sciences, Paris, 1985.

98. **Petersen, G. M., Rotter, J., McCraken, J., Raelson, J., New, M., Terasaki, P., Park, M., Sparkes, R., and Ward, J.,** Selective advantage of the 21-hydroxylase deficiency (21-OH-DEF) gene in Alaskan eskimos: use of a linked marker to identify heterozigotes, *Am. J. Hum. Genet.,* 36, 1775, 1984.

99. **Bjorkman, P. J., Saper, M. A., Samraoui, B., Bennett, W. S., Strominger, J. L., and Wiley, D. C.,** Structure of the human class I histocompatibility antigen, HLA-A2, *Nature (London),* 329, 506, 1987.

100. **Bjorkman, P. J., Saper, M. A., Samraoui, B., Bennett, W. S., Strominger, J. L., and Wiley, D. C.,** The foreign antigen binding site and T cell recognition regions of class I histocompatibility antigens, *Nature (London),* 329, 512, 1987.

101. **Maryanski, J., Abastado, J.-P., and Kourilsky, P.,** Specificity of peptide presentation by a set of hybrid mouse class I MHC molecules, *Nature (London),* 330, 660, 1987.

102. **Billington, W. D.,** Influence of immunologic dissimilarity of mother and foetus on size of placenta in mice, *Nature (London),* 202, 317, 1964.

103. **Ober, C., Simpson, J. L., Ward, M., Radvany, R. M., Andersen, R., Elias, S., Sabbagha, R., and The Diep study group,** Prenatal effects of maternal-fetal HLA compatibility, *AJRIM,* 15, 141, 1987.

104. **Yamasaki, K., Boyse, E. A., Mike, V., Thalen, T. H., Mathieson, B. J., Abbot, J., Boyse, J., Zayas, Z., and Thomas, L.,** Control of mating preferences in mice by genes in the major histocompatibility complex, *J. Exp. Med.,* 144, 1324, 1976.

105. **Chaouat, G.,** Reproductive immunology: materno-fetal relationship. Fundamental investigations and future applied prospects, Chaouat, G., Ed., Colloque International INSERM-CNRS, 1986, 154.

106. **Kourilsky, P. and Claverie, J.-M.,** The peptidic self model: a hypothesis on the molecular nature of the immunological self, *Ann. Inst. Pasteur/Immunol.,* 137D, 3, 1986.

107. **Kourilsky, P., Chaouat, G., Rabourdin-Combe, C., and Claverie, J.-M.,** Working principles in the immune system implied by the "peptidic self" model, *Proc. Natl. Acad. Sci. U.S.A.,* 84, 3400, 1987.

108. **Güssow, D. and Ploegh, H.,** Soluble class I antigens: a conundrum with no solution?, *Immunol. Today,* 8, 220, 1987.

109. **Shiroishi, T., Sagai, T., Natsuume-Sakai, S., and Moriwaki, K.,** Lethal deletion of the complement component C4 and steroid 21-hydroxylase genes in the mouse H-2 class III region, caused by meiotic recombination, *Proc. Natl. Acad. Sci. U.S.A.,* 84, 2819, 1987.

110. **Jones, M., Wagner, R., and Radman, M.,** Mismatch repair and recombination in *E. coli, Cell,* 50, 621, 1987.

111. **Mostachfi, P. and Sicard, A. M.,** Polarity of localised conversion in *Streptococcus pneumoniae* transformation, *Mol. Gen. Genet.,* 208, 361, 1987.

112. **Claverys, J. P. and Lacks, S. A.,** Heteroduplex deoxyribonucleic acid base mismatch repair in bacteria, *Microbiol. Rev.,* 50, 133, 1986.

113. **Selker, E. U., Cambareri, E. B., Jensen, B. C., and Haack, K. R.,** Rearrangement of duplicated DNA in specialized cells of neurospora, *Cell,* 51, 741, 1987.

114. **Figueroa, F., Golubic, M., Nizetic, D., and Klein, J.,** Evolution of mouse major histocompatibility complex genes borne by t chromosomes, *Proc. Natl. Acad. Sci. U.S.A.,* 82, 2819, 1985.

115. **Seeman, G. H. A., Rein, R. S., Brown, C. S., and Ploegh, H. L.,** Gene conversion like mechanisms may generate polymorphism in human class I genes, *EMBO J.,* 5, 547, 1986.

116. **Cochet, M., et al.,** in preparation.

117. **Abastado, J.-P., et al.,** unpublished data.

Chapter 11

DIVERSITY AND REGULATION OF MHC CLASS II GENES

B. Mach, C. Berte, J.-M. Tiercy, and W. Reith

TABLE OF CONTENTS

I. INTRODUCTION

The immune response is controlled by the expression of specific major histocompatibility complex (MHC) class II molecules on distinct subsets of cells. Qualitatively, the diversity of the different MHC class II molecules capable of stimulating specific T cells determines the triggering of an immune response to specific antigens. In this sense, MHC class II diversity corresponds functionally to a diversity of immune performance, and MHC class II genes were, in fact, originally described as Immune responses (Ir) genes.[1-2] Quantitatively, the regulation of expression of MHC class II molecules is directed by very strict and specific control mechanisms. The level of expression of MHC class II molecules on the surface of antigen-presenting cells directly determines T-cell recognition and stimulation.[3] It follows that the immune response is also controlled by Ir genes in a quantitative sense. This quantitative control concerns the amount of class II molecules expressed, the cell type where class II expression takes place and the timing of induction of class II molecules on specific cells. The reasons for linking together *diversity* and *regulation* of human leucocyte antigen (HLA) class II genes in this chapter is that both qualitatively and quantitatively, these genes and their products exert a direct control over the immune response. This is not only true in the case of a normal immune response, but also for abnormal responses, as in the case of autoimmune disorders, where both qualitative and quantitative features of HLA class II genes, namely, structural characteristics and level of expression, are thought to play a critical role.[4-7]

In addition to their well-known function in antigen presentation to T cells throughout the body, MHC class II molecules also play an essential role in the acquisition of tolerance during thymic education of T lymphocytes. Expression of class II antigens in the thymus serves a dual function in the acquisition of T-cell restriction and T-cell tolerance, respectively. A positive selection of T cells specific to self MHC molecules is thought to operate in the thymic epithelium.[8-11] In addition, T cells recognizing self antigens presented by class II molecules on cells of bone marrow origin in the thymus medulla undergo either clonal deletion or silencing.[12-13] T cells reacting to autologous MHC alone are also selected against. Consequently, the structural diversity of class II molecules, as well as the regulation of the level of their expression, are both directly responsible for the generation of T-cell restriction and T-cell tolerance. It is important to realize that the qualitative and quantitative features of MHC class II expression in specific thymic cells are not necessarily identical to those observed in the periphery on antigen-presenting cells and that any difference could have important physiological consequences. For instance, the lack of expression of certain MHC class II isotypes in the thymus, or simply a reduced level of expression, would prevent the acquisition of T-cell tolerance for self antigens that are presented efficiently by these particular isotypes and this could ultimately result in a situation of autoimmune reactivity.

This chapter is *not* conceived as a systematic review of the diversity and the regulation of HLA class II genes. It reports on certain selected and important issues and on new developments, primarily focused on work from our own laboratory.

II. MOLECULAR BASIS FOR HLA CLASS II DIVERSITY

A. THE DIFFERENT LEVELS OF DIVERSITY OF HLA CLASS II ANTIGENS

The remarkable diversity of MHC class II molecules results from a number of specific features altogether unique to this genetic system. One can recognize different levels which determine this diversity.

1. Genetic Complexity and Isotypic Diversity

Since the original cloning of DR, DQ, and DP α and β chain genes,[14-22] it became clear that the HLA-D region consists of a number of nonallelic class II genes, grouped into highly related subregions, HLA-DR, -DQ, and -DP (Figure 1). Two striking features emerged from these structural studies: first, there are multiple α and β chain genes, and direct evidence for recent gene duplication, even within a given subregion, has been documented.[23] Second, these multiple genes are not all functional and in each of the three subregions, a number of nonexpressed class II loci were discovered.[21-25] In some cases one deals with obvious pseudogenes (DPA2, DPB2, or DRB2), while in others the lack of expression is a puzzle, such as in the case of DQA2 and DQB2, also referred to as DX genes.[26-27]

In addition to the HLA-DR, -DQ, and -DP genes, whose product and biological function are known, there exist at least two other expressed genes, DOβ[28] and DZα.[29] Both have been mapped in the HLA-D region[30] but no product has ever been identified. Furthermore, the DOβ gene does not show the normal pattern of expression of HLA class II genes.[28]

As indicated in Figure 1, four HLA class II heterodimers are produced, with in particular two distinct DR molecules. The finding of multiple DRβ chain genes[16,31] and the understanding of their organization[23] lead to the observation that not all individuals express the same DRβ loci! Indeed, the second DRβ chain is encoded in DR3, DR5, DRw6 haplotypes by locus DRB3, while it is encoded by DRB4 in DR4 and DR7 haplotypes.[32] In the case of DR2 haplotypes, it is not yet clear which DRB gene encodes the second β chain. It follows that one unique feature of HLA class II gene complexity are polymorphic variations in the choice of the specific DRB locus expressed.

The existence of multiple class II genes, generated during evolution by gene duplication, together with the finding of several nonexpressed class II genes represents an interesting paradox.[33] The pressure for multiple loci, in terms of a greater potential for diversity at the level of the individual, must have been balanced by deleterious consequences of the expression of too many class II products. Somehow, there must be a penalty for expressing too many different class II antigens in the same individual, either as the result of holes created in the T-cell repertoire or from the development of autoimmunity.

2. Molecular Basis For An Extensive Allelic Polymorphism

By far the greatest contribution to MHC class II diversity is allelic polymorphism.[33] Originally recognized as differences in the serological reactivity of class II antigens,[34] HLA class II polymorphism is now characterized at the level of DNA sequences. A striking feature of this genetic system is that the different HLA class II loci exhibit very different degrees of polymorphism, as illustrated in Figure 2. More than 30 alleles have been identified at the sequence level for DRB1 alone, while DRA is invariant[14,16,35-38] It is likely that sequence data from other ethnic groups will extend this picture, but the trend is that illustrated in Figure 2. The number of alleles discovered at the DNA sequence level is considerably greater than was anticipated earlier on the basis of the polymorphism detected by serology.

As described elsewhere,[33] HLA class II polymorphism is not evenly distributed in class II genes, but restricted to the exons encoding the first extracellular domain. Furthermore, within this domain, one observes highly conserved regions, including a number of strictly invariant segments, and three hypervariable regions.[33] These distinguish different alleles and must therefore determine the restricted specificity of T-cell recognition. It is likely that, throughout evolution, class II polymorphism has evolved as the result of gene duplication, recombination, and in a number of cases from gene conversion events, as first demonstrated in the case of DR3 and DRw6 genes.[35] This issue was discussed in more detail elsewhere.[33,34]

3. *Trans*-Complementation

An additional contribution to HLA class II diversity results from the possibility of α/β dimer formation from the products of each of the two haplotypes in heterozygous individuals. Since in HLA-DR, the α chain is not polymorphic, *trans*-complementation can of course only operate for HLA-DQ and -DP, where both the α and the β chains are polymorphic.[39]

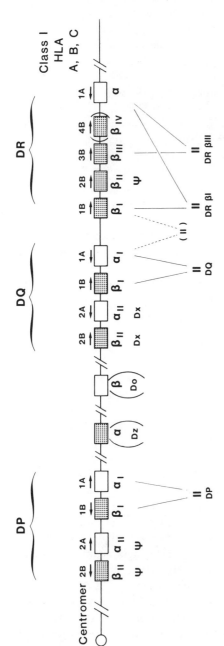

FIGURE 1. Schematic map of the HLA-D region. The relative position of the individual genes is not to scale. The α/β dimers encoded by the HLA-DP, DQ, and DR subregions are indicated. Nomenclature is according to Reference 67.

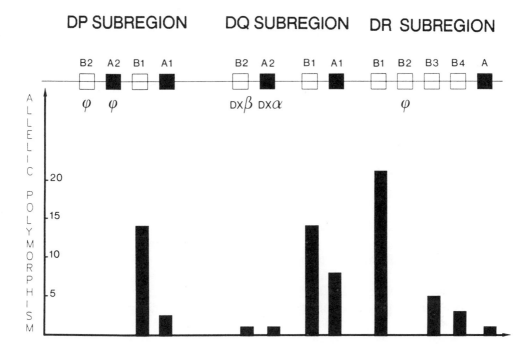

FIGURE 2. Extent of allelic polymorphism at the different HLA class II loci. The number of alleles known to date reflects primarily sequence data obtained in Caucasians. The extent of allelic polymorphism varies in the different HLA class II loci. (From Berdoz, J. et al., *Immunogenetics,* 29, 241, 1989. With permission.)

Experiments in mice have indicated that *trans*-complementation was not the rule and that there was in fact a selective pressure for α/β pairing among products of the same haplotype.[40] This has not yet been analyzed in humans with a large range of different haplotypes.

4. Combinatorial Diversity with Illegitimate *Trans*-Isotypic α/β Pairing

Finally, the possibility exists that α and β chains encoded in two distinct subregions can form an ''illegitimate'' α/β pair and thus produce a class II molecule with a distinct structure and, consequently, a novel specificity.[33,41] It is clear from experiments with transfected cells that such pairing can occur, both in mice[42] and in man.[126,127] It is also evident from a fairly systematic study of cells transfected with various mice α and β genes, that this *trans*-isotypic pairing might only take place within certain combinations of haplotypes and that it occurs with a very much lower frequency than the legitimate isotypic pairing.[43] We have argued elsewhere that the formation of such novel and illegitimate HLA class II molecules could take place as the result of changes in the level of expression of individual class II loci and that, under certain circumstances, the resulting expression of new class II specificities could lead to an autoimmune reaction.[33] Indeed, if not expressed in the thymus, an illegitimate *trans*-isotypic dimer would be considered as foreign. In addition, certain self antigens might be presented much more efficiently by these novel class II molecules and corresponding T lymphocytes would still exist in the T-cell repertoire. Table 1 summarizes the four levels of HLA class II diversity.

B. A HIGHLY SPECIFIC HLA TYPING PROCEDURE BASED ON OLIGONUCLEOTIDE HYBRIDIZATION

Soon after the cloning of the polymorphic HLA-DRβ chain gene,[15,16] it was possible to demonstrate that HLA class II polymorphism can be recognized and analyzed directly at the DNA level.[44] We initially showed that the pattern of DNA fragments containing class II

TABLE 1
The Different Levels of HLA Class II Diversity

1. Genetic complexity: multiple class II loci (isotypes)	Four distinct products expressed: HLA DRB1,DRB3/ 4,DQ and DP.
2. Allelic polymorphism	Multiple alleles at most loci; about 60 different alleles now recognized. Diversity at the level of the individual (heterozygotes); diversity at level of the population.
3. *Trans*-complementation	α/β dimer formation with products of both chromosomes (for DQ and DP).
4. Formation of *trans*-isotypic illegitimate dimers	Pairing of α/β chains encoded by two different subregions: DQ/DP, DR/DQ or DR/DP. Can also occur in *trans*.

genes, obtained following digestion with restriction enzymes (RFLP), could represent a drastically novel approach to HLA typing, referred to as "DNA typing". This led to a large number of studies performed with that procedure and informative results have been generated.[45-49] As more DNA sequence data became available, however, we realized that most restriction sites responsible for RFLP typing are in the noncoding region, and are, therefore, phenotypically irrelevant. More importantly, specific amino acid differences which distinguish different alleles could not be analyzed directly. For these reasons, we proposed a procedure based on the hybridization of HLA class II allele and locus-specific oligonucleotides[50] and showed that this could represent a large scale routine method, capable of identifying HLA class II alleles differing by a single amino acid residue.[50] As expected, oligonucleotide typing (or "oligo typing") was then shown to allow the identification of "hidden" alleles, not detectable by the classical serological or cellular methodology.[51] Oligonucleotide typing can also be performed on RNA samples, with the advantage of an analysis by a simple dot blot hybridization.[52] Recently, the large scale use of "oligo typing" from DNA samples has been considerably simplified by the development of a DNA amplification step.[53-55] This amplification step, referred to as polymerase chain reaction or PCR,[53] greatly shortens the time required compared to the original "oligo typing" procedure[50] and has allowed a generalization and simplification of DNA typing.

An important advantage of oligonucleotide typing, in addition to its simplicity, is that it allows a much more accurate identification of any allele not easily typable by serology. For instance the three alleles of locus HLA-DRB3, 52a, 52b, and 52c, can now be easily and rapidly identified by dot blot hybridization of amplified DNA with allele-specific oligonucleotide probes, as illustrated in Figure 3. This procedure is especially useful for all the newly discovered HLA class II alleles and subtypes which cannot be identified by routine methods, not only for DR but also for DQ and DP, for which oligo typing is becoming the technique of choice. Oligonucleotide typing can also replace cellular typing, as shown recently in the case of different Dw subtypes of the group of DRw6 haplotypes,[56] which cannot be resolved by routine methods. In the case of matching of unrelated donor-recipient pairs for transplantation, in particular, we have been able to identify allelic differences in so called HLA "matched" pairs and to select the best donors on the basis of DNA sequence identity. In certain cases, this has allowed us to predict a positive or a negative MLR.

Table 2 lists the different alleles of HLA-DR and -DQ currently recognized. For the alleles of DQA and DQB genes, the prevalent association with DR alleles is indicated in parentheses. This association results from the known extensive linkage disequilibrium between the DR and the DQ subregion. With an extensive panel of specific oligonucleotide probes based on DNA sequence data, we can now identify most of the alleles listed in Table 2 by oligonucleotide typing.

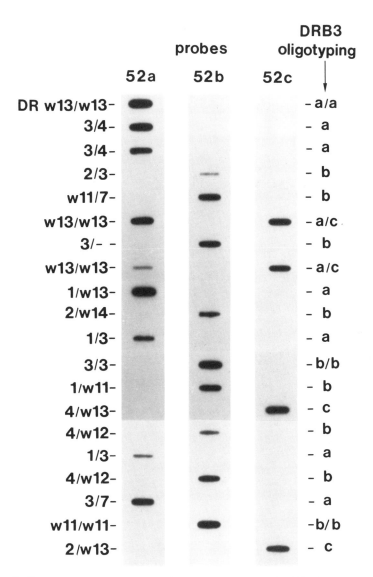

FIGURE 3. Example of "DNA typing" by oligonucleotide hybridization to dot blots of amplified DNA. DNA from a number of DRw52 individuals was amplified (PCR),[53] spotted on a membrane and subjected to "oligo typing" according to the published procedure.[50] The oligonucleotide probes used were specific for each of the three alleles of locus DRB3, 52a, b, and c,[51] now referred to as DRB3*0101, DRB3*0201 and DRB3*0301.[67]

C. IDENTIFICATION OF SEROLOGICAL AND T-CELL RECOGNITION EPITOPES ON HLA-DR MOLECULES

The function of MHC class II molecules in antigen presentation to T cells depends on their fine structure. The specificity of class II molecules, either isotypic or allelic, is generally recognized by the reactivity of specific T lymphocytes or by antibodies. Information about the amino acid sequence of a number of class II molecules has allowed structure-function correlations to be performed, with the objective of identifying the fine structure of class II molecules involved in specific interaction.

A key step in this direction was the elucidation of the three-dimensional structure of HLA class I[57] and the finding that contact with the T-cell receptor concerns two α helices

TABLE 2
DR and DQ Alleles Now Recognized at the DNA Sequence Level

Loci

DQB1	DQA1	DRB1	DRB3	DRB4	DRA
1.1 DQw5	1.1	1.1	0101 52a	53	DRA
(DR1)	(DR1)	1.2 BON			
		2.1 Dw2	0201 52b1		
1.2 DQw6	1.2	2.2 Dw12	0202 5222		
(DR2/Dw2)	(DR2,DR6)	2.3 Dw21			
		3.1 DRw17	0301 52c		
1.3 DQw6	1.3	3.1 DRw18			
(DR2/Dw12)	(DR5,DRw6,DR8)	0401 Dw4			
		0402 Dw10			
1.4 DQw5		0403 Dw13			
(DR2/Dw21)	2	0404 Dw14			
	(DR3,DR5)	0505 Dw15			
1.5 DQw5		7	Locus unknown:		
(DRw6/Dw9)		8			
	3.1	9	2.1 Dw2		
1.6 DQw6	(DR7)	10			
(DRw6/Dw18)		11.1	2.2 Dw12		
	3.2	11.2			
1.7 DQw6	(DR4,DR9)	12.1	2.3 Dw21		
(DRw6/Dw19)		12.2			
	3.3	13.1 Dw18			
2 DQw2	(DR8,DQ3.1)	13.2 Dw19			
(DR3,DR7)		14.1 Dw9			
	3.4	14.2 Dw16			
3.1 TA10+	(DR8,DQWA)				
DQw7					
(DR3,DR5,DR8)					
3.2 TA10−					
DQw8					
(DR4)					
3.3 DQw9					
(DR9)					
3.4 DQWA DQw4					
(DR4,DR8)					

Note: This table must now be updated according to Reference 126.

separated by a groove which most likely contains the peptide antigen.[58] There is a great similarity in the structure of HLA class I and class II chains with respect to the position of conserved and variable residues. In particular the α2 domain of class I is similar in sequence to the β1 domain of class II and the α1 domain of class I matches the α1 domain of class II. On that basis, a theoretical three-dimensional model of HLA class II molecules can be derived, based on the structure of class I molecules.[59] Figure 4A represents the case of HLA-DR, with a comparison of the β1 domain of DRβ chains with the α2 domain of HLA class I chains. When represented on the theoretical three-dimensional model of the HLA-DR molecule, the positions of the hypervariable allelic residues (black and shaded circles) correspond to both the bottom of the groove and the α helical edges (Figure 4B). These hypervariable amino acids are responsible for the allelically restricted recognition of antigens by T cells, for allo-recognition by class II-specific alloreactive T cells and for the specific serological recognition of class II alleles by antibodies.

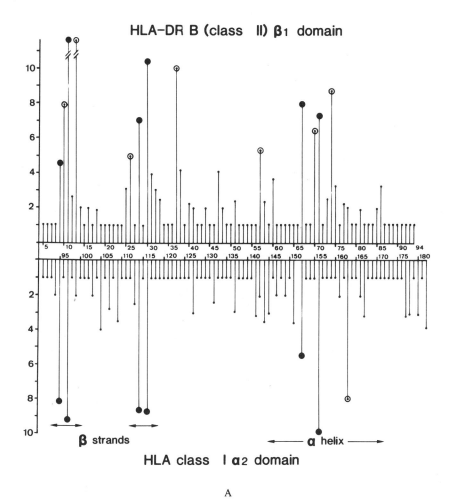

HLA-DR B (class II) β₁ domain

HLA class I α2 domain

A

FIGURE 4. Position of allelic amino acid variations on a putative three dimensional model of HLA-DR. A. Correlation between the constant and variable positions of the α2 domain of HLA class I α chains and of the β1 domain of HLA-DRβ chains. B. Three dimensional representation of HLA-DR based on the analogy with the structure of HLA class I (References 57, 59 and A above). The DR molecule is viewed from above the antigen presenting "platform", as seen by the T-cell receptor. Dark circles represent residues variable in class I and class II and shaded circles residues variable only in class II. Numbers refer to the amino acid position of DRβ chains (top) and of HLA α chains (bottom) according to the alignment in A above.

Three structures are important in these recognition events and in the interaction between an antigen-presenting cell and a specific T lymphocyte. The portion of class II molecules that binds foreign peptide antigens, primarily in the groove, is referred to as desotope.[9] The residues of the peptide involved in this binding represent the agretope.[9] Finally, the specific structures of the MHC class II-peptide complex that are recognized by the T-cell receptor are the epitopes. Specific epitopes can involve residues of the foreign MHC-bound peptide and of the exposed portions of the class II molecule.

In order to identify specific serological epitopes on class II molecules, we have made use of mouse cells transfected with HLA-DR α and β chain genes and which express a single class II molecule.[60,61,62] A series of DR-specific monoclonal antibodies were tested for their reactivity and correlations could be made between the specificity of individual monoclonal antibodies and the amino acid sequence of the variable β chain of the DR molecules. On that basis, the reactivity of a number of antibodies was assigned to individual

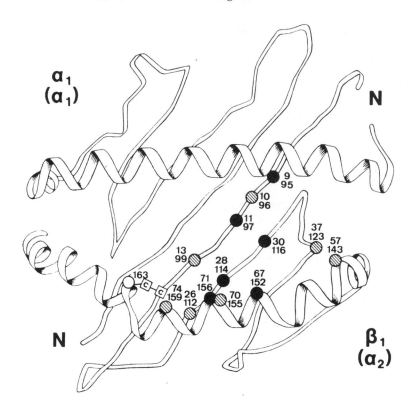

FIGURE 4B.

amino acid positions.[63] As expected, some of these epitopes are located on the highly exposed α helical portion. More interestingly, we observed two cases where the epitope was mapped in the β sheet portion in the grove of the class II molecule.[63] This indicates either that these amino acid variations induce conformational changes on the exposed portion of the α helix or that the T-cell receptor can "see" the bottom of the groove or a bound peptide.

In another set of experiments, we have analyzed the reactivity of T-cell clones toward specific alleles of HLA-DR of known sequence. Certain T cells were found specific for alleles of DRB3 (52a, 52b, and 52c), for either allo-recognition or antigen-specific restricted recognition.[64,65] On the basis of the differences in amino acid sequence, it was possible to identify and to locate the fine structures that determine T-cell recognition to small amino acid clusters on the three dimensional structure of HLA-DR.[66] These clusters involve amino acid residues located in the groove (potential desotopes) as well as on the α helix. We concluded from this analysis that different T cells recognize distinct clusters of amino acids located on different regions of the class II molecule.

III. REGULATION AND DISREGULATION OF HLA CLASS II GENES

A. BIOLOGY OF MHC CLASS II REGULATION

The regulation of MHC class II genes is of special interest for three main reasons. The expression of MHC class II molecules on the surface of a cell confers the capacity to stimulate specific T lymphocytes[68] and, consequently, the functional significance of the control of these genes is enormous. The regulated expression of MHC class II antigens determines and controls the immune response, is required for the induction of tolerance, and is very likely also involved in the pathogenesis of autoimmunity.[6,69] Second, class II genes represent an example of very tightly regulated genes, with a strict developmental control, a narrow tissue-specific pattern of expression, and a sensitive inducibility in response to various stimuli. Third, they represent a family of closely linked and structurally related genes generally regulated as a group in a coordinate fashion,[33] with rare examples of dissociation in their respective expression.[70] MHC class II genes represent, therefore, an interesting model system for a better understanding of gene regulation and, at the same time, the control of the expression of Ir genes is one of the essential phenomena in immunology.

HLA class II genes are expressed constitutively on a very limited number of cell types: B lymphocytes, macrophages, dendritic cells, epidermal Langerhans cells and microglial cells.[71-76] Activated T lymphocytes also express class II antigens in man, but not in mice.[77] Certain epithelial cells, in particular, in the gut, are class II positive. The expression of class II genes can be induced in a large number of class II-negative cells, in particular, by interferon gamma, which also enhances class II expression in macrophages.[78] The mechanisms controlling class II gene must, therefore, account for both constitutive and inducible expression.

In most situations, one observes a coordinate regulation of class II genes and the gene encoding the invariant (In) chain,[79-82] a structurally unrelated and nonpolymorphic polypeptide which associates with MHC class II dimers intracellularly. The complex of class II and the In chain forms in the endoplasmic reticulum and dissociates during transport of class II molecules, probably in the *trans*-golgi. We have shown that the In chain is not required for the correct presentation of HLA class II molecules at the cell surface.[60,83] Its role could well be in the control of interactions between processed peptide antigens and class II dimers. Coexpression of class II genes and the In gene in most biological circumstances is a striking feature of class II gene regulation.

We have studied different experimental models and, in particular, constitutive expression in B-cell lines and interferon γ-induced expression in fibroblasts. An essential aspect of this work was the study of HLA class II regulatory mutants, found to be affected in *trans*-acting factors which control HLA class II gene expression. Our studies concerned different levels of gene expression, namely analysis of steady-state mRNA levels, measure of gene transcription, identification of DNase I hypersensitive sites in class II promoters and, more importantly, the identification of protein factors that bind to specific target sequences in the controlling region of HLA class II promoters.

B. MHC CLASS II PROMOTERS AND DNA BINDING PROTEINS
1. HLA Class II Promoters and Conserved Regions

The regulation of expression of genes transcribed by RNA polymerase II is a complex phenomenon which involves multiple protein interactions with the promoter region.[84] In the case of HLA class II genes, a region of about 150 base pairs directly upstream from the start of transcription has been shown by transfection experiments to be necessary and sufficient for both B-cell-specific and interferon-γ-induced expression.[85-87] These transfection data do not exclude the role of sequences located further upstream for tissue-specific expression, but they point to the essential role of the first 150 nucleotides.

It was known for some time that certain segments of MHC class II promoters were highly conserved in sequence, both in man and mouse.[88-90] In particular, two conserved boxes, called the X and the Y boxes, are located at positions −95 to −108 and −63 to −74, respectively, separated by exactly 20 bases (Figure 5). These sequences appear to function as *cis*-acting elements in the control of class II genes.[85,86,91]

2. Identification of Four Proteins Binding to HLA Class II Promoters

Novel procedures have been developed for the analysis of protein factors that specifically bind to control DNA regions and for the identification of the specific binding sites. These

		X			Y		
		−125		−112	−92	−83	
Consensus α & β :		CCYAGNRACNGATG			CTGATTGGYY		bp to

		X box	spacer	Y box		bp to
DR 3	β I	A-C--CA--T----	ATGCTATTGAACTCAGATG	------C-TT	CTCC	141
DR 3	β II	AGC--CA--T----	ATGATACTGAACTCAGATG	--------TT	CTCC	142
DR 3	β III	AGC--CA--T----	ATGCTGTTGAACTCAGATG	--------TT	CTCC	142
DR6B	β III	AGC--CA--T----	ATGCTGTTGAACTCAGATG	--------TT	CTCC	142
DR 4	β (Ψ)	A-C--CA--T----	ATGCTATTGAACTCAGACG	----CATT	CTCC	142
I-E	β	A-T--CA--T----	ATGCTGGACTCCTTTGATG	-------CT	CCCA	135
I-E	β 2	A-C--CA--T----	CTGCTGGTGCTCTTGGCTG	--------TT	CTTC	106
DQw3	β	--T--AG--A---T	AGGTCCTTCAGCTCCAGTG	--------TT	CCTT	127
DQw2	β	--C--AG--A----	AGGTCCTTCAGCTCCAGTG	--------TT	CCTT	124
I-A	β	--C--AG--A----	ACAGACTTCAGGTCCAATG	--------TT	CCTC	119
I-A	β 2	--T--CA--AA---	ATGCAAA GT CTTCGCTC	T-------TT	AACA	113
DX	β	--C--AGG-A----	AGGTCCTTCAGCTCCAGTG	--------TT	CCTT	
DP	β	--T--TG-GCA---	ACTCATACAAAGCTCAGTG	TCC-----TT	CTTT	140
SX	β	--C--TG-GCA---	ACTCATACAAAGCTCAGTG	-CC-----TT	CTTT	140

Consensus β only :   A------Y-AR^G_CYY-RRTG   CYYY

		X box	spacer	Y box		bp to
DR	α	--T--CA--A----	CGTCATCTCAAAATATTTTT	-------CC	AAAG	124
I-E	α	--T--CA--A----	TGTCAGTCTGAAACATTTTT	-------TT	AAAA	93
DP	α	--C--CA--A--GA	ATGTCAGCTCTATGATTTCT	-----A--TG	AATC	133
DZ	α	--C--CA--A---A	CATTCACTCAGAGAATTTCT	G-------CT	GAAG	111
DQ	α	G-T--TA--T--GA	TGTCACCATGGGGGATTTTT	--A-----CC	AAAA	110
DX	α	A-TG-CA-ACA--A	TGTCACCATAGGGGATTTTT	-------CC	AAAA	

Consensus α only :   YRTY--Y-YRRRR-ATTTYT   RAAR

INV.		--C--AA--AAG--	44 bp, no homology with α or β	---CC---GG	AGCC	173

FIGURE 5.   Two highly conserved regions in the promoters of human and mouse MHC class II genes. Two conserved short stretches are referred to as the X box and the Y box.

rely on the formation of DNA-protein complexes with nuclear extracts[92] and an analysis by gel retardation assays[93,94] and by DNase I footprinting[95] or methylation interference assays.[96] Proteins that bind to MHC class II promoters have been identified with these procedures.[86,97-99] In a systematic study of protein binding to the promoters of HLA-DR, -DQ and -DP, we have now identified four specific factors and their respective binding sites. The octamer binding protein (OBP) and the Y box binding protein (NF-Y) are ubiquitous proteins, described earlier,[96,98] and which are known to bind to the promoters of a number of other genes. A specific protein binding to the X box of HLA DR genes was recently described.[99] With appropriate DNA fragments, it could be shown that both NF-Y and RF-X can bind simultaneously.[99] As discussed below, binding of RF-X seems to be essential for class II gene expression, hence the name "regulatory factor-X".[99] A fourth protein, which binds to the 5′ portion of the spacer region between the X and the Y box, was recently identified (NF-S).[100] For each of these factors the exact binding site on the promoter sequence has been determined and a schematic representation of the four factors found in class II positive B cells on HLA class II promoters is given in Figure 6.

3. Different Relative Binding Affinities for the Promoters of HLA-DR, -DQ, and -DP Genes

RF-X was observed with DNA fragments containing the X box of the HLA-DRA promoter.[99] Its binding requires an unusually long oligonucleotide (54 mer). Interestingly, the binding of RF-X to the X box of HLA-DP is reduced and binding to HLA-DQ is barely detectable.[100] In contrast, NF-S binds strongly to the DQ promoter, less to DP, and almost not all to the DR promoters. This unexpected difference in the relative affinity of RF-X and

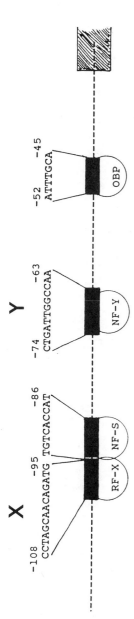

FIGURE 6. Four DNA -binding proteins identified on HLA class II promoters. OBP and NF-Y are known to be ubiquitous proteins also binding to promoters of other unrelated genes. [96,98] RF-X binds specifically to the X box[99] and binds to the X box of DR (shown) much better than to the X box of DQ.[100] NF-S binds to the spacer region in between X and Y and binds much better to the DQ S region (shown) than to DR.[100] Binding of RF-X is specifically lacking in HLA class II deficient regulatory mutants. [99]

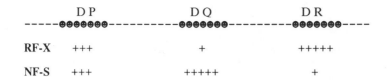

	D P	D Q	D R
RF-X	+++	+	+++++
NF-S	+++	+++++	+

FIGURE 7.   DNA binding factors RF-X and NF-S exhibit different relative affinities for the promoters of HLA-DR, -DQ, and -DP. See text.

NF-S for the promoters of the three subregions of HLA class II is illustrated in Figure 7. Since RF-X binds only very poorly to the DQ promoter, and, as also expected, very poorly to the mouse IE promoter,[100] it explains why this factor was not identified elsewhere with the use of DQ or IE DNA fragments. It will be of great interest to establish if the different relative affinities of the two factors for DR, DP, and DQ genes play a role in the rare situations of dissociated expression of DR without DQ.[70]

In certain situations, the transcriptional activity of a gene is associated with specific features of chromatin structure manifested by hypersensitivity to DNase at specific sites.[101] Two such sites can be found in the same region of HLA class II promoters and, as shown below, their presence correlates with the binding of RF-X.[102]

C. INDUCTION OF HLA CLASS II GENES BY INTERFERON γ INVOLVES A *TRANS*-ACTING ACTIVATION FACTOR

Interferon γ induces the expression of MHC class II genes in most class II negative cells *in vitro* and *in vivo* and enhances their expression in macrophages.[78,82,105,106] This induction process is characterized by an unusually long lag phase and by the activation of class II genes at the level of transcription.[105,106] From a study of the kinetics of transcription of class II genes in a human class II negative fibroblastic line, following exposure to interferon γ, we have concluded that transcription is only initiated after about 6 h.[105] This lag suggested that the effect of interferon is not direct but is mediated by earlier events. This was shown to be the case by experiments that established the requirement for *de novo* synthesis of a protein(s) in the mechanism of class II activation.[105,106] Both the transcriptional level of activation and the requirement for protein synthesis in the activation of class II genes by interferon γ are now firmly established, although some earlier data had suggested different mechanisms.[107,108]

From these experiments we conclude that the mechanism of interferon γ induction of class II genes is a two-step process, which first involves a *trans*-acting activator protein, whose synthesis is directly induced by interferon.[106,109] We have also established that this synthesis occurs very early after exposure of cells to interferon γ and that once the activator is made, protein synthesis is no longer required for class II gene activation.[109]

Transcriptional activation of class II genes by interferon γ was studied at the level of promoters. Interestingly, the process does not involve any novel protein binding to the promoter region, even though this segment (150 base pairs) is sufficient for interferon γ responsiveness.[110] The four DNA-binding proteins observed in extracts of class II positive B cells were detected in extracts of uninduced, class II negative fibroblasts, as well as in induced cells.[110] The two DNase I hypersensitive sites are also observed independently of the transcriptional activation. We have concluded that the activator protein induced by interferon γ can activate a regulatory protein (such as RF-X) which is already bound to the promoter in an inactive state prior to the induction of transcription.[102]

D. HLA CLASS II REGULATORY MUTANTS LACK A SPECIFIC CLASS II PROMOTER BINDING PROTEIN, RF-X

The use of mutants is frequently the best way to discover normal biological structures

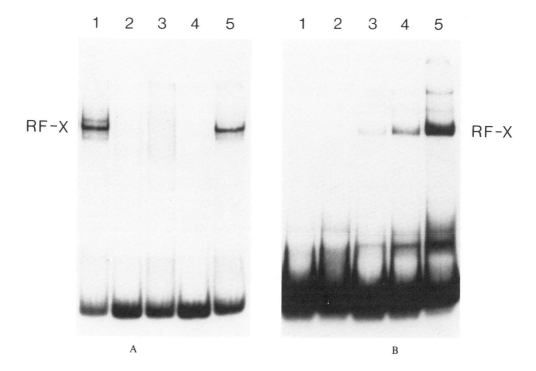

FIGURE 8.   DNA-protein binding assays ("band shift") with natural (A) and recombinant (B) RF-X factor and a DR-X box oligonucleotide. Band shift assays are performed as described[99] with a 54 mer oligonucleotide centered on the DR X box. (A) binding with nuclear extracts from normal B cells (1), SCID regulatory mutants (2—4) and an *in vitro* mutant (5). (B) Following cloning of RF-X cDNA,[123] the recombinant protein was expressed in *E. coli* and used for binding assays at two concentrations (4—5). Controls included the same oligonucleotide with the X box mutated and control bacterial extracts with the normal X oligonucleotide.

and functions. Certain HLA class II-deficient regulatory mutants have indeed been very informative for our understanding of class II gene regulation.[99] An inherited disease characterized by severe immunodeficiency and due to the lack of expression of HLA class II antigens (see review in Reference 111) was studied in detail.[112,113] The genetic defect in HLA class II deficient SCIDS is not in class II genes but in a *trans*-acting regulatory factor essential for HLA class II gene expression.[113] All cells of the body including non-lymphoid cells are defective and both constitutive and interferon-γ-induced expression of class II genes is affected.[114] An interesting feature of class II-deficient severe combined immunodeficiencies (SCIDs) is that the gene for the In chain, normally coregulated with class II genes, is expressed, although at reduced levels.[112-114]

Established B-cell lines from these patients with a class II regulatory defect have no HLA-DR, -DQ and -DP mRNA and are blocked at the level of transcription.[99] At the level of class II promoters, SCID cells were found to be totally deficient in the binding of RF-X to the class II promoter[99] and to lack the two DNase I hypersensitive sites normally seen on the promoter.[102] Figure 8 A shows a DNA-binding experiment with an X box oligonucleotide. Lane 1 is with a normal B-cell extract and lanes 2 to 4 are with extracts from B cells obtained from three different SCID patients. The lack of binding of RF-X in these regulatory mutants strongly suggests that this represents the molecular defect responsible for their HLA class II-negative phenotype and that RF-X is an essential factor in the regulation of HLA class II gene expression.[99] It is not yet established if the SCID defect is due to the absence of the RF-X protein or to a structural defect in the sequence of the DNA binding domain of RF-X. From preliminary experiments that followed the cloning of the RF-X gene, we tend to favor the latter mechanism.

Although RF-X normally binds very poorly to the HLA-DQ promoter,[100] the defect is sufficient to abolish expression of all class II genes. The much stronger binding of RF-X to DR than to DQ promoters also explains why others, using DQ X oligonucleotides, had failed to observe the lack of RF-X in similar mutants.[97] With the finding of a defect in RF-X binding in patients with HLA class II deficiency, this represents the first disease due to a DNA-binding regulatory factor.[99]

Other HLA class II regulatory mutants have been generated *in vitro* from B cell lines.[115,116] They are also due to a defect in a *trans*-acting regulatory gene[117,118] and have the same characteristic phenotype with a lack of class II mRNA but expression of In mRNA.[119] With one such *in vitro* mutant, RJ5.2.5, we have observed that, contrary to the SCID mutants, RF-X binds normally (Figure 8A, lane 5) and the two DNase I sites are also normal.[120] All four proteins binding to HLA class II promoters are normally present. This implies that two distinct molecular defects must be involved in these two kinds of regulatory mutants and that the binding of RF-X to the X box of class II promoters is necessary but not sufficient for gene expression. It also implies that at least another factor, yet unidentified, is necessary for the control of HLA class II gene expression. It could be a cofactor, required for RF-X activity, but not detectable by direct binding to the promoter.

E. MOLECULAR CLONING OF THE HLA CLASS II REGULATORY FACTOR RF-X

Because of the biological relevance of RF-X in the control of class II gene regulation, as indicated by the study of the SCID mutants, the cloning of the RF-X gene had become an important priority. Nuclear regulatory factors are only present in low amounts and the purification of sufficient quantities of the protein for microsequencing appeared an unrealistic task. We made use of the DNA binding specificity of RF-X for the X box of the HLA-DR promoter to screen a cDNA expression library in lambda gt11, arguing that a fusion protein with the DNA binding domain of RF-X would be specifically recognized by the X box oligonucleotide. Conditions for such a screening have been described.[121,122] Out of about 900,000 clones, one clone was consistantly positive following several rounds of screening.[123] As a control, we employed an identical X oligonucleotide where the X box alone had been mutated. The cDNA clone has a 4.2-kb insert which encodes a recombinant protein that displays the same binding specificity as natural RF-X (Figure 8B). It binds to the same position within the X box and exhibits the same gradient of affinity for the X box of the promoters of HLA-DR, DP, and DQ.[123] This gene could not have been cloned by screening with a DQ X box DNA fragment. Recombinant RF-X is smaller than the natural protein, indicating that this cDNA clone is not full length. An interesting property of the RF-X gene is that it is highly conserved by Southern blot hybridization across a range of DNA from other species.[123] It is interesting to speculate on the role of an MHC class II regulatory factor in these species. Finally, one can expect that, if the defect in SCIDs is in the RF-X gene itself, cloning of the gene might open the way to the perspective of gene therapy for this generally fatal disease.

F. THE COMPLEXITY OF MHC CLASS II REGULATION

The binding of RF-X is necessary for HLA class II gene expression, as shown by the regulatory mutants. It is, however, not sufficient. RF-X was found normally bound in the case of HLA class II negative fibroblasts, prior to stimulation with interferon γ,[110] and in the case of the class II negative *in vitro* mutant RJ5.2.5, where no class II transcription can be detected.[120] Interestingly, one observes a perfect correlation between the binding of RF-X and the presence of the two DNase I hypersensitive sites[102,120] (Figure 9). This suggests that binding of RF-X affects the chromatin structure in a way that generates these DNase sites.

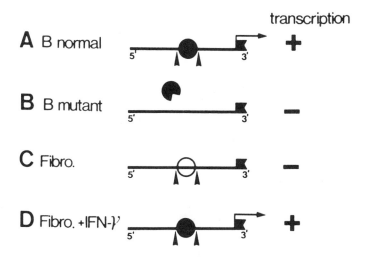

FIGURE 9. Schematic representation of the role of RF-X as a necessary but not sufficient factor for MHC class II gene transcription. See text. (From Gonzcy, P., Reith, W., Barras, E., Hadam, M. R., Lisowska-Grospierre, B., Griscelli, C., and Mach, B., *Mol. Cell. Biol.,* 9, 296, 1989. With permission.)

Figure 9 illustrates the state of HLA class II promoters in four distinct situations. Lack of RF-X binding abolishes transcription (second lane) but the presence of RF-X is not always sufficient. We propose that RF-X can be bound in an inactive form in a fairly ubiquitous manner in many cell types and that it can be activated on the promoter by specific factors. One such factor could be the activator directly induced by interferon γ. Alternatively, the activation of RF-X might involve the synthesis of a novel form of the regulatory protein through alternative splicing. Both forms would have the same DNA-binding domain and therefore bind to the promoter; but another domain of the regulatory protein, responsible for activity, would only be expressed upon splicing of the corresponding exon.

Class II-deficient regulatory mutants illustrate the effect of the lack of RF-X binding, with a resulting lack of class II expression and severe immunodeficiency. It is possible that other mutations might affect RF-X in a different manner, leading to a disregulation of MHC class II gene expression, with, for instance, an enhanced expression of class II genes[6,69] or an increased responsiveness to interferon γ.[7] Modifications in either the structure or the regulation of RF-X are expected to affect the control MHC class II gene expression. Since HLA class II expression in the thymus is of great importance in the generation of restriction and tolerance, it is important to establish if RF-X is also involved or if one deals with thymus-specific factors. The use of anti-RF-X antibodies, produced from recombinant RF-X will facilitate studies on the cellular specificity of the factor.

RF-X is the only identified factor clearly involved in class II gene regulation. Yet it is obvious that several other proteins also play a role. Figure 10 illustrates the known and less unknown aspects of class II gene regulation. On the left, the activator induced early by interferon γ is probably not binding to DNA itself but affects the activity of promoter factors. In addition to RF-X, other nonidenfitied *trans*-acting factors must include the factor defective in *in vitro* mutants[118,120] and possibly a factor responsible for the dominant "extinction" of HLA class II expression.[124] Certain class II negative cells, such as the end product of B-cell differentiation, plasmocytes, have been shown by fusion experiments to act as dominant suppressors of the expression of class II genes.[124,125] One can anticipate that the genes for those regulatory factors which do not bind to DNA will only be cloned by direct expression in an appropriate selection system.

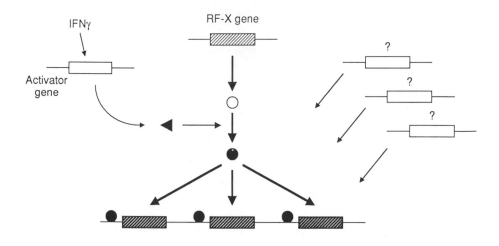

FIGURE 10.   Current view of the complexity of MHC class II gene regulation. Rectangles represents genes. Dark triangle represents the class II transactivator, whose synthesis is induced de novo by INF γ. Circle represents class II regulatory factor RF-X, which binds to class II promoters. Other unidentified factors included the product of a gene affected in an *in vitro* generated class II deficient mutant and factors involved in the dissociation of DR, DQ, and DP gene expression.

ACKNOWLEDGMENTS

We are grateful to all members of the laboratory for discussion and for unpublished information, in particular to I. Amaldi, C. Herrero-Sanchez, and M. Kobr. We also acknowledge the collaboration of Drs. B. Grospierre, C. Griscelli, and M. R. Hadam on the work with SCID cell lines. Work described in this chapter was supported by the Swiss National Science Foundation.

REFERENCES

1. **Benaceraff, B.,** Role of MHC gene products in immune regulation, *Science,* 212, 1229, 1981.
2. **Benaceraff, B. and McDevitt, H. O.,** Histocompatibility-linked immune response genes, *Science,* 175, 273, 1972.
3. **Matis, L. A., Glimscher, L. H., Paul, W. E., and Schwartz R. H.,** Magnitude of response of histocompatibility-restricted T cell clones is a function of the concentrations of antigen and Ia molecule, *Proc. Natl. Acad. Sci. U.S.A.,* 80, 6019, 1983.
4. **Todd, J. A., Acha-Orbea, H., Bell, J. I., Chao, N., Fronek, Z., Jacob, C. O., McDermott, M., Sinha, A. A., Timmermann, L., Steinman, L., and McDevitt, H. O.,** A molecular basis for MHC class II-associated autoimmunity, *Science,* 240, 1003, 1988.
5. **Scharf, S. J., Friedmann, A., Brautbar, C., Szafer, F., Steinman, L., Horn, G., Gyllensten, U., and Erlich, H. A.,** HLA class II allelic variation and susceptibility to pemphigus vulgaris, *Proc. Natl. Acad. Sci. U.S.A.,* 85, 3504, 1988.
6. **Bottazzo, G. F., Pujol-Borrell, R., and Hanafusa, T.,** Role of aberrant HLA-DR expression and antigen presentation in induction of endocrine autoimmunity, *Lancet,* 2, 1115, 1983.
7. **Massa, P. T., Ter Meulen, V., and Fontana, A.,** Hyperinductibility of Ia antigen on astrocytes correlates with strain-specific susceptibility to experimental autoimmune encephalomyelitis, *Proc. Natl. Acad. Sci. U.S.A.,* 84, 4219, 1987.
8. **Schwartz, R.,** T-lymphocyte recognition of antigen in association with gene products of the major histocompatibility complex, *Annu. Rev. Immunol.,* 3, 237, 1985.

9. **Schwartz, R.,** T-lymphocyte recognition of antigen in association with gene products of the major histocompatibility complex, *Annu. Rev. Immunol.,* 3, 237, 1985.

10. **Reinherz, E. L.,** A molecular basis for thymic selection: regulation of T11 induced thymocyte expansion by the T3-Ti antigen/MHC receptor pathway, *Immunol. Today,* 6, 75, 1985.

11. **Lo, D. and Sprent, J.,** Identity of cells that imprint H-2 restricted T-cell specificity in the thymus. *Nature (London),* 319, 672, 1986.

12. **Reinherz, E. L. and Schlossman, S. F.,** The differentiation and function of human T lymphocytes, *Cell,* 19, 821, 1980.

13. **Marrack, P. and Kappler, J.,** The T cell receptor, *Science,* 238, 1073, 1987.

14. **Lee, J., Trowsdale, J., and Bodmer, W.,** cDNA clones coding for the heavy chain of human HLA-DR antigen, *Proc. Natl. Acad. Sci. U.S.A.,* 79, 545, 1982.

15. **Long, E. O., Wake, C. T., Strubin, M., Gross, N., Accolla, R. S., Carrel, S., and Mach, B.,** Isolation of distinct cDNA clones encoding HLA-DR β chains by use of an expression assay, *Proc. Natl. Acad. Sci. U.S.A.,* 79, 7465, 1982.

16. **Long, E. O., Wake, C. T., Gorski, J., and Mach, B.,** Complete sequence of an HLA-DR β chain deduced from a cDNA clone and identification of multiple non-allelic DR β chain genes, *EMBO J.,* 2, 389, 1983.

17. **Larhammar, D., Schenning, L., Gustafsson, K., Wiman, K., Claesson, L., Rask, L., and Peterson, P. A.,** Complete amino acid sequence of an HLA-DR antigen-like β chain as predicted from the nucleotide sequence: similarities with immunoglobulins and HLA-A, -B, and -C antigens, *Proc. Natl. Acad. Sci. U.S.A.,* 79, 3687-3691, 1982.

18. **Auffray, C., Korman, A. J., Roux-Dosseto, M., Bono, R., and Strominger, J. L.,** cDNA clone for the heavy chain of the human B cell alloantigen DC1: strong sequence homology to the HLA-DR heavy chain, *Proc. Natl. Acad. Sci. U.S.A.,* 79, 6337, 1982.

19. **Long, E. O., Gorski, J., and Mach, B.,** Structural relationship of the SB β-chain gene to HLA-D-region genes and murine I-region genes, *Nature (London),* 310, 233, 1984.

20. **Gorski, J., Rollini, P., Long, E. O., and Mach, B.,** Molecular organization of the HLA-SB region of the human major histocompatibility complex and evidence for two SB-β chain genes, *Proc. Natl. Acad. Sci. U.S.A.,* 81, 3934, 1984.

21. **Trowsdale, J., Kelly, A., Lee, J., Carson, S., Austin, P., and Travers, P.,** Linkage map of two HLA-SBβ and two HLA-SBα-related genes: an intron in one of the SBβ genes contains a processed pseudogene, *Cell,* 38, 241, 1984.

22. **Servenius, B., Gustafsson, K., Widmark, E., Emmoth, E., Andersson, G., Larhammar, D., Rask, L., and Peterson, P. A.,** Molecular map of the human HLA-SB (HLA-DP) region and sequence of an SB α (DP α) pseudogene, *EMBO J.,* 3, 3209, 1984.

23. **Rollini, P., Mach, B., and Gorski, J.,** Linkage map of three HLA-DR β-chain genes: evidence for a recent duplication event, *Proc. Natl. Acad. Sci. U.S.A.,* 82, 7197, 1985.

24. **Rollini, P., Mach, B., and Gorski, J.,** Characterization of an HLA-DR β pseudogene in the DRw52 supertypic group, *Immunogenetics,* 25, 336, 1987.

25. **Larhammar, D., Servenius, B., Rask, L., and Peterson, P. A.,** Characterization of an HLA DR β pseudogene, *Proc. Natl. Acad. Sci. U.S.A.,* 82, 1475, 1985a.

26. **Auffray, C., Lillie, J. W., Arnot, D., Grossberger, D., Kappes, D., and Strominger, J. L.,** Isotypic and allotypic variation of human class II histocompatibility antigen α-chain genes, *Nature (London),* 308, 327, 1984.

27. **Jonsson, A. K., Hyldig-Nielsens, J. J., Servenius, B., Larhammar, D., Andersson, G., Jörgensen, F., Peterson, P. A., and Rask, L.,** Class II genes of the human major histocompatibility complex, *J. Biol. Chem.,* 262, 8767, 1987.

28. **Tonnelle, C., DeMars, R., and Long, E. O.,** DOβ: a new β chain in HLA-D with a distinct regulation of expression, *EMBO J.,* 4, 2839, 1985.

29. **Trowsdale, J. and Kelly, A.,** The human HLA class II α chain gene DZ α is distinct from genes in the DP, DQ and DR subregions, *EMBO J.,* 4, 2231, 1985.

30. **Hardy, D. A., Bell, J. I., Long, E. O., Lindsten, T., and McDevitt, H. O.,** Mapping of the class II region of the human major histocompatibility complex by pulsed-field gel electrophoresis, *Nature (London),* 323, 453, 1986.

31. **Böhme, J., Andersson, M., Andersson, G., Möller, E., Peterson, P. A., and Rask, L.,** HLA-DRβ genes vary in number between different DR specificities, whereas the number of DQ βgenes is constant, *J. Immunol.,* 135, 2149, 1985.

32. **Gorski, J., Rollini, P., and Mach, B.,** Structural comparison of the genes of two HLA-DR supertypic groups: the loci encoding DRw52 and DRw53 are not truly allelic, *Immunogenetics,* 25, 397, 1987.

33. **Mach, B., Gorski, J., Rollini, P., Berte, C., Amaldi, I., Berdoz, J., and Ucla, C.,** Polymorphism and regulation of HLA Class II genes of the major histocompatibility complex, *Cold Spring Harbor Symp. Quant. Biol.,* 51, 67, 1986.

34. **Albert, E. D., Baur, M. P., and Mayr, W. R.,** *Histocompatibility Testing 1984,* Springer-Verlag, Berlin, 1984.
35. **Gorski, J. and Mach, B.,** Polymorphism of human Ia antigens: gene conversion between two DR β loci results in a new HLA-D/DR specificity, *Nature (London),* 322, 67, 1986.
36. **Gregersen, P. K., Shen, M., Song, Q. L., Merryman, P., Degar, S., Seki, T., Maccari, J., Goldberg, D., Murphy, H., Schwenzer, J., Wang, C. Y., Winchester, R. J., Nepom, G. T., and Silver, J.,** Molecular diversity of HLA-DR4 haplotypes, *Proc. Natl. Acad. Sci. U.S.A.,* 83, 2642, 1986.
37. **Bell, J. I., Denney, D., Jr., Foster, L., Belt, T., Todd, J. A., and McDevitt, H. O.,** Allelic variation in the DR subregion of the human major histocompatibility complex, *Proc. Natl. Acad. Sci. U.S.A.,* 84, 6234, 1987.
38. **Wake, C. T., Long, E. O., Strubin, M., Gross, N., Accolla, R., Carrel, R., and Mach, B.,** Isolation of cDNA clones encoding HLA-A-DR α chains, *Proc. Natl. Acad. Sci. U.S.A.,* 79, 6979, 1982.
39. **Charron, D. J., Lotteau, V., and Turmel, P.,** Hybrid HLA-DC antigens provide molecular evidence for gene transcomplementation, *Nature (London),* 312, 157, 1984.
40. **Germain, R. N., Bentley, D. M., and Quill, H.,** Influence of allelic polymorphism on the assembly and surface expression of class II MHC (Ia) molecules, *Cell,* 43, 233, 242, 1985.
41. **Schwartz, R. H., Chen, C., and Paul, W. E.,** Gene complementation in the T lymphocyte proliferative response to poly (Glu^{56}Lys^{35}Phe^9)r. Functional evidence for a restriction element coded for by both the I-A and I-E subregions, *Eur. J. Immunol.,* 10, 708, 1980.
42. **Germain, R. N. and Quill, H.,** Unexpected expression of a unique mixed-isotype class II MHC molecule by transfected L-cells, *Nature (London),* 320, 72, 1986.
43. **Sant, A. J., Maloy, W. L., Coligan, J., and Germain, R. N.,** The efficiency of interisotypic Ia α/β dimer expression assayed by *in vivo* competition, 6th HLA Cloning Workshop, Airlie House, Virginia, May 1988.
44. **Wake, C. T., Long, E. O., and Mach, B.,** Allelic polymorphism and complexity of the genes for HLA-DR β-chains-direct analysis by DNA-DNA hybridization, *Nature (London),* 300, 372-374, 1982.
45. **Kim, S. E., Holbeck, S. L., Nisperos, B., Hansen, J. A., Maeda, H., and Nepom, G. T.,** Identification of a polymorphic variant associated with HLA-DQw3 and characterized by specific restrictions sites within the DQ β-chain gene, *Proc. Natl. Acad. Sci. U.S.A.,* 82, 8139, 1985.
46. **Inoko, H., Ando, A., Tsuji, K., Matsuki, K., Juji, K., Honda, Y.,** HLA-DQβ chain DNA restriction fragments can differentiate between healthy and narcoleptic individuals with HLA-DR2, *Immunogenetics,* 23, 126, 1986.
47. **Cohen, D., Paul, P., Le Gall, I., Marcadet, A., Font, M. P., Cohen-Haguenauer, O., Sayagh, B., Cann, H., Lalouel, J. M., and Dausset, J.,** DNA polymorphism of HLA class I and class II regions, *Immunol. Rev.,* 85, 87, 1985.
48. **Tilanus, M. G. J., Morolli, B., Van Eggermond, M. C. J. A., Schreier, G. M. T., de Vries, R. R. P., and Giphart, M. J.,** Dissection of HLA class II haplotypes in HLA-DR4 homozygous individuals, *Immunogenetics,* 23, 333, 1986.
49. **Bidwell, J. L., Bidwell, E. A., Savage, D. A., Middleton, D., Klouda, P. R., and Bradley, B. A.,** A DNA-RFLP typing system that positively identifies serologically well-defined and ill-defined HLA-DR and DQ alleles, including DRw10, *Transplantation,* 45, 640, 1988.
50. **Angelini, G., de Préval, C., Gorski, J., and Mach, B.,** High-resolution analysis of the human HLA-DR polymorphism by hybridization with sequence-specific oligonucleotide probes, *Proc. Natl. Acad. Sci. U.S.A.,* 83, 4489, 1986.
51. **Tiercy, J. M., Gorski, J., Jeannet, M., and Mach, B.,** Identification and distribution of three serologically undetected alleles of HLA-DR by oligonucleotide DNA-typing analysis, *Proc. Natl. Acad. Sci. U.S.A.,* 85, 198, 1988.
52. **Ucla, C., Van Rood, J. J., Gorski, J., and Mach, B.,** Analysis of HLA-D micropolymorphism by a simple procedure. RNA oligonucleotide hybridization, *J. Clin. Invest.,* 80, 1155, 1987.
53. **Mullis, K. B. and Faloona, F. A.,** Specific synthesis of DNA *in vitro* via a polymerase catalysed chain reaction, *Methods Enzymol.,* 155, 335, 1987.
54. **Saiki, R. K., Bugawan, T. L., Horn, G. T., Mullis, K. B., and Erlich, H. A.,** Analysis of enzymatically amplified β-globin and HLA-DQα DNA with allele-specific oligonucleotide probes, *Nature (London),* 324, 163, 1986.
55. **Saiki, R. K., Gelfand, D. H., Stoffel, S., Scharf, S. J., Higuchi, R., Horn, G. T., Mullis, K. B., and Erlich, H. A,** Primer-directed enzymatic amplification of DNA with a thermostable DNA polymerase, *Science,* 239, 487, 1988.
56. **Tiercy, J. M., Gorski, J., Bétuel, H., Freidel, A. C., Gebuhrer, L., Jeannet, M., and Mach, B.,** DNA typing of DRw56 subtypes: correlation with DRB1 and DRB3 allelic sequences by hybridization with oligonucleotide probes, *Human Immunol.,* 24, 1, 1989.
57. **Bjorkman, P. J., Saper, M. A., Samraoui, B., Bennett, W. S., Strominger, J. L., and Wiley, D. C.,** Structure of the human class I histocompatibility antigen, HLA-A2, *Nature (London),* 329, 506, 1987.

58. **Bjorkman, P. J., Saper, M. A., Samraoui, B., Bennett, W. S., Strominger, J. L., and Wiley, D. C.,** The foreign antigen binding site and T cell recognition regions of class I histocompatibility antigens, *Nature (London),* 329, 512, 1987.

59. **Brown, J. H., Jardetzky, T., Saper, M. A., Samraoui, B., Bjorkman, P., and Wiley, D. C.,** A hypothetical model of the foreign antigen binding site of class II histocompatibility molecules, *Nature (London),* 332, 845, 1988.

60. **Rabourdin-Combe, C. and Mach, B.,** Expression of HLA-DR antigens at the surface of mouse L cells co-transfected with cloned human genes, *Nature (London),* 303, 670, 1983.

61. **Gorski, J., Tosi, R., Strubin, M., Rabourdin-Combe, C., and Mach, B.,** Serological and immuno-chemical analysis of the products of a single HLA DR-α and DR-β chain gene expressed in a mouse cell line after DNA-mediated co-transformation reveals that the β chain carries a known supertypic specificity, *J. Exp. Med.,* 162, 105, 1985.

62. **Berte, C. C., Tanigaki, N., Tosi, R., Gorski, J., and Mach, B.,** Serological recognition of HLA-DR allodeterminant corresponding to DNA sequence involved in gene conversion, *Immunogenetics,* 27, 167, 1988.

63. **Berte, C. C., Gorski, J., Reith, W., and Mach, B.,** Epitope mapping of HLA-DR antigens with the use of DNA-transfected cells, *Immunobiology of HLA,* Vol. 2, *Immunogenetics and Histocompatibility,* Dupont, B., Ed., Springer Verlag, New York, 1989, 245.

64. **Sheehy, M. J., Rowe, J. R., Koning, F., and Jorgensen, L.,** Functional polymorphism of the HLA-DR beta-III chain, *Hum. Immunol.,* 21, 49, 1987.

65. **Irle, C., Jaques, D., Tiercy, J. M., Fuggle, S. V., Gorski, J., Termijtelen, A., Jeannet, M., and Mach, B.,** Functional polymorphism of each of the two HLA-DR beta chain loci demonstrated with antigen specific DR3 and DRw52-restricted T cell clones, *J. Exp. Med.,* 167, 853, 1988.

66. **Gorski, J., Irle, C., Mickelson, E. M., Sheehy, M. J., Termijtelen, A. M., Ucla, C., and Mach, B.,** Correlation of structure with T cell responses of the three members of the HLA-DRw52 allelic series, *J. Exp. Med.,* 170, 1027, 1989.

67. WHO nomenclature report: Nomenclature for factors of the HLA system 1984, *Tissue Antigens,* 24, 73, 1984.

68. **Pober, J. S., Collins, T., Gimbrone, Jr., M. A., Cotran, R. S., Gitlin, G. D., Fiers, W., Clayberger, C., Krensky, A. M., Burakoff, S. J., and Reiss, C. S.,** Lymphocytes recognize human vascular endothelial and dermal fibroblast Ia antigens induced by recombinant immune interferon, *Nature (London),* 305, 726, 1983.

69. **Bottazzo, G. F., Todd, I., Mirakian, R., Belfiore, A., and Pujol-Borrell, R.,** Organ-specific autoim-munity: a 1986 overview, *Immunol. Rev.,* 94, 137, 1986.

70. **Fakenburg, J. H. F., Fibbe, W. E., Goselink, H. M., Van Rood, J. J., Jansen, J.,** Human hematopoetic progenitor cells in long-term cultures express HLA-DR antigens and lack HLA-DQ antigens, *J. Exp. Med.,* 162, 1359, 1985.

71. **Hammerling, G. J.,** Tissue distribution of Ia antigens and their expression on lymphocyte subpopulations, *Transplant. Rev.,* 30, 64, 1976.

72. **Nadler, P. I., Klingenstein, R. J., Richman, L. K., and Ahmann, G. B.,** The murine Kuppfer cell, *J. Immunol.,* 125, 2521, 1980.

73. **Natali, P., Bigotti, A., Cavalieri, R., Nicotra, M. R., Tecce, R., Manfredi, D., Chen, Y. X., Nadler, L. M., and Ferrone, S.,** Gene products of the HLA-D region in normal and malignant tissues of non-lymphoid origin, *Hum. Immunol.,* 15, 220, 1986.

74. **Stingl, G. S., Katz, I., Clement, L., Green, I., and Shevach, E. M.,** Immunologic functions of Ia-bearing epidermal langerhans cells, *J. Immunol.,* 121, 2005, 1978.

75. **De Tribolet, N., Hamou, M. F., Mach, J.-P., Carrel, S., and Schreyer, M.,** Demonstration of HLA-DR antigens in normal human brain, *J. Neurol. Neurosurg. Psychiat.,* 47, 417, 1984.

76. **Ting, J. P. Y., Shigekawa, B. L., Linthicom, D. S., Weiner, L. S., and Frelinger, J. A.,** Expression and synthesis of murine immune response-associated (Ia) antigens by brain cells, *Proc. Natl. Acad. Sci. U.S.A.,* 78, 3170, 1981.

77. **Flavell, R. A., Allen, H., Burkly, L. C., Sherman, D. H., Waneck, G. L., and Widera, G.,** Molecular biology of the H-2 histocompatibility complex, *Science,* 233, 437, 1986.

78. **Unanue, E. R. and Allen, P. M.,** The basis for the immunoregulatory role of macrophages and other accessory cells, *Science,* 236, 551, 1987.

79. **Strubin, M., Mach, B., and Long, E. O.,** The complete sequence of the mRNA for the HLA-DR-associated invariant chain reveals a polypeptide with an unusual transmembrane polarity, *EMBO J.,* 3, 869, 1984.

80. **Strubin, M., Long, E. O., and Mach, B.,** Two forms of the Ia antigen-associated invariant chain results from alternative initiations at two In-phase AUGs, *Cell,* 47, 619, 1986.

81. **Strubin, M., Berte, C., and Mach, B.,** Alternative splicing and alternative initiation of translation explain the four forms of the Ia antigen-associated invariant chain, *EMBO J.,* 5, 3483, 1986.

82. **Collins, T., Korman, A. J., Wake, C. T., Boss, J. M., Kappes, D. J., Fiers, W., Ault, K. A., Gimbrone, M. A., Jr., Strominger, J. L., and Pober, J. S.,** Immune interferon activates multiple class II major histocompatibility complex genes and the associated invariant chain gene in human endothelial cells and dermal fibroblasts, *Proc. Natl. Acad. Sci. U.S.A.,* 81, 4917, 1984.

83. **Sekaly, R. P., Tonnelle, C., Strubin, M., and Mach, B.,** Cell surface expression of class II histocompatibility antigens occurs in the absence of the invariant chain, *J. Exp. Med.,* 164, 1490, 1986.

84. **Maniatis, T., Goodbourn, S., and Fischer, J. A.,** Regulation of inducible and tissue-specific gene expression, *Science,* 236, 1237, 1987.

85. **Boss, J. M. and Strominger, J. L.,** Regulation of a transfected human class II major histocompatibility complex gene in human fibroblasts, *Proc. Natl. Acad. Sci. U.S.A.,* 83, 9139, 1986.

86. **Sherman, P. A., Basta, P. V., and Ting, J. P. Y.,** Upstream DNA sequences required for tissue-specific expression of the HLA-DRa gene, *Proc. Natl. Acad. Sci. U.S.A.,* 84, 4254, 1987.

87. **Sullivan, K. E., Calman, A. F., Nakanishi, M., Tsang, S. Y., Wang, Y., and Peterlin, B. M.,** A model for the transcriptional regulation of MHC class II genes, *Immunol. Today,* 8, 289, 1987.

88. **Mathis, D. J., Benoist, C. O., Williams II, V. E., Kanter, M. R., and McDevitt, H. O.,** The murine Eα immune response gene, *Cell,* 32, 745, 1983.

89. **Saito, H., Maki, R. A., Clayton, L. K., and Tonegawa, S.,** Complete primary structures of the Eβ chain and gene of the mouse major histocompatibility complex, *Proc. Natl. Acad. Sci. U.S.A.,* 80, 5520, 1985.

90. **Kelly, A. and Trowsdale, J.,** Complete nucleotide sequence of a functional HLA-DPβ gene and the region between the DPβ1 and DPgrα1 genes: comparison of the 5′ ends of HLA class II genes, *Nucleic Acids Res.,* 13, 1607, 1985.

91. **Dorn, A., Durand, B., Marfing, C., Le Meur, M., Benoist, C., and Mathis, D.,** The conserved MHC class II boxes-X and -Y are transcriptional control elements and specifically bind nuclear proteins, *Proc. Natl. Acad. Sci. U.S.A.,* 84, 6249, 1987a.

92. **Dignam, J. D., Lebovitz, R. M., and Roeder, R. G.,** Accurate transcription initiation by RNA polymerase II in a soluble extract from isolated mammalian nuclei, *Nucleic Acids Res.,* 11, 1475, 1983.

93. **Fried, M. and Crothers, D. M.,** Equilibria and kinetics of lac repressor-operator interactions by polyacrylamide gel electrophoresis, *Nucleic Acids Res.,* 9, 6505, 1981.

94. **Carthew, R. W., Chodosh, L. A., and Sharp, P. A.,** An RNA polymerase II transcription factor binds to an upstream element in the adenovirus major late promoter, *Cell,* 43, 439, 1985.

95. **Galas, D. J. and Schmitz, A.,** DNAase footprinting: a simple method for the detection of protein-DNA binding specificity, *Nucleic Acids Res.,* 9, 3047, 1978.

96. **Staudt, L. M.,, Singh, H., Sen, R., Wirth, T., Sharp, P. A., and Baltimore, D.,** A lympoid-specific protein binding to the octamer motif of immunoglobulin genes, *Nature (London),* 323, 640, 1986.

97. **Miwa, K., Doyle, C., and Strominger, J. L.,** Sequence-specific interactions of nuclear factors with conserved sequences of human class II major histocompatibility complex genes, *Proc. Natl. Acad. Sci. U.S.A.,* 84, 4939, 1987.

98. **Dorn, A., Bollekens, J., Staub, A., Benoist, C., and Mathis, D.,** A multiplicity of CCAAT box-binding proteins, *Cell,* 50, 863, 1987.

99. **Reith, W., Satola, S., Herrero Sanchez, C., Amaldi, I., Lisowska-Grospierre, B., Griscelli, C., Hadam, M. R., and Mach, B.,** Congenital immunodeficiency with a regulatory defect in MHC class II gene expression lacks a specific HLA-DR promoter binding protein, RF-X, *Cell,* 53, 897, 1988.

100. **Kobr, M., Reith, W., and Mach, B.,** Two distinct factors binding to MHC class II promoters with different relative affinities for individual HLA class II loci, submitted.

101. **Elgin, S. C. R.,** DNAse I hypersensitive sites of chromatin, *Cell,* 27, 413, 1981.

102. **Gönzcy, P., Reith, W., Barras, E., Hadam, M. R., Lisowska-Grospierre, B., Griscelli, C., and Mach, B.,** Inherited immunodeficiency with a transregulatory defect in MHC class II gene expression differs in the chromatin structure of the HLA-DRA gene, *Mol. Cell Biol.,* 9, 296, 1989.

103. **Basham, T. Y. and Merigan, T. C.,** Recombinant interferon-c increases HLA-DR synthesis and expression, *J. Immunol.,* 130, 1492, 1983.

104. **Basham, T., Smith, W., Lanier, L., Morhenn, V., and Merigan, T.,** Regulation of expression of class II major histocompatibility antigens on human peripheral blood monocytes and Langerhans cells by interferon, *Hum. Immunol.,* 10, 83, 1984.

105. **Amaldi, I., Reith, W., Berte, C., and Mach, B.,** Induction of HLA class II genes by interferon gamma is transcriptional and requires protein synthesis, *J. Immunol.,* 142, 999, 1989.

106. **Reith, W., Satola, S., Amaldi, I., Herrero-Sanchez, C., Berte, C., Ulevitch, R., and Mach, B.,** Regulation of HLA class II genes: identification of a regulatory promoter binding protein missing in class II-deficient congenital immunodeficiency, *Immunobiology of HLA,* Vol. 2, *Immunogenetics and Histocompatibility,* Dupont, B., Ed., Springer-Verlag, New York, 1989, 345.

107. **Rosa, F. and Fellous, M.,** The effect of gamma interferon on MHC antigens, *Immunol. Today,* 5, 261, 1984.

108. **Collins, T., Lapierre, L. A., Fiers, W., Strominger, J. L., and Pober, J. S.,** Recombinant human tumor necrosis factor increases mRNA levels and surface expression of HLA-A,B antigens in vascular endothelial cells and dermal fibroblasts, *in vitro, Proc. Natl. Acad. Sci. U.S.A.,* 83, 446, 1986.

109. **Reith, W., Amaldi, I., Berte, C., Ulevitch, R., and Mach, B.,** The induction of MHC class I and class II genes by interferon gamma involves distinct transacting factors, submitted.

111. **Griscelli, C., Lisowska-Grospierre, B., and Mach, B.,** Combined immunodeficiency with defective expression in MHC class II genes, *Immunodeficiency Rev.,* 1, 135, 1989.

112. **Lisowska-Grospierre, B., Charron, D., De Préval, C., Durandy, A. Griscelli, C., and Mach, B.,** A defect in the regulation of major histocompatibility complex class II gene expression in human HLA-DR negative lymphocytes from patients with combined immunodeficiency syndrome, *J. Clin. Invest.,* 76, 381, 1985.

113. **De Préval, C., Lisowska-Gropierre, B., Loche, M., Griscelli, C., and Mach, B.,** A *trans*-acting class II regulatory gene unlinked to the MHC controls expression of HLA class II genes, *Nature (London),* 318, 291, 1985.

114. **De Préval, C., Hadam, M. R., and Mach, B.,** Regulation of genes for HLA class II antigens in cell lines from patients with severe combined immunodeficiency, *N. Engl. J. Med.,* 318, 1295, 1988.

115. **Gladstone, P. and Pious, D.,** Stable variants affecting B cell alloantigens in human lymphoid cells, *Nature (London),* 271, 459, 1978.

116. **Accolla, R. S.,** Human B cell variants immunoselected against a single Ia antigen subset have lost expression of several Ia antigen subsets, *J. Exp. Med.,* 157, 1053, 1983.

117. **Gladstone, P. and Pious, D.,** Identification of a *trans*-acting function regulating HLA-DR expression in a DR-negative B cell variant, *Somatic Cell Gen.,* 6, 285, 1980.

118. **Accolla, R. S., Carra, G., and Guardiola, J.,** Reactivation by a *trans*-acting factor of human major histocompatibility complex Ia gene expression in interspecies hybrids between an Ia-negative human B-cell variant and an Ia-positive mouse B-cell lymphoma, *Proc. Natl. Acad. Sci. U.S.A.,* 82, 5145, 1985.

119. **Long, E. O., Mach, B., and Accolla, R. S.,** Ia-negative B-cell variants reveal a coordinate regulation in the transcription of the HLA class II gene family, *Immunogenetics,* 19, 349, 1984.

120. **Satola, S., Reith, W., Amaldi, I., Gönczy, P., Accolla, R., Grospierre, B., Griscelli, C., Hadam, M., and Mach, B.,** Distinct transacting regulatory factors are involved in two forms of HLA class II regulatory mutants, submitted.

121. **Singh, H., LeBowitz, J. H., Baldwin, Jr., A. S., and Sharp, P. A.,** Molecular cloning of an enhancer binding protein: isolation by screening of an expression library with a recognition site DNA, *Cell,* 52, 415, 1988.

122. **Vinson, C. R., LaMarco, K. L., Johnson, P. F., Landschulz, W. H., and McKnight, S. L.,** *In-situ* detection of sequence-specific DNA binding activity specified by a recombinant bacteriophage, *Genes Dev.,* 2, 801, 1988.

123. **Reith, W., Barras, E., Satola, S., Kobr, M., Reinhart, D., Herrero-Sanchez, C., and Mach, B.,** Cloning of the major histocompatibility complex class II promoter binding protein affected in a hereditary defect in class II gene regulation, *Proc. Natl. Acad. Sci. U.S.A.,* 86, 4200, 1989.

124. **Benham, F. J., Quintero, M. A., and Goodfellow, P. N.,** Human-mouse hybrids with an embryonal carcinoma phenotype continue to transcribe HLA-A,B,C, *EMBO J.,* 2, p.1963, 1983.

125. **Latron, F., Jotterand-Bellomo, M., Maffei, A., Scarpellino, L., Bernard, M., Strominger, J. L., and Accolla, R. S.,** Active suppression of major histocompatibility complex class II gene expression during differentiation from B cells to plasma cells, *Proc. Natl. Acad. Sci. U.S.A.,* 85, 2229, 1988.

126. **Bodmer, J. G. et al.,** Nomenclature for factors of the HLA system, *Immunol. Today,* in press.

127. **Long, E. O.,** personal communication.

Chapter 12

TRANSCRIPTIONAL REGULATION OF HLA CLASS II GENES: *CIS*-ACTING ELEMENTS AND *TRANS*-ACTING FACTORS

Andrew F. Calman and B. Matija Peterlin

TABLE OF CONTENTS

I. INTRODUCTION

The regulation of human leucocyte antigen (HLA) class II gene expression is intimately related to the function of class II proteins as antigen-presenting molecules. Whereas class I molecules are found on virtually all nucleated cells, expression of class II molecules is normally restricted to antigen-presenting cells, such as macrophages and B cells.[1] Certain lymphokines augment class II gene expression in these cells, increasing their ability to present antigens to CD4$^+$ T cells.[2-4] *Cis*-acting transcriptional promoter and enhancer sequences, which enable class II genes to respond appropriately to tissue-specific and lymphokine-induced *trans*-acting transcription factors, have coevolved with class II protein coding sequences to enable these molecules to function optimally in host defense. The severe combined immune deficiency seen in children who congenitally fail to express class II antigens (the class II bare lymphocyte syndrome) attests to the importance of these molecules in the immune response.[5]

Constitutive high-level expression of class II molecules is confined to B cells and thymic epithelial cells. Expression of class II in B cells facilitates their antigen-specific interactions with CD4$^+$ T cells.[6] The level of class II determinants on murine B cells is increased by IL-4, but this effect has not yet been demonstrated with human cells.[7] T cells, unlike B cells, do not express class II in the resting state. However, class II synthesis is induced in human T cells upon cellular activation via the T-cell antigen receptor.[8] Class II expression is also seen in T cells transformed by the leukemogenic retroviruses HTLV-I and HTLV-II.[9]

The low level of class II expression in resting macrophages is greatly increased by the lymphokine IFN-γ, produced by helper T cells.[10,11] The time course of induction is slow, with maximal cell-surface class II expression occurring 4 to 6 d after IFN-γ administration.[11] Many other cell types also express class II in response to IFN-γ under certain circumstances. Primary fibroblast cultures and many epithelial tumor cells respond to physiological doses of IFN-γ,[12-14] while administration of high (pharmacologic) doses of IFN-γ to mice results in class II expression in many other tissue types.[15] Other substances, such as tumor necrosis factor-α, lymphotoxin, vitamin D_3 and some prostaglandins, can also affect class II expression.[16-18] Finally, inappropriate expression of class II in tissues undergoing autoimmune destruction has been observed.[19] In both tissue-specific and IFN-γ inducible expression, the levels of class II mRNA and protein are generally proportionate, and the rate of transcription is probably the major determinant of the level of class II biosynthesis.[11] Although post-transcriptional effects may also contribute, they have not been extensively studied.

The genes encoding human class II α and β chains, clustered on chromosome 6, have recently been mapped by pulse-field gel electrophoresis (Figure 1).[20-22] The corresponding genes from the DR3 and DR4 haplotypes have been cloned and sequenced.[23-29] Most of these genes are coordinately regulated, although quantitative differences exist in their levels of expression. The DOβ gene, whose corresponding α gene is unknown, is an exception. It is not IFN-γ inducible in antigen-presenting cells, and is expressed as mRNA only at low levels in B cells.[13] Several class II genes are not expressed. Of these, the DPαII and DPβII (or SXαII and SXβII) genes contain several mutations (including stop codons) in their coding regions, as well as promoter mutations, and are definitely pseudogenes in the haplotypes studied.[26] The DRβII gene is a pseudogene in the DRw52 group of haplotypes, as are two DRβ genes in the DRw53 group.[23,30] No transcripts from the DXα and DXβ (or DQαII and DQβII) genes have been detected, even though they contain no aberrant mutations in their coding sequences.[25] Mutations within the promoters of these genes may have rendered them transcriptionally inactive. The invariant (Ii or γ chain gene, whose product associates with class II α and β chains during their biosynthesis, is expressed in most cell types; levels of Ii mRNA are also increased by administration of IFN-γ.[31]

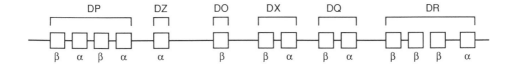

FIGURE 1.   Molecular map of class II MHC genes of the DR3 haplotype. Scale is approximate.

This review will focus on our studies of the regulation of human class II genes in two contexts which are relevant to their biological function: constitutive high-level expression of class II in B cells, and IFN-γ inducible expression of class II in macrophages and other cell types. We have primarily studied the DRα gene, whose DNA sequence is invariant between haplotypes. We have also examined some aspects of DQα and DQβ gene expression. The chromatin structure of class II genes in different cell types was analyzed by mapping DNase I hypersensitive sites.[32] Methylation-sensitive restriction endonucleases were used to examine the degree of cytosine methylation within and near the DRα gene.[33] Using a transient transfection assay, we dissected the B cell-specific and IFN-γ inducible *cis*-acting DNA sequences in class II promoters, from which transcription initiates and transcriptional enhancers, which act at a distance to facilitate increased transcription from homologous or heterologous promoters.[14,34,35] The direct binding of tissue-specific DNA-binding proteins to short sequences within class II promoter and enhancer elements was examined by gel mobility shift and DNase I footprinting assays.[36] Finally, the generation of class II-negative mutant B cell lines which lack class II-specific *trans*-acting transcription factors has enabled us to begin a genetic analysis of these factors, and may allow for isolation and characterization of their genes.[37] Taken together, these data form the basis for a model of the transcriptional regulation of class II genes.

II. CHROMATIN STRUCTURE

In order to identify sequences which might regulate DRα expression, our initial studies focused on the chromatin structure of the DRα gene in different functional states of class II gene expression.[32] Transcriptional regulatory elements, such as promoter and enhancer sequences, represent sites of interaction between DNA and sequence-specific DNA-binding proteins. Such interactions often perturb the local chromatin structure, for example, by disrupting histone H1 bridges or altering the phasing of nucleosomal arrays. These loci of DNA-protein interactions are more sensitive to digestion by DNase I than is bulk chromatin, and can be visualized as distinct bands by Southern blot analysis.[38-40] Thus, this technique can provide clues to the location of *cis*-acting regulatory sequences.

The entire DRα gene, along with 2.4 kilobase pairs (kb) of 5' flanking and 0.4 kb of 3' flanking sequences, was examined for DNase I hypersensitive sites in several cell lines.[32] In cells which constitutively express high levels of DRα, such as the B cell line BJAB and the T cell line HUT78 (which has the phenotype of an activated, class II⁺ T cell), three distinct hypersensitive sites (HS) were observed (Figure 2). HS I is located approximately 100 bp upstream of the transcription initiation site; this is the location of the conserved class II promoter sequences (see below). HS II maps slightly downstream of exon 1, while HS III is located near the center of intron 1. These latter two sites are contained within a tissue-specific transcriptional enhancer in the first intron (see below).[34] The presence of intronic DNase I hypersensitive sites in class II-expressing cells is not unique to the DRα gene; a similar site is found within a transcriptional enhancer element in the first intron of the DQα gene.[35]

In U-937 and HL-60 monocytoid cells and HeLa cervical carcinoma cells, which express class II in response to IFN-γ, HS I is present, but HS II and HS III are absent. Administration

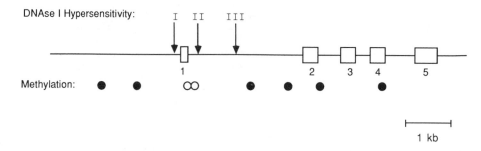

FIGURE 2.   DNase I hypersensitivity and cytosine methylation analysis of the HLA-DRα gene. Exons 1 through 5 are denoted by open boxes. Arrows indicate DNase I hypersensitive sites. Open and closed circles indicate cytosine residues which are demethylated and methylated, respectively, in most cell types studied.

of IFN-γ has no effect on any of these sites. Surprisingly, class II-negative, noninducible Jurkat CD4[+] T cells similarly exhibit HS I, but not HS II or HS III. Thus, in all cells studied, the region of the DRα promoter is DNase I-hypersensitive, and thus presumably accessible to DNA-binding proteins, whereas HS II and III, which are associated with the transcriptional enhancer, are present only in cells which constitutively express high levels of DRα.[32] IFN-γ induction of DRα gene expression does not result in the appearance of new DNase I hypersensitive sites,[32] suggesting that B cell-specific transcription of the DRα gene may involve both the promoter and sequences from intron 1, while IFN-γ induction is probably mediated by promoter sequences alone.

Another parameter related to the transcriptional activity of eukaryotic genes is the degree of *in vivo* methylation of cytosine nucleotides within and near genes. Methylation of cytosines within CpG dinucleotides, especially near regulatory sequences, often correlates negatively with transcriptional activity, and inhibitors of cytosine methylation, such as 5-azacytidine, often results in expression of previously silent genes (including DRα).[41-43] Using restriction enzymes which are sensitive to methylation of cytosines within their recognition sites, we examined the methylation status of ten CpG dinucleotides in and near the DRα gene in cells whose expression of DRα was constitutively high, inducible by IFN-γ, or noninducible. We found no consistent correlation between DRα expression and the degree of cytosine methylation at the sites examined.[33] Others have reported similar results.[44-47] However, we found a striking difference between the sites in and near the first exon, which were demethylated in all cells studied, and the sites further upstream and downstream (including those in the first intron), which were generally methylated (Figure 2).[33] Thus, although methylation patterns did not differ significantly between DRα expressing and nonexpressing cells, sequences near the promoter were essentially free of cytosine methylation in all cells examined. It has been reported that methylated DNA is a better substrate for formation and tight packing of nucleosomes.[42] Thus, DNA demethylation near the promoter may influence local chromatin structure, and could facilitate binding of regulatory proteins to the DRα promoter.[42,48]

III. PROMOTER ELEMENTS

Class II promoters are complex structures whose DNA sequences have evolved to respond to a large set of intracellular cues, including *trans*-acting transcription factors specific for cell type, developmental stage, and the presence of lymphokines, such as IL-4 and IFN-γ. The study of the *cis*-acting elements which compose these promoters is facilitated by the promoters' compactness (approximately 100 to 150 bp), and by comparison of the sequences conserved among the various class II promoters.[1,24,49-51]

Class II promoters (including the Ii promoter) contain several shared sequences ("boxes") interspersed with nonconserved sequences. From the 3′ to 5′ direction, these are the TATA

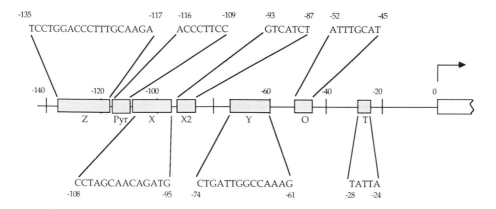

FIGURE 3. Conserved regions of the HLA-DRα promoter. DNA sequences of the Z box, pyrimidine tract, X box, X2 box, Y box, octamer, and TATA box are shown.

box, CCAAT box (or, in the case of DRα, Ig octamer), Y box, X box, and Z (or W) box.[1,24,49-51] Additional conserved sequences which we have shown to be important for promoter function or protein binding (see below) include the pyrimidine tract and the X2 box, which lie immediately 5' and 3' of the X box, respectively (Figure 3). All class II genes contain these motifs, though some are more conserved than others. The sequence spanning the Z box, pyrimidine tract, and X box is very similar to a consensus sequence found in genes which are inducible by IFN-α and IFN-β.[52] The class II α and β gene families each contain specific conserved sequences between the X and Y boxes.[28] Each of the boxes appears to interact with specific *trans*-acting transcription factors, and each plays a role either as a basal promoter element or as a tissue-specific or lymphokine-responsive element.[14,51,53] However, combinations of the elements are more transcriptionally active than is each element individually.[14] The TATA box is found in most eukaryotic promoters, and appears to direct the proper positioning of the transcription complex.[54] The other boxes constitute upstream promoter elements which are essential for efficient, correctly initiated transcription, and are largely responsible for the tissue-specificity and IFN-γ inducibility of class II genes.[12,14,53]

Plasmids containing 5' deletions of the DRα promoter linked to the bacterial chloramphenicol acetyltransferase (CAT) reporter gene were transfected into Raji and BJAB B cells, Jurkat T cells, and resting and IFN-γ induced HeLa cells. Raji and BJAB cells express large quantities of class II mRNA and protein. Jurkat T cells express no detectable class II mRNA or protein. Although Jurkat cells have the phenotype of resting T cells by most criteria, they cannot be induced to express class II by activation via their antigen receptors, nor can any other known T-cell tumor line. HeLa cells were used because they express no detectable class II mRNA in the resting state, and synthesize relatively large amounts of DRα and DRβ MRNA and protein after IFN-γ administration.[14] Two days after transfection, the specific CAT activity in cellular lysates was determined by an enzymatic assay. Because CAT mRNA stability and translation efficiency are constant, this assay measures the transcription rate of the test promoters.

Transfected plasmids containing promoter sequences beginning at positions −268, −150, or −136 (the approximate 5' boundary of the Z box) were all expressed at high levels in B cells, low levels in T cells, and were inducible by IFN-γ in HeLa cells (Figure 4). Thus, sequences between positions −136 and +29 were sufficient for B cell-specific and IFN-γ inducible expression. However, when a further 5 bp were deleted, to position −131 within the Z box, expression both in B cells and in IFN-γ treated HeLa cells was decreased. Paradoxically, increased expression of the −131 construct was seen in Jurkat cells and, to

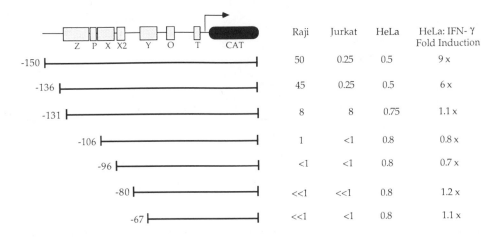

	Raji	Jurkat	HeLa	HeLa: IFN-γ Fold Induction
-150	50	0.25	0.5	9 x
-136	45	0.25	0.5	6 x
-131	8	8	0.75	1.1 x
-106	1	<1	0.8	0.8 x
-96	<1	<1	0.8	0.7 x
-80	<<1	<<1	0.8	1.2 x
-67	<<1	<1	0.8	1.1 x

FIGURE 4. Deletion analysis of the HLA-DRα promoter. Plasmids containing various 5' deletions of the DRα promoter, linked to the CAT gene, were transfected into different cell types. Results of a standardized CAT assay (expressed as percentage of chloramphenicol converted to acetylated forms) are shown. Results in HeLa cells treated with IFN-γ are expressed as "fold induction" relative to untreated HeLa cells.

a lesser extent, in resting HeLa cells.[14] This may represent the removal of a sequence around position −136 which functions as a negative regulatory element in cells which do not express class II. Such a negative element has also been found in analogous position in the DQβ gene.[12]

When the DRα promoter deletion was extended to position −106, within the X box, CAT activity in all cells was very low. Further deletions, to position −96 within the X box, position −80 between X and Y, or position −67 within the Y box, resulted in even lower levels of CAT activity in all cells.[14] Deletion analysis of the DQβ gene in transfected human fibroblasts generated similar results, except that evidence was found for an additional negative regulatory element between the X and Y boxes.[12] Our failure to find such an element in the DRα gene may be due to sequence divergence between class II α and β genes in this region, or to differences in the endpoints of the deletions.

To assess further the contributions of the various conserved boxes to B cell-specific and IFN-γ inducible transcription of the DRα gene, restriction fragments and synthetic oligonucleotides corresponding to individual boxes and combinations thereof were linked to the tk promoter, directing the transcription of the CAT gene, and analyzed by the transient transfection assay (Figure 5).[14] Oligonucleotides containing the X box (position −109 to −92), Y box (−85 to −60), or Z box and pyrimidine tract (−136 to −107) did not increase the transcriptional activity of the tk promoter in B cells, nor did they confer IFN-γ inducibility in HeLa cells. A combination of the X, X2 and Y boxes (positions −110 to −60) enhanced transcription from the tk promoter in B cells, T cells, and HeLa cells, but did not respond to IFN-γ.[14,35] This combination functioned in an orientation- and position-independent fashion, and hence fulfilled the criteria for a transcriptional enhancer.[35] Most significantly, the combination of the Z box, pyrimidine tract, X and X2 boxes (−136 to −80) conferred upon the tk promoter extremely high CAT activity in B cells, much lower activity in T cells and resting HeLa cells, and IFN-γ inducibility in HeLa cells. A smaller fragment, containing only the pyrimidine tract and X box (positions −116 to −92), retained most of this B cell-specific transcriptional activity.[14]

Gel mobility shift studies were used to identify DNA-binding proteins capable of interacting with class II promoter elements. In this assay, binding of proteins from crude nuclear extracts to a radiolabeled DNA fragment decreases the fragment's electrophoretic mobility on low ionic-strength polyacrylamide gels; each DNA-protein complex has a char-

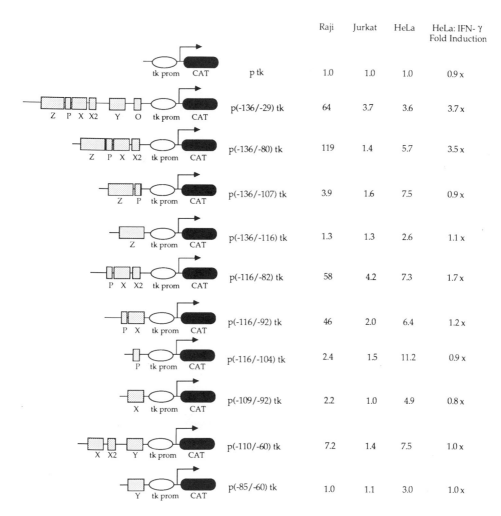

	Raji	Jurkat	HeLa	HeLa: IFN-γ Fold Induction
p tk	1.0	1.0	1.0	0.9 x
p(-136/-29) tk	64	3.7	3.6	3.7 x
p(-136/-80) tk	119	1.4	5.7	3.5 x
p(-136/-107) tk	3.9	1.6	7.5	0.9 x
p(-136/-116) tk	1.3	1.3	2.6	1.1 x
p(-116/-82) tk	58	4.2	7.3	1.7 x
p(-116/-92) tk	46	2.0	6.4	1.2 x
p(-116/-104) tk	2.4	1.5	11.2	0.9 x
p(-109/-92) tk	2.2	1.0	4.9	0.8 x
p(-110/-60) tk	7.2	1.4	7.5	1.0 x
p(-85/-60) tk	1.0	1.1	3.0	1.0 x

FIGURE 5. Transfection analysis of conserved DRα promoter elements. Conserved class II boxes, singly and in combination, were linked to the tk promoter and CAT gene and transfected into various cells. CAT results for all plasmids were expressed as relative CAT activity, with the activity of the tk promoter alone set equal to 1.0 for each cell type.

acteristic mobility.[55] An oligonucleotide containing the DRα promoter octamer and its flanking sequence displayed the typical gel-shift pattern previously seen with immunoglobulin octamer elements (Figure 6).[56,57] A DNA-protein complex of relatively low mobility was formed with crude nuclear extracts from B, T, and HeLa cells, and a faster-migrating band was observed only in B cells. These bands correspond to the DNA-binding proteins NF-A1 and NF-A2.[58] The Y box formed two DNA-protein complexes of identical mobility in all cells studied.[51,81] In the I-Eα gene, the murine homolog of DRα, these complexes have been shown to contain the protein NF-Y, one of a family of at least four different factors which interact with sequences containing CCAAT motifs (in the Y box, this motif is present as its reverse complement, ATTGG).[59,60]

The X box formed a complex of three slowly migrating bands, which we collectively term complex X, in all cells studied (Figure 6). The relative amounts and mobilities of the three bands differed subtly among the different cell types, but it is not clear whether this is functionally significant. HeLa cells contained an additional, faster-migrating band, which we designate complex X2. Using smaller oligonucleotides, the binding site of factor X was

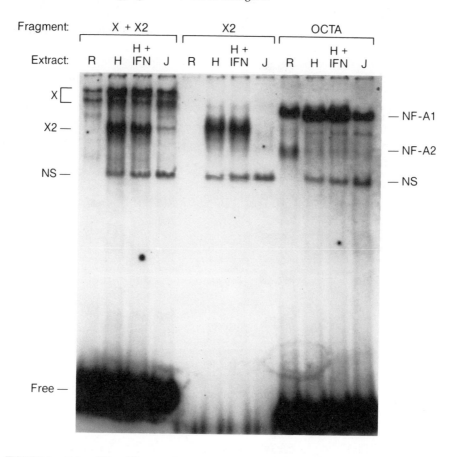

FIGURE 6.   Gel-mobility shift assay for proteins binding to X box, X2 box and octamer from the DRα promoter. ^{32}P-labeled oligonucleotides containing the X and X2 boxes, X2 box alone, or octamer sequence were incubated with 1 μg of poly(dI-dC) and 4 μg of crude nuclear protein from Raji cells (R), Jurkat Cells (J), HeLa cells (H), or HeLa cells treated for 48 h with 100 units/ml recombinant human IFN-γ (H + IFN), then electrophoresed on a low-ionic strength polyacrylamide gel essentially as described.[56] The specificity and mapping of all of the complexes shown was confirmed by competition with unlabeled oligonucleotides (not shown). NS indicates a band which was formed with many different labeled oligonucleotides, and whose formation was not inhibited by unlabeled specific oligonucleotides: thus, it represents a nonsequence-specific interaction.

localized to positions −108 to −93, whereas factor X2 binds to sequences between −94 and −82 (Figure 6). None of the bands observed with the X box, Y box or octamer was grossly altered in mobility or concentration when HeLa cells were stimulated with IFN-γ for 12 to 48 h (Figure 6 and data not shown). Proteins binding to the Z box and pyrimidine tract were much more difficult to detect using this assay. Although we have been able to identify faint sequence-specific DNA-protein complexes with oligonucleotides containing the Z box and pyrimidine tract, definitive results will require either protein purification or further optimization of binding conditions.

To assess the sites of interaction of the various *cis*-acting promoter elements with DNA-binding proteins in the context of the entire promoter, we performed DNase I footprint analysis. Crude nuclear protein extracts were incubated with an end-labeled DRα promoter DNA fragment, then dilute DNase I was added. Gaps in the ladder of DNA fragments produced by this limited DNase I digestion indicate sites of specific DNA-protein interactions.[61] With protein extracts from two different B cells (Raji and BJAB), virtually the entire promoter showed signs of DNA-protein interaction (Figure 7 a,b and data not shown). The octamer, Y and X boxes were protected from DNase I digestion on both DNA strands. Z binding factors did not produce a clear footprint. However, DNase I hypersensitive sites (indicated by bands of increased cleavage compared to the control) were found on both strands, within and just 5' of the Z box; such hypersensitive sites are often associated with DNA-protein binding domains.[62]

HeLa cell extracts yielded a footprint which differed from those of B cells in several respects (Figure 7, a,c). The octamer and Y box were less well protected from digestion, despite the presence of abundant NF-A1 and NF-Y activity in HeLa cell extracts in the gel-shift assay. Apparently these factors can bind to the isolated octamer and Y box oligonucleotides, but do not interact as efficiently with these sequences in the context of the entire promoter. Gel-shift experiments clearly demonstrated that factors X2 and NF-Y could bind simultaneously to an oligonucleotide containing both the X2 and Y boxes, so NF-Y binding in HeLa extracts was not simply blocked by binding of factor X2.[81] The X box was protected by the HeLa extract, but the 3' boundary of the interaction extended to position -82, compared to -95 in the other cell types. This extended footprint corresponded precisely to the binding site of factor X2, and inclusion of an excess of X2 oligonucleotide in the footprint reaction inhibited formation of the footprint over this X2 box.[81] This region of factor X2 binding, containing the consensus sequence GTCATCT (Figure 3), is conserved in most human class II genes 2 to 4 bp 3' of the classical X box. As in Raji cells, no footprint was seen in the Z box, but hypersensitive sites within and flanking these sequences were found. In extracts prepared from HeLa cells treated with IFN-γ for 48 h, no clear differences were seen on the footprint assay (data not shown). With Jurkat cells, the footprints of the Z and X boxes resembled those of B cells, while the Y and octamer footprints were similar to those of HeLa cells (Figure 7 a,d).

In summary, all of the previously identified conserved regions of the DRα promoter (octamer, Y, X, and Z boxes), as well as a newly identified conserved sequence, the X2 box, bound to sequence-specific DNA-binding proteins, although the Z proteins have been difficult to characterize in crude extracts. Purified proteins may help provide a more complete picture of DNA-protein interactions in the Z box. In the context of the whole promoter, significant tissue-specific differences were seen in the footprint assay in the octamer, Y, and X2 boxes. In the octamer and Y box, protection was strongest with B cell extracts, despite the presence of NF-A1 and NF-Y activity in the other cell types. The X2 box was only footprinted by the HeLa cell extract. These differences may be interpreted in light of the transfection data to give a tentative, testable model of the two modes of DRα expression.[1]

IV. TRANSCRIPTIONAL ENHANCERS

Transcriptional enhancer elements are important in the regulation of many genes. These cis-acting DNA sequences function by increasing the rate of transcriptional initiation from a promoter which may be located far apart on the chromosome. The function of enhancers is relatively independent of distance and orientation.[63] Their activity may be lineage-dependent (as in immunoglobulin heavy and κ light chain enhancers) or may be modified by signaling events caused by cellular activators, hormones, or lymphokines.[64-67] To look for class II transcriptional enhancers, we first subdivided the DRα gene into several restriction fragments. Each fragment was inserted into a plasmid containing the tk promoter directing the transcription of the CAT gene and analyzed by transient transfection and CAT assay.[34]

Two nonoverlapping DRα restriction fragments were found which significantly enhanced transcription from the tk promoter (Figure 8). The first, a 750 bp SacI-BglII fragment, contains sequences from exon 1 and intron 1. This fragment, designated fragment H, increased transcription from the tk promoter two- to threefold. The second, a 750 bp BglII-ClaI fragment designated fragment I, caused a three- to sevenfold increase in CAT activity.

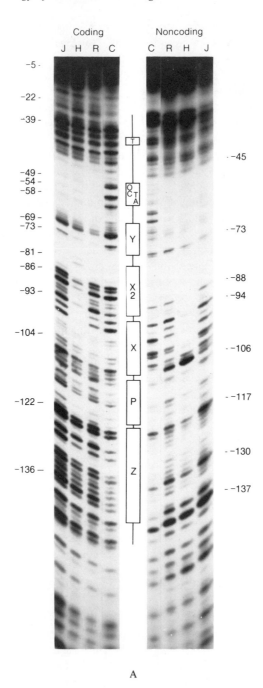

A

FIGURE 7. (A) DNase I footprint analysis of the HLA-DRα promoter. A restriction fragment containing the DRα promoter was end-labeled on either the coding or noncoding strand, incubated with 40 μg of crude nuclear protein from Raji cells (R), Jurkat cells (J), HeLa cells (H), or with no protein (C), digested for one minute with DNase I, then electrophoresed on a polyacrylamide sequencing gel essentially as described.[78] Bands which appear darker in the control (C) lane than in the other lanes represent areas where the DNA is protected from DNase digestion by binding of proteins. Bands which appear lighter in the control lane than in the other lanes represent DNase I hypersensitive sites where DNA cleavage is facilitated by DNA-protein interactions. (B—D) Summary of DNase footprint analysis in Raji (B), HeLa (C), and Jurkat (D) cells. Solid boxes indicate region protected from DNase I digestion; hatched boxes indicate weakly protected regions. Arrows denote DNase I hypersensitive sites. Boxes and arrows above the printed nucleotide sequence represent results for the coding strand; those below the sequence represent results for the noncoding strand.

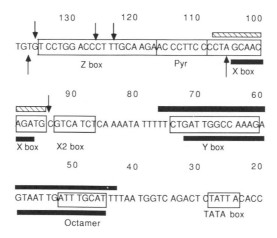

FIGURE 7B.

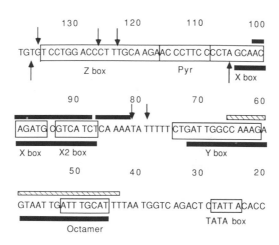

FIGURE 7C.

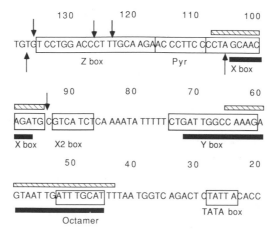

FIGURE 7D.

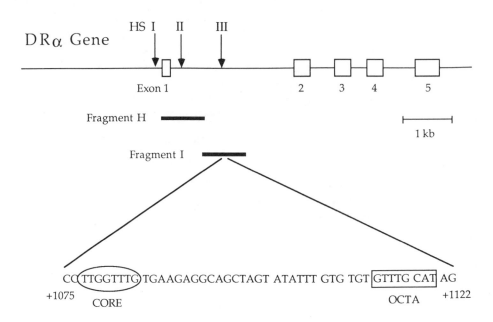

FIGURE 8.   The HLA-DRα transcriptional enhancer. Fragments H and I, which behaved as lymphoid-specific transcriptional enhancers in a transient transfection assay, are indicated, as are DNase I hypersensitive sites I, II, and III.[32,34] A sequence within fragment I, containing a core enhancer sequence and an imperfect immunoglobulin octamer, is shown. Fragment H also contains a core enhancer sequence (not shown).

Both fragments were active only in T and B cells, and functioned in either orientation; additionally, fragment I functioned whether it was placed 5′ or 3′ of the CAT gene. RNA analysis demonstrated that these transcriptional enhancers functioned by increasing the quantity of correctly initiated transcripts.[34] No other fragments, from 2.8 kb upstream of the transcription start site to 3.2 kb downstream (within intron 4) were able to significantly increase CAT expression from the tk promoter.

Significantly, the DNase I hypersensitive sites in the first intron of the DRα gene, seen only in cells which constitutively express high levels of DRα, mapped to regions possessing transcriptional enhancer activity in the transfection assay.[32,34] HS II maps into fragment H, while HS III maps into fragment I. The absence of HS II and III in class II-negative T cells suggests that the DRα enhancer may be inactive in the endogenous DRα gene, even though it displays transcriptional enhancer activity in transient transfection assays.

When the DNA sequence from the first intron was examined, regions of sequence similarity to regulatory elements in other genes were found. Within a 43 bp sequence in fragment I (positions +1077 to +1120 bp) is found a core transcriptional enhancer element similar to those found in viral enhancers, as well as a sequence identical at seven of eight positions to the immunoglobulin octamer (Figure 8).[34,63,68] Oligonucleotides containing either of these sequence motifs showed strong activity in transient transfection assays.[81] To determine whether proteins NF-A1 and NF-A2 interact with the imperfect octamer element, a radiolabeled oligonucleotide containing these sequences was incubated with crude nuclear protein extracts from Raji cells, then electrophoresed on low-ionic strength polyacrylamide gels. Bands corresponding to NF-A1 and NF-A2 were indeed detected; their mobility was identical to that of NF-A1 and NF-A2 complexes of the DRα promoter octamer, and their formation was inhibited by an excess of cold oligonucleotide containing the complete octamer sequence (Figure 9). Thus, octamer-binding proteins involved in B cell-specific expression of Ig genes may also participate in B cell-specific DRα gene expression by binding to octamer sequences in both the promoter and the enhancer. However, because this transcrip-

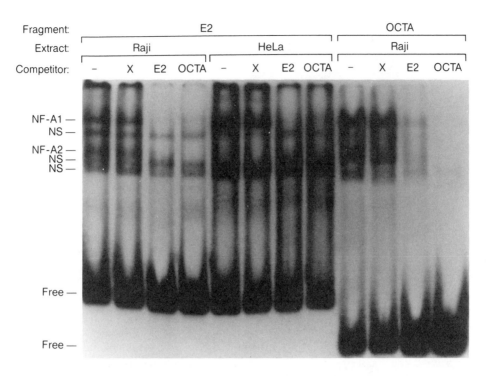

FIGURE 9.   Gel-mobility shift assay of the imperfect octamer from the DRα enhancer. Oligonucleotides containing either the octamer from the DRα promoter (OCTA fragment) or the imperfect octamer from the DRα enhancer (E2 fragment) were synthesized and end-labeled. Labeled oligonucleotides were incubated with nuclear extracts from Raji or HeLa cells, in the presence of poly(dI-dC) and an excess of unlabeled competitor (either OCTA, E2, or an oligonucleotide containing the DRα X box). DNA-protein complexes NF-A1 and NF-A2 had identical mobilities and tissue distribution with both E2 and OCTA, and formation of these complexes was inhibited by an excess of unlabeled OCTA or E2, but not by an excess of the X oligonucleotide. The bands labeled NS were not inhibited by an excess of the specific oligonucleotide (E2), and hence do not represent true sequence-specific binding.

tional enhancer functions not only in B cells but also in T cells, which contain NF-A1 but lack NF-A2, and because other transcriptionally active enhancer fragments also bind to proteins in the gel mobility shift assay,[81] the precise roles of NF-A1 and NF-A2 in the function of the DRα transcriptional enhancer are not clear.

To determine whether the existence of lymphoid-specific transcriptional enhancers is a general feature of class II gene expression, we scanned the DQα and DQβ genes for sequences which could increase transcription from a linked tk promoter, using the same transient transfection assay.[35] A systematic survey of 34 kb of DNA encompassing these two genes, as well as their 5' and 3' flanking regions, revealed the existence of several regulatory elements (Figure 10). A restriction fragment from the 5' flanking region of the DQα gene increased transcription from the tk promoter in an orientation-independent manner, but functioned only when placed upstream of the promoter. A sequence from the first intron of the DQα gene, near the location of a DNase I hypersensitive site in DQ-expressing cells, fulfilled the criteria of a transcriptional enhancer since it functioned in either orientation, whether placed upstream or downstream of the tk promoter. A third element, which maps to the second and third introns of the DQβ gene, also functioned as a transcriptional enhancer.

Although all of the DRα, DQα, and DQβ transcriptional enhancer elements functioned in both T and B cells, the absence of HS II and III in the DRα gene in T cells[32] and the inactivity of class II promoters in resting T cells (see below) suggest that enhancer activity is only functionally significant in B cells and possibly activated T cells. Perhaps the tran-

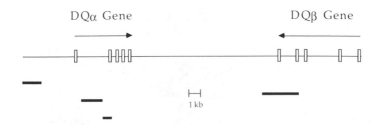

FIGURE 10. Transcriptional enhancers in the HLA-DQ subregion. The DQα and DQβ genes are shown; exons are depicted by open boxes, and arrows indicate the direction of transcription. Solid boxes indicate restriction fragments having transcriptional enhancer activity in a transient transfection assay.

scriptional enhancer interacts with lymphoid-specific *trans*-acting factors only when it is linked to an active promoter. Transfection studies using resting and activated peripheral T cells might help to clarify this point.

V. GENETIC APPROACHES TO *TRANS*-ACTING TRANSCRIPTION FACTORS

Biochemical assays such as DNase I footprinting can give invaluable information about those *trans*-acting transcription factors which are DNA-binding proteins. However, other transcription factors might exist which do not themselves bind DNA. For example, certain proteins may recognize combinations of DNA-binding proteins bound to DNA, while other may interact with RNA polymerase II to form a mobile subunit of the transcription complex. Although such factors are likely to contribute to tissue-specific and inducible gene expression, they have been refractory to study in mammalian systems due to the difficulty of performing *in vitro* transcription assays. A genetic approach to the identification of *trans*-acting transcription factors, through the creation, mutagenesis, and rescue of cell lines specifically deficient in class II transcription, offers the possibility of identifying and cloning the genes for transcription factors which may or may not themselves be DNA-binding proteins.

The existence of *trans*-acting transcription factors which specifically activate class II genes in B lymphocytes is supported by studies on a congenital immune deficiency known variously as class II-negative severe combined immunodeficiency (SCID) or class II bare lymphocyte syndrome (BLS).[69] BLS patients' lymphocytes fail to express either class I determinants, class II determinants, or both, despite the presence of grossly intact MHC genes.[5] In the case of class II BLS, the absence of cell surface class II on B cells is apparently due to decreased steady-state class II mRNA levels.[5] This defect, which behaves as an autosomal recessive trait, does not segregate on chromosome 6, demonstrating that the syndrome is caused by a defective or absent class II-specific transcription factor whose gene is not linked to the MHC.[5]

Because lymphocytes from BLS patients often revert to a class II-positive phenotype after viral transformation, they are difficult to study *in vitro*. Following pioneering studies by Pious and co-workers, several groups have produced B-cell lines which are stably class II-negative, by mutagenesis and selection for class II-negative variants of B-cell tumor lines.[37,70,71] We isolated two such class II-negative cell lines, designated RM2 and RM3, derived from the Burkitt lymphoma Raji (Figure 11).[37] Raji cells were mutagenized with ethylmethanesulfonate, which generates point mutations, then subjected to repeated rounds of complement-mediated cytolysis using anti-class II monoclonal antibodies directed against monomorphic determinants. Finally, class II-negative cells were isolated by fluorescence-activated cell sorting and cloned by limiting dilution.

The RM2 and RM3 cell lines, isolated from independent experiments, displayed very

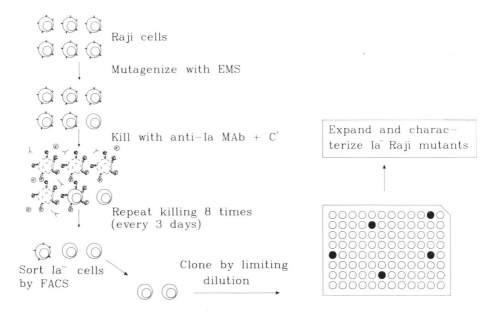

FIGURE 11. Selection scheme for the isolation of class II-negative B cell mutants. Double lines at cell surface indicate surface class II molecules.

TABLE 1
Class II Expression in RM2 and RM3 Cells

	Raji	RM2	RM3
Cell surface protein			
DR	100	0.11	<0.09
DQ	100	<2.7	<2.7
Class I	100	136	151
Steady-state mRNA			
DRα	100	8	<2
DRβ	100	4	<2
DQα	100	<2	<2
DQβ	100	6	<2
DPβ	100	6	<2
DOβ	100	31	37
Ii	100	46	43
Class I	100	85	84
Transcription rate			
DRα	100	7	<5
DQα	100	<5	<5
Class I	100	89	112
mRNA half-life			
DRα	5 h	5 h	N.D.
CAT activity of transfected *cis*-acting elements			
DRα enhancer (pE1, pE2, pE3)	100	~100	~100
DRα promoter (p 5'-136)	100	21	10
Pyrimidine tract + X box (p -116/-91 tk)	100	14	21

low and undetectable levels of cell surface class II molecules, respectively. Decreases in steady-state mRNA levels encoding DP, DQ, and DRα and β paralleled the low cell surface protein levels (Table 1).[37] In contrast, normal levels of class I mRNA and protein, and of other B cell markers, such as Ig κ light chain mRNA, CD19; and CD21, were found in the mutant cell lines.[37,81] *In vitro* nuclear run-on transcription experiments, which directly measure transcription rates, demonstrated that the defect in class II expression in RM2 and RM3 cells was attributable to decreased class II transcription (Table 1).[37] Finally, the class II genes in the mutants had undergone no gross deletions or rearrangements, as judged by low-stringency Southern hybridization analysis. Thus, these cells fail to express class II determinants due to a mutation which affects a class II-specific *trans*-acting transcription factor.

To prove that the transcriptional defect in the mutants was due to a *trans*-acting factor, we stably transfected RM2, RM3, and Raji cells with a plasmid containing an Epstein-Barr virus (EBV) origin of replication (which enables the plasmid to replicate extrachromosomally in these cells), a DRα minigene (constructed by deletion of exons 2, 3 and 4), and a hygromycin resistance gene. While levels of the hygromycin resistance gene mRNA were comparable in all three cell lines, the amount of mRNA from the transfected DRα minigene was high in Raji, very low in RM2, and undetectable in RM3.[36] In transient transfection using the DRα promoter linked to the CAT gene, CAT activity was substantially higher in Raji than in the mutant cells. This effect mapped to the pyrimidine cluster and X box: sequences from positions -116 to -92 of the DRα promoter, linked to the tk promoter and the CAT gene, were much more active in Raji than in RM2 or RM3 cells.[36] In contrast, fragments from the enhancer showed comparable activity in all three cells.

Fusion of RM2 or RM3 to another B cell line (of different HLA haplotype) restored expression of DR determinants of the Raji haplotype.[36] However, fusion of RM2 to RM3 did not appear to result in reexpression of class II, suggesting that the defects in the two cell lines map to the same complementation group. Thus, RM2 and RM3 appear to have lost class II expression due to a recessive allele encoding a defective *trans*-acting transcription factor which interacts (directly or indirectly) with the pyrimidine tract and X box. However, we observed no difference between Raji and mutant cells in protein binding to these sequences (or to any of the other conserved promoter boxes) in the gel mobility shift or footprint assays. Furthermore, the points of protein-DNA contact within the X box were identical in Raji and mutant cells, as shown by a methylation interference assay.[36] This leads to two possible explanations: either the defective factor is not a DNA-binding protein, or the mutations in RM2 and RM3 are point mutations which inactive transcriptional function, but not DNA binding, in a DNA-binding protein which interacts with the pyrimidine cluster or X box. The resolution of this issue, and precise identification of the defective or missing polypeptide, may require cloning of the gene encoding this *trans*-activator (Figure 12). Preliminary studies in this direction in other laboratories have shown promising results: the gene encoding a class II-specific transcription factor has been mapped to murine chromosome 16, and a human class II-negative B cell mutant has been rescued by transfection of mouse genomic DNA.[72,73]

VI. MODEL OF CLASS II GENE EXPRESSION

Any model of class II gene expression must explain the transcription of these genes under diverse circumstances, including resting B cells, activated T cells, and IFN-γ induced macrophages and fibroblasts. At a gross level, we may consider expression to be controlled by two *cis*-acting elements: the promoter and enhancer. In B cells both the promoter and the enhancer are constitutively active, resulting in high levels of class II expression. The effect of IL-4 on these elements is unknown. In resting T cells, the enhancer is active but the promoter is not. Cellular activation is presumed to induce *trans*-acting factors which activate the promoter, but no good tumor cell model exists to study this phenomenon. In IFN-γ inducible cells, the DRα enhancer appears to play no role. The enhancer has no activity in transfection assays in these IFN-γ induced cells, and its associated DNase I hypersensitive sites are present neither before nor after induction.[32,34] Rather, the promoter switches from a transcriptionally silent to a transcriptionally active state as a result of IFN-γ treatment.[14]

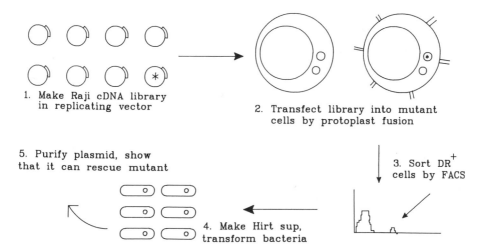

1. Make Raji cDNA library in replicating vector

2. Transfect library into mutant cells by protoplast fusion

5. Purify plasmid, show that it can rescue mutant

3. Sort DR$^+$ cells by FACS

4. Make Hirt sup, transform bacteria

FIGURE 12. Cloning regulatory genes by mutant rescue. In this proposed method for isolating the genes encoding specific *trans*-acting transcription factors, a cDNA library from class II-positive cells is constructed using a plasmid vector containing an Epstein-Barr virus origin of replication. Such plasmids replicate extrachromosomally to high copy number in cells which express Epstein-Barr nuclear antigen (such as RAji derivatives).[79] This permits high transfection efficiency, because integration into chromosomal DNA (a low-efficiency event) is not required. Additionally, the extrachromosomal plasmid can be readily separated from chromosomal DNA by the Hirt supernatant technique,[80] thus facilitating isolation of the gene of interest after cells expressing the gene have been identified by fluorescence-activated cell sorting.

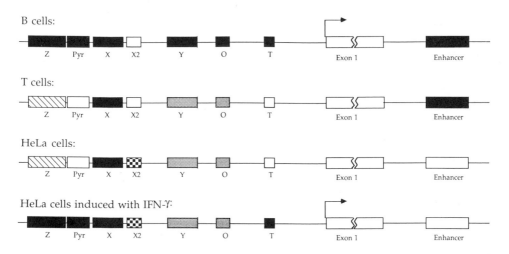

B cells:

Z   Pyr   X   X2   Y   O   T   Exon 1   Enhancer

T cells:

Z   Pyr   X   X2   Y   O   T   Exon 1   Enhancer

HeLa cells:

Z   Pyr   X   X2   Y   O   T   Exon 1   Enhancer

HeLa cells induced with IFN-γ:

Z   Pyr   X   X2   Y   O   T   Exon 1   Enhancer

FIGURE 13. Model of HLA-DRα gene regulation. The contributions of the different promoter and enhancer elements to DRα transcription in different phenotypic states are shown schematically; their functions are inferred from transfection experiments, DNA-protein binding assays, and by analogy with other eukaryotic promoters. Solid boxes depict sequences which function as positive transcriptional elements; hatched boxes depict negative transcriptional elements. Shaded boxes indicate regions which show relatively weak binding to regulatory proteins, and may function as inefficient positive transcriptional elements. Checkered boxes indicate sequences whose function is unknown but which bind to proteins. Open boxes represent regions which function as neither positive nor negative elements, and do not appear to bind to proteins.

The promoter sequences involved in transcriptional regulation have been analyzed in detail by transfection analysis. The roles of the DNA-binding proteins which interact with them may be inferred from these data and from direct studies of DNA binding by crude nuclear extracts from various cell types. Figure 13 depicts a model for HLA-DRα transcriptional regulation based on these data and by analogy with other well-studied eukaryotic genes. The TATA box probably interacts with ubiquitous RNA polymerase II transcription factors, and the octamer probably has maximal activity in B cells, which contain NF-A2 as well as NF-A1. The Y box is involved in efficiency and accuracy of transcription initiation, as shown most clearly by studies in transgemic mice.[53] The X box and its pyrimidine extension (to position −116 in the DRα gene) constitute the core transcription regulatory element responsible for B cell-specific expression; in contrast, IFN-γ inducible expression requires the Z box, pyrimidine tract, X box and possibly X2 box.[14] Finally, Z box sequences between positions −136 and −131 function as a positive upstream element in B cells and IFN-γ induced cells, and as a negative element in class II-negative T cells and uninduced HeLa cells.[14]

In spite of the co-localization of some *cis*-acting elements involved in class II expression in B cells and IFN-γ inducible cells, several lines of evidence indicate that these two modes of expression involve distinct *trans*-acting factors. First, certain mutations within the Z box inactivate IFN-γ induction but not B cell-specific expression.[82] Second, direct assays of DNA-protein interactions such as gel-shift electrophoresis and DNase I footprinting demonstrate clear differences between B cell and HeLa extracts in the regions of the octamer, X and Y boxes. Finally, results from fusion of human class II-negative mutant B cells to murine IFN-γ inducible cells provide genetic evidence for two distinct modes of expression.[74] Treatment of these hybrids with murine IFN-γ induced synthesis of murine class II mRNA and protein. However, human class II antigens were not induced by this treatment, although levels of human class I antigens did increase. In contrast, fusion of the class II-negative mutant cells with murine B cells did restore human class II expression.[75] This suggests that factors present in normal B cells, but not those induced by IFN-γ in inducible cells, can complement the defective B cell-specific factor in the mutant, even across species boundaries.

Although we and others have defined *cis*-acting elements within the X and Z boxes which are involved in IFN-γ induction, the precise mechanism of IFN-γ induction of class II genes is still unknown.[12,14] Undoubtedly the DNA-binding proteins which interact with the X box, pyrimidine tract, Z box and perhaps X2 box are involved, but they may or may not be quantitatively or qualitatively modified in response to IFN-γ administration. Alternative models can be envisioned in which IFN-γ induces modifications of *trans*-acting transcription factors which interact with these DNA-binding proteins but do not themselves bind DNA. It is clear, however, that IFN-γ induction of class II genes requires protein synthesis, as demonstrated by the ability of cycloheximide to block induction of class II mRNA.[83] This requirement for *de novo* synthesis of some *trans*-acting factor probably accounts for the slow kinetics of class II induction by IFN-γ.

VII. FUTURE PROSPECTS

The model of class II expression presented above leads to a number of hypotheses which can be tested by transfection assays of plasmids in which the DRα promoter has been systematically mutated. Mutations within putative positive elements should reduce transcription, whereas mutations in negative elements would be expected to result in increased expression. The precise DNA sequences required for binding of specific proteins, as well as the relationship between protein binding and transcriptional activity, can be assessed by examining the binding of nuclear proteins to DRα promoter sequences which have sustained specific mutations. In this way, the precise function and sequence-specificity of each DNA-binding protein can be eludicated. Studies in transgenic mice may yield additional insights into the elements which control tissue-specific expression.

Recently, purification of DNA-binding proteins by affinity chromatography has been described.[76] Cloning of genes encoding tissue-specific *trans*-acting factors, by sequencing purified DNA-binding proteins, is now in progress in several laboratories. The factors which bind to the X, X2, and Y boxes of the DRα promoter may be amenable to this approach. However, *trans*-acting transcription factors which are not DNA-binding proteins cannot be isolated in this way. The rescue of class II-negative B cells by transfection of libraries of DNA clones provides another means of cloning genes encoding *trans*-acting factors, whether or not they are DNA-binding proteins. These genes could be tested for their ability to complement *in vitro* the class II expression defect in cells from bare lymphocyte syndrome patients, and might eventually be useful for gene therapy of these patients.

The ultimate goal of studies on class II gene regulation is to isolate and characterize all of the *trans*-acting factors which influence transcription of class II genes. The power of this approach would be immeasurably enhanced by the development of an appropriately-regulated class II *in vitro* transcription assay. *In vitro* transcription of other mammalian genes has been achieved, but tissue-specificity has been extremely elusive in such systems.[77] Reconstitution *in vitro* of active transcriptional complexes which accurately mimic the different physiological modes of class II gene expression will have to await technological improvements in *in vitro* transcription.

REFERENCES

1. **Sullivan, K. E., Calman, A. F., Nakanishi, M., Tsang, S. Y., Wang, Y., and Peterlin, B. M.,** A model for the transcriptional regulation of MHC class II genes, *Immunol. Today,* 8, 289, 1987.
2. **Pober, J. S., Collins, T., Gimbrone, M. A., Jr., Cotran, R. S., Gitlin, J. D., Fiers, W., Clayberger, C., Krensky, A. M., Burakoff, S. J., and Reiss, C. S.,** Lymphocytes recognize human vascular endothelial and dermal fibroblast Ia antigens induced by recombinant immune interferon, *Nature (London),* 305, 726, 1983.
3. **Zlotnik, A., Shimonkevitz, R. P., Gefter, M. L., Kappler, J., and Marrack, P.,** Characterization of the γ-interferon-mediated induction of antigen-presenting ability in P388D1 cells, *J. Immunol.,* 131, 2814, 1983.
4. **Walker, E. B., Lanier, L. L., and Warner, N. L.,** Concomitant induction of the cell surface expression of Ia determinants and accessory cell function by a murine macrophage tumor cell line, *J. Exp. Med.,* 155, 629, 1982.
5. **de Preval, C., Lisowska-Grospierre, B. Loche, M., Griscelli, C., and Mach, B.,** A *trans*-acting class II regulatory gene unlinked to the MHC controls expression of HLA class II genes, *Nature (London),* 318, 291, 1985.
6. **Lanzavecchia, A.,** Antigen uptake and accumulation in antigen-specific B cells, *Immunol. Rev.,* 99, 39, 1987.
7. **Noelle, R., Krammer, P. H., Ohara, J., Uhr, J. W., and Vitetta, E. S.,** Increased expression of Ia antigens on resting B cells: additional role of B-cell growth factor, *Proc. Natl. Acad. Sci. U.S.A.,* 81, 6149, 1984.
8. **Reinherz, E. L., Kung, P. C., Pesando, J. A., Ritz, J., Goldstein, G., and Schlossman, S. F.,** Ia determinants on human T-cell subsets defined by monoclonal antibody: activation stimuli required for expression, *J. Exp. Med.,* 150, 1472, 1979.
9. **Suciu-Foca, N., Rubinstein, P., Popovic, M., Gallo, R. C., and King, D. W.,** Reactivity of HTLV-transformed human T-cell lines to MHC class II antigens, *Nature (London),* 312, 275, 1984.
10. **Rosa, F., Hatat, D., Abadie, A., Wallach, D., Revel, M., and Fellous, M.,** Differential regulation of HLA-DR mRNAs and cell surface antigens by interferon, *EMBO J.,* 2, 1585, 1983.
11. **Collins, T., Korman, A. J., Wake, C. T., Boss, J. M., Kappes, D. J., Fiers, W., Ault, K. A., Gimbrone, M. A., Jr., Strominger, J. L., and Pober, J. S.,** Immune interferon activates multiple class II major histocompatibility complex genes and the associated invariant chain gene in human endothelial cells and dermal fibroblasts, *Proc. Natl. Acad. Sci. U.S.A.,* 81, 4917, 1984.
12. **Boss, J. M. and Strominger, J. L.,** Regulation of a transfected human class II major histocompatibility complex gene in human fibroblasts, *Proc. Natl. Acad. Sci. U.S.A.,* 83, 9139, 1986.
13. **Tonnelle, C., DeMars, R., and Long, E. O.,** DOβ: a new β chain in HLA-D with a distinct regulation of expression, *EMBO J.,* 4, 2839, 1985.

14. **Tsang, S. Y., Nakanishi, M., and Peterlin, B. M.,** B-cell-specific and IFN-γ inducible regulation of the HLA-DRα genes, *Proc. Natl. Acad. Sci. U.S.A.,* 85, 8598, 1988.

15. **Skoskiewicz, M. J., Colvin, R. B., Schneeberger, E. E., and Russell, P. S.,** Widespread and selective induction of major histocompatibility complex-determined antigens in vivo by γ interferon, *J. Exp. Med.,* 162, 1645, 1985.

16. **Pujol-Borrell, R., Todd, I., Doshi, M., Bottazzo, G. F., Sutton, R., Gray, D., Adolf, G. R., and Feldmann, M.,** HLA class II induction in human islet cells by interferon-γ plus tumor necrosis factor or lymphotoxin, *Nature (London),* 326, 304, 1987.

17. **Morel, P. A., Manolagas, S. C., Provvedini, D. M., Wegmann, D. R., and Chiller, J. M.,** Interferon-γ-induced Ia expression in WEHI-3 cells is enhanced by the presence of 1,25-dihydroxyvitamin D_3, *J. Immunol.,* 136, 2181, 1986.

18. **Snyder, D. S., Beller, D. I., and Unanue, E. R.,** Prostaglandins modulate macrophage Ia expression, *Nature (London),* 299, 163, 1982.

19. **Bottazzo, G. F., Dean, B. M., McNally, J. M., MacKay, E. H., Swift, P. G. F., and Gamble, D. R.,** *In situ* characterization of autoimmune phenomena and expression of HLA molecules in the pancreas in diabetic insulinitis, *N. Engl. J. Med.,* 313, 353, 1985.

20. **Carroll, M. C., Katzmann, P., Alicot, E. M., Koller, B. H., Geraghty, D. E., Orr, H. T., Strominger, J. L., and Spies, T.,** Linkage map of the human major histocompatibility complex including the tumor necrosis factor genes, *Proc. Natl. Acad. Sci. U.S.A.,.,* 84, 8535, 1987.

21. **Hardy, D. A., Bell, J. I., Long, E. O., Lindsten, T., and McDevitt, H. O.,** Mapping of the class II region of the human major histocompatibility complex by pulsed-field gel electrophoresis, *Nature (London),* 323, 453, 1986.

22. **Lawrance, S. K., Smith, C. L., Srivastava, R., Cantor, C. R., and Weissman, S. M.,** Megabase-scale mapping of the HLA gene complex by pulsed field gel electrophoresis, *Science,* 235, 1387, 1987.

23. **Andersson, G., Larhammar, D., Widmark, E., Servenius, B., Peterson, P. A., and Rask, L.,** Class II genes of the major histocompatibility complex: organization and evolutionary relationship of the DRβ genes, *J. Biol. Chem.,* 262, 8748, 1987.

24. **Servenius, B., Rask, L., and Peterson, P. A.,** Class II genes of the major histocompatibility complex: the DOβ gene is a divergent member of the class II β gene family, *J. Biol. Chem.,* 262, 8759, 1987.

25. **Jonsson, A-K., Hyldig-Nielsen, J.-J., Servenius, B., Larhammar, D., Andersson, G., Joergensen, F., Peterson, P. A., and Rask, L.,** Class II genes of the major histocompatibility complex: Comparisons of the DQ and DXα and β genes, *J. Biol. Chem.,* 262, 8767, 1987.

26. **Gustafsson, K., Widmark, E., Jonsson, A-K., Servenius, B., Sachs, D. H., Larhammar, D., Rask, L., and Peterson, P. A.,** Class II genes of the major histocompatibility complex: evolution of the DP region as deduced from nucleotide sequences of the four genes, *J. Biol. Chem.,* 262, 8778, 1987.

27. **Spies, T., Sorrentino, R., Boss, J. M., Okada, K., and Strominger, J. L.,** Structural organization of the DR subregion of the human major histocompatibility complex, *Proc. Natl. Acad. Sci. U.S.A.,* 82, 5165, 1985.

28. **Okada, K. Boss, J. M., Prentice, H., Spies, T., Mengler, R., Auffray, C., Lillie, J., Grossberger, D., Strominger, J. L.,** Gene organization of DC and DX subregions of the human major histocompatibility complex, *Proc. Natl. Acad. Sci. U.S.A.,* 82, 3410, 1985.

29. **Rollini, P., Mach, B., and Gorski, J.,** Linkage map of three HLA-DR beta-chain genes: evidence for a recent duplication event, *Proc. Natl. Acad. Sci. U.S.A.,* 82, 7197, 1985.

30. **Rollini, P., Mach, B., and Gorski, J.,** Characterization of an HLA-DRβ pseudogene in the DRw52 supertypic group, *Immunogenetics,* 25, 336, 1987.

31. **O'Sullivan, D. M., Larhammar, D., Wilson, M. C., Peterson, P. A., and Quaranta, V.,** Structure of the human Ia-associated invariant (γ)-chain gene: identification of 5′ sequences shared with major histocompatibility complex class II genes, *Proc. Natl. Acad. Sci. U.S.A.,* 83, 4484, 1986.

32. **Peterlin, B. M., Hardy, K. J., and Larsen, A. S.,** Chromatin structure of the HLA-DRα gene in different functional states of major histocompatibility complex class II gene expression, *Mol. Cell. Biol.,* 7, 1967, 1987.

33. **Wang, Y. and Peterlin, B. M.,** Methylation patterns of HLA-DRα genes in six mononuclear cell lines, *Immunogenetics,* 24, 298, 1986.

34. **Wang, Y., Larsen, A. S., and Peterlin, B. M.,** A tissue-specific transcriptional enhancer is found in the body of the HLA-DRα gene, *J. Exp. Med.,* 166, 625, 1987.

35. **Sullivan, K. E. and Peterlin, B. M.,** Transcriptional enhancer in the HLA-DQ subregion, *Mol. Cell. Biol.,* 7, 3315, 1987.

36. **Calman, A. F. and Peterlin, B. M.,** Evidence for a *trans*-acting factor that regulates the transcription of class II major histocompatibility complex genes: genetic and functional analysis, *Proc. Natl. Acad. Sci. U.S.A.,* 85, 8830, 1988.

37. **Calman, A F. and Peterlin, B. M.,** Mutant human B cell lines deficient in class II major histocompatibility complex transcription, *J. Immunol.,* 139, 2489, 1987.

38. **Weintraub, H. and Groudine, M.,** Chromosomal subunits in active genes have an altered conformation, *Science,* 193, 848, 1976.

39. **Mills, F. C., Fisher, L. M., Kuroda, R., Ford, A. M., and Gould, H. J.,** DNAse I hypersensitive sites in the chromatin of human μ immunoglobulin heavy-chain genes, *Nature (London),* 306, 809, 1983.

40. **Parslow, T. G. and Granner, D. K.,** Structure of a nuclease-sensitive region inside the immunoglobulin kappa gene: evidence for a role in gene regulation, *Nucleic Acids Res.,* 11, 4775, 1983.

41. **Doerfler, W.,** DNA methylation and gene activity, *Annu. Rev. Biochem.,* 52, 93, 1983.

42. **Keshet, I., Lieman-Hurwitz, J., and Cedar, H.,** DNA methylation affects the formation of active chromatin, *Cell,* 44, 535, 1986.

43. **Peterlin, B. M., Gonwa, T. A., and Stobo, J. D.,** Expression of HLA-DR by a human monocyte cell line is under transcriptional control, *J. Mol. Cell. Immunol.,* 1, 191, 1984.

44. **Carrington, M. N., Salter, R. D., Cresswell, P., and Ting, J. P.-Y.,** Evidence for methylation as a regulatory mechanism in HLA-DRα gene expression, *Immunogenetics,* 22, 219, 1985.

45. **Levine, F. and Pious, D.,** Different roles for cytosine methylation in HLA class II gene expression, *Immunogenetics,* 22, 427, 1985.

46. **Reitz, M. S., Mann, D. L., Eiden, M., Trainor, C. D., and Clarke, M. F.,** DNA methylation and expression of HLA-DRα, *Mol. Cell. Biol.,* 4, 890, 1984.

47. **Sano, H., Compton, L. J., Shiomi, N., Steinberg, A. D., Jackson, R. A., and Sasaki, T.,** Low expression of human histocompatibility leukocyte antigen-DR is associated with hypermethylation of human histocompatibility leukocyte antigen-DRα gene region in B cells from patients with systemic lupus erythematosis, *J. Clin. Invest.,* 76, 1314, 1985.

48. **Bird, A. P.,** CpG-rich islands and the function of DNA methylation, *Nature (London),* 321, 209, 1986.

49. **Kelly, A. and Trowsdale, J.,** Complete nucleotide sequence of a functional HLA-DPβ gene and the region between the DPβ1 and DPα1 genes: comparison of the 5′ ends of HLA class II genes, *Nucleic Acids Res.,* 13, 1607, 1985.

50. **Mathis, D. J., Benoist, C. O., Williams, V. E., Kanter, M. R., and McDevitt, H. O.,** The murine Eα immune response gene, *Cell,* 32, 745, 1983.

51. **Miwa, K., Doyle, C., and Strominger, J. L.,** Sequence-specific interactions of nuclear factors with conserved sequences of human class II major histocompatibility complex genes, *Proc. Natl. Acad. Sci. U.S.A.,* 84, 4939, 1987.

52. **Friedman, R. L. and Stark, G. R.,** α-interferon-induced transcription of HLA and metallothionein genes containing homologous upstream sequences, *Nature (London),* 314, 637, 1985.

53. **Dorn, A., Durand, B., Marfing, C., Le Meur, M., Benoist, C., and Mathis, D.,** Conserved major histocompatibility complex class II boxes — X and Y — are transcriptional control elements and specifically bind nuclear proteins, *Proc. Natl. Acad. Sci. U.S.A.,* 84, 6249, 1987.

54. **Breathnach, R. and Chambon, P.,** Organization and expression of eucaryotic split genes coding for proteins, *Annu. Rev. Biochem.,* 50, 349, 1981.

55. **Fried, M. and Crothers, D. M.,** Equilibria and kinetics of lac repressor-operator interactions by polyacrylamide gel electrophoresis, *Nucleic Acids Res.,* 9, 6505, 1981.

56. **Singh, H., Sen, R., Baltimore, D., and Sharp, P. A.,** A nuclear factor that binds to a conserved sequence motif in transcriptional control elements of immunoglobulin genes, *Nature (London),* 319, 154, 1986.

57. **Sherman, P. A., Basta, P. V., and Ting, J. P.-Y.,** Upstream DNA sequences required for tissue-specific expression of the HLA-DRα gene, *Proc. Natl. Acad. Sci. U.S.A.,* 84, 4254, 1987.

58. **Staudt, L. M., Singh, H., Sen, R., Wirth, T., Sharp, P. A., and Baltimore, D.,** A lymphoid-specific protein binding to the octamer motif of immunoglobulin genes, *Nature (London),* 323, 640, 1986.

59. **Hooft van Huijsduijnen, R. A., Bollekens, J., Dorn, A., Benoist, C., and Mathis, D.,** Properties of a CCAAT box-binding protein, *Nucleic Acids Res.,* 15, 7265, 1987.

60. **Dorn, A., Bollekens, J., Staub, A., Benoist, C., and Mathis, D.,** A multiplicity of CCAAT box-binding proteins, *Cell,* 50, 863, 1987.

61. **Galas, D. and Schmidtz, A.,** DNase footprinting: a simple method for the detection of protein-DNA binding specificity, *Nucleic Acids Res.,* 5, 3157, 1978.

62. **Jones, K. A., Yamamoto, K. R., and Tjian, R.,** Two distinct transcription factors bind to the HSV thymidine kinase promoter in vitro, *Cell,* 42, 559, 1985.

63. **Khoury, G. and Gruss, P.,** Enhancer elements, *Cell,* 33, 313, 1983.

64. **Emorine, L., Kuehl, M., Weir, L., Leder, P., and Max, E. E.,** A conserved sequence in the immunoglobulin Jκ-Cκ intron: possible enhancer element, *Nature (London),* 304, 447, 1983.

65. **Gillies, S. D., Morrison, S. L., Oi, V. T., and Tonegawa, S.,** A tissue-specific transcription enhancer element is located in the major intron of a rearranged immunoglobulin heavy chain gene, *Cell,* 33, 717, 1983.

66. **Payvar, F., DeFranco, D., Firestone, G. L., Edgar, B., Wrange, O., Okret, S., Gustafsson, J. A., and Yamamoto, K. R.,** Sequence-specific binding of glucocorticoid receptor to MTV DNA at sites within and upstream of the transcribed region, *Cell,* 35, 381, 1983.

67. **Goodbourn, S., Zinn, K., and Maniatis, T.,** Human β-interferon gene expression is regulated by an inducible enhancer element, *Cell,* 41, 509, 1985.

68. **Parslow, T. G., Blair, D. L., Murphy, W. J., and Granner, D. K.,** Structure of the 5' ends of immunoglobulin genes: a novel conserved sequence, *Proc. Natl. Acad. Sci. U.S.A.,* 81, 2650, 1984.

69. **Hadam, M. R., Dopfer, R., Dammer, G., Peter, H-H., Schlesier, M., Mueller, C., and Niethammer, D.,** Defective expression of HLA-D region determinants in children with congenital agammaglobulinemia and malabsorption: a new syndrome, in *Histocompatibility Testing,* Albert, E. D., Baur, M. P., and Mayr, W. R., Eds., Springer-Verlag, New York, 1984, 35.

70. **Gladstone, P. and Pious, D.,** Identification of a *trans*-acting function regulating HLA-DR expression in DR-negative B-cell variant, *Somatic Cell Genet.,* 6, 285, 1980.

71. **Accolla, R. S.,** Human B cell variants immunoselected against a single Ia antigen subset have lost expression of several Ia antigen subsets, *J. Exp. Med.,* 157, 1053, 1983.

72. **Accolla, R. S., Jotterand-Bellomo, M., Scarpellino, L., Maffei, A., Carra, G., and Guardiola, J.,** aIr-1, a newly found locus on mouse chromosome 16 encoding a *trans*-acting activator factor for MHC class II gene expression, *J. Exp. Med.,* 164, 369, 1986.

73. **Guardiola, J., Scarpellino, L., Carra, G., and Accolla, R. S.,** Stable integration of mouse DNA into Ia-negative human B-lymphoma cells causes reexpression of the human Ia-positive phenotype, *Proc. Natl. Acad. Sci. U.S.A.,* 83, 7415, 1986.

74. **Maffei, A., Scarpellino, L., Bernard, M., Carra, G., Jotterand-Bellomo, M., Juardiola, J., and Accolla, R. S.,** Distinct mechanisms regulate MHC class II gene expression in B cells and macrophages, *J. Immunol.,* 139, 942, 1987.

75. **Accolla, R. S., Carra, G., and Guardiola, J.,** Reactivation by a *trans*-acting factor of human major histocompatibility complex Ia gene expression in interspecies hybrids between an Ia-negative human B-cell variant and an Ia-positive mouse B-cell lymphoma, *Proc. Natl. Acad. Sci. U.S.A.,* 82, 5145, 1985.

76. **Kadonaga, J. T. and Tjian, R.,** Affinity purification of sequence-specific DNA binding proteins, *Proc. Natl. Acad. Sci. U.S.A.,* 83, 5889, 1986.

77. **Sen, R. and Baltimore, D.,** *In vitro* transcription of immunoglobulin genes in a B-cell extract: effect of enhancer and promoter sequences, *Mol. Cell. Biol.,* 7, 1989, 1987.

78. **Ohlsson, H. and Edlund, T.,** Sequence-specific interactions of nuclear factors with the insulin gene enhancer, *Cell,* 45, 35, 1985.

79. **Yates, J. L., Warren, N., and Sugden, B.,** Stable replication of plasmids derived from Epstein-Barr virus in various mammalian cells, *Nature (London),* 313, 812, 1985.

80. **DeLorbe, W. J., Luciw, P. A., Goodman, H. M., Varmus, H. E., and Bishop, J. M.,** Molecular cloning and characterization of avian sarcoma virus circular DNA molecules, *J. Virol.,* 36, 50, 1980.

81. **Calman, A. and Peterlin, B. M.,** unpublished results.

82. **Tsang, S. Y., Nakanishi, M., and Peterlin, B. M.,** unpublished results.

83. **Tsang, S. Y. and Peterlin, B. M.,** unpublished results.

APPENDIX

FIRST DOMAIN SEQUENCES OF DRβ, DQα, AND DQβ CHAIN ALLELES

DRβ First Domain Amino Acid Sequence

```
                          6        10            20            30            40            50
                 G D T R P R F L W Q  L K F E C H F F N G  T E R V R L L E R C  I Y N Q E E S V R F  D S D V G E Y R A V
DR1βI (1)        - - - - - - - - - -  - - - - - - - - - -  - - - - - - - - - -  - - - - - - - - - -  - - - - - - - - - -
DR1 NASC (2,3)   - - - - - - - - - -  - - - - - - - - - -  - - - - - - - - - -  - - - - - - - - - -  - - - - - - - - - -
DR1 CETUS (4)    - - - - - - - - - -  - - - - - - - - - -  - - - - - - - - - -  - - - - - - - - - -  - - - - - - - - - -
DR2 Dw2βI (5)    - - - - - - - - - Q  - D - Y - - H - - -  - - - - - F - D - Y  - - F - H - Y - - -  - - - - - - F - - -
DR2 Dw12βI (5)   - - - - - - - - - Q  - D - Y - - H - - -  - - - - - F - D - Y  - - F - H - Y - N -  - L - - - - - - - -
DR2 AZHβI (6)    - - - - - - - - - -  - D - Y - - H - - -  - - - - - F - D - Y  - - F - H - Y - N -  - - - - - - - - - -
DR3βI (7)        - - - - - - - - - -  E Y - S T S - - - -  - - - - - F - Y - D  - - F H H - Y - - -  - F - - - - - - - -
DR3 DQw4 (8)     - - - - - - - - - -  E Y - S T S - - - -  - - - - - F - D - Y  - - F - H - Y - - -  - - - - - - - - - -
DR4 Dw4βI (9,10) - - - - - - - - - -  E - - V - H - - - -  - - - - - F - D - Y  - - F H H - Y - Y -  - - - - - - F - - -
DR4 Dw10βI (10)  - - - - - - - - - -  E - - V - H - - - -  - - - - - F - D - Y  - - F H H - Y - Y -  - - - - - - F - - -
DR4 Dw13βI (11)  - - - - - - - - - -  E - - V - H - - - -  - - - - - F - D - Y  - - F H H - Y - Y -  - - - - - - - - - -
DR4 Dw14βI
    (10,11)      - - - - - - - - - -  E - - V - H - - - -  - - - - - F - D - Y  - - F - H - Y - Y -  - - - - - - - - - -
DR4 Dw15βI (10)  - - - - - - - - - -  E - - V - H - - - -  - - - - - F - D - Y  - - F H H - Y - Y -  - - - - - - - - - -
DR4 KT2 (12)     - - - - - - - - - -  E - - V - H - - - -  - - - - - F - D - Y  - - F H H - Y - Y -  - - - - - - - - - -
DR4 JHA (13)     - - - - - - - - - -  E - - V - H - - - -  - - - - - F - D - Y  - - F H H - Y - Y -  - - - - - - - - - -
DR4 CETUS (4)    - - - - - - - - - -  E - - V - H - - - -  - - - - - F - D - Y  - - F H H - Y - Y -  - - - - - - - - - -
DR4 WARAO (13)   - - - - - - - - - -  E - - V - H - - - -  - - - - - F - D - Y  - - F H H - Y - Y -  - - - - - - - - - -
DR5 Dw5βI (14)   - - - - - - - - - -  E Y - S T S - - - -  - - - - - F - D - Y  - - F H H - Y - Y -  - - - - - - - - - -
DR5 JVM (15)     - - - - - - - - - -  E Y - S T S - - - -  - - - - - F - D - Y  - - F H H - Y - Y -  - - - - - - - - - -
DRw6aβI (16)     - - - - - - - - - -  E Y - S T S - - - -  - - - - - F - D - Y  - - F H H - Y - Y -  - - - - - - - - - -
DRw6bβI (16)     - - - - - - - - - -  E Y - S T S - - - -  - - - - - F - D - Y  - - F H H - Y - Y -  - - - - - - - - - -
DRw6 AMALA (17)  - - - - - - - - - -  E Y - S T S - - - -  - - - - - F - D - Y  - - F H H - N - N -  - - - - - - - - - -
DR7βI (18,19)    - - - Q - - - - - -  E Y - G - Y K - - -  - - - - - Q F - L -  - F - L - - F - - -  - - - - - - F - - -
DRw8βI (15)      - - - - - - - - - -  E Y - S T G - Y - -  - - - - - F - D - Y  - - F - H - Y - N -  - - - - - - - - - -
DRw8 TAB (4)     - - - - - - - - - -  E Y - S T G - Y - -  - - - - - F - D - Y  - - F H H - Y - Y -  - - - - - - - - - -
DRw8 SPL (4)     - - - Q - - - - - -  E Y - S T G - Y - -  - - - - - F - D - Y  - - F H H - Y - Y -  - - - - - - - - - -
DRw9βI (15,19)   - - - - - - - - - K  E - - D - - - - - -  - - - - - Y - H G H  - - - - - - Y A - -  - - - - - - - - - -
DRw10βI (2,7)    - - - - - - - - - -  E - - V - - - - - -  - - - - - - - R - -  R V H R - - Y A - Y  - - - - - - - - - -
DRw12 (20)       - - - - - - - - - -  E Y - S T G - - - L  - - - - - - - H - -  - H - - - - L L - -  - - - - - - - - - -

DR2 Dw12βIII(5)  - - - - - - - - - -  - - - P R - - - - -  - - - - - F - D - Y  - F - D - - F - - F  L - - - - - F - - -
DR2 Dw2βIII(5)   - - - - - - - - - -  - - - P R - - - - -  - - - - - F - D - Y  - F - D - - Y A - -  - - - - - - F - - -
DR2 AZHβIII (6)  - - - Q - - - - - -  E L - P R - - - - -  - - - - - Y - D - Y  - F - D - - L A - Y  - - - - - - F - - -
DRw52a (16)      - - - - - - - - - -  E L - R S - - - - -  - - - - - Y - D - Y  - H Y H - - F L - -  - - - - - - F - - -
DRw52b (16)      - - - - - - - - - -  E L - - S - - - - -  - - - - - H - Y - Y  - H Y H - - Y A - -  - - - - - - F - - -
DRw52c (21)      - - - Q - - - - - -  - E - - S - - - - -  - - - - - F - Y - Y  - H Y - - - L A - -  - - - - - - - - - L
DRw53 (9,22)     - - - - - - - - - -  - - - A - C - - - L  - - - - - W N - I Y  - - - - - - Y N - Y  - - L - - - - - - Q
```

DRβ First Domain Amino Acid Sequence (continued)

```
                        58                60            70              80              90
                        T E L G R P D A E Y W N S Q K D L L E Q R R A A V D T Y C R H N Y G V G E S F T V Q R R
DR1β1                   - - - - - - - - - - - - - - - - - - - - - - - - - - - - - - - - - - - - - - - - - - - -
DR1 NASC                - - - - - - - - - - - - - - - - - - - - - - - - - - - - - - - - - - - - - - - - - - - -
DR1 CETUS               - - - - - - - - - - - - - - - - - - - - E - - - - - - - - - - - - - - - - - - - - - - -
DR2 Dw2β1               - - - - - - - - - - - - - - - - F - - D - - - - - - - - - - - - - - - - - - - - - A V - -
DR2 Dw12β1              - - - - - - - - - - - - - - - - F - - D - - - - - - - - - - - - - - - - - - - - - - V - -
DR2 AZHβ1               - - - - - - - - - - - - - - - - I - - - - - A - - - - - - - - - - - - - - - - - - - V - -
DR3β1                   - - - - - - - - - - - - - - - - - - - - K - - - - N - - - - - - - - - - A - V - - - - - - -
DR3 DQw4                - - - - - - - - - - - - - - - - - - - - K - G R - N - - - - - - - - - - - - V - - - - - - -
DR4 Dw4β1               - - - - - - - - - - - - - - - - - - - - K - G R - - - - - - - - - - - - - - V - - - - - - -
DR4 Dw10β1              - - - - - - - - - - - - - - - - - I - D E - - E - - - - - - - - - - - - - - V - - - - - - -
DR4 Dw13β1              - - - - - - - - - - - - - - - - - I - - - - - E - - - - - - - - - - - - - - V - - - - - - -
DR4 Dw14β1              - - - - - - - - - - - - - - - - - I - - - - - E - - - - - - - - - - - - - - V - - - - - - -
DR4 Dw15β1              - - - - - - - - - - - - - - - - - I - - - - - E - - - - - - - - - - - - - - V - - - - - - -
DR4 KT2                 - - - - - - S - - - - - - - - - - - - - - - - - - - - - - - - - - - - - - - V - - - - - - -
DR4 JHA                 - - - - - - - - - - - - - - - - - - - - - - - - - - - - - - - - - - - - - - - - - - - - -
DR4 CETUS               - - - - - - - - - - - - - - - - - - - - - - E E - - - - - - - - - - - - - - - - - - - - -
DR4 WARAO               - - - - - E H - - - - - - - - - - - - - - - E E - - - - - - - - - - - - - - - - - - - - -
DR5 Dw5β1               - - - - - E - - - - - - - - - - F - - D E - - E - - - - - - - - - - - - - V - - - - - - -
DR5 JVM                 - - - - - - - - - - - - - - - - - - - D - - - - - V - - - - - - - - - - - - V - - - - - - -
DRw6aβ1                 - - - - - - - - - - - - - - - - I - - D - - - - - - - - - - - - - - - - - - V - - - - - - -
DRw6bβ1                 - - - - - - A - - - - - - - - - F - - D E - G E - V - - - - - - - - - - - - V - - - - - - -
DRw6 AMALA              - - - - - - - - H - - - - - - - I - - R E - - E - - - - - - - - - - - - - - - - - - - - -
DR7β1                   - - - - - - V - S - - - - - - - I - - D R - - E - - - - - - - - - - - - - - V - - - - - - -
DRw8β1                  - - - - - - - - - - - - - - - - I - - D - - G L - - - - - - - - - - - - - - - - - - - - - -
DRw8 TAB                - - - - - - - - - - - - - - - - I - - D - - - L - - - - - - - - - - - - - - - - - - - - - -
DRw8 SPL                - - - - - - - - - - - - - - - - F - - D - - G L - - - - - - - - - - - - - - - - - - - - - -
DRw9β1                  - - - - - - V - S - - - - - - - F - - D R - - E - V - - - - - - - - - - A V - - - - - - - -
DRw10β1                 - - - - - - V - S - - - - - - - I - - R - - - - - - - - - - - - - - - - - - - - - - - - - -
DRw12                   - - - - - - - - S - - - - - - - I - - D - - - - - - - - - - - - - - - - - - - - - - - - - -
                                                                                                                
DR2 Dw2βIII             - - - - - - - - - - - - - - - - I - - - - - - - - - - - - - - - - - - - - - - - - - - - -
DR2 Dw12βIII            - - - - - - V - - - - - - - - - I - - - A - - - - - - - - - - - - - - - - - - A V - - - - -
DR2 AZHβIII             - - - - - - - - - - - - - - - - F - - - A - - - - - - - - - - - - - - - - - - - V - - - - -
DRw52a                  R - - - - - - - - - - - - - - - - - - D - - K R - N - - - - - - - - - - - - - - - - - - - -
DRw52b                  - - - - - - V - - - - - - - - - - - - - - - R G - N - - - - - - - - - - - - V - - - - - - -
DRw52c                  - - - - - - V - S - - - - - - - - - - Q - - K G - N - - - - - - - - - - - - - - - - - - - -
DRw53                   - - - - - - - - - - - - - - - - - - - R - - - E - - - - - - - - Y - - - - - - V - - - - - -
```

DQα First Domain Amino Acid Sequences

```
                               1        10         20         30         40
DR1  DQw1.1 (23-25)       E D I V A D H V A S  C G V N L Y Q F Y G  P S G Q Y T H E F D  G D E E F Y V D L E   N
DR2  DQw1.2 (23-26)       - - - - - - - - - -  - - - - - - - - - -  - - - - - - - Q - -  - - - Q - - - - - -   -
DRw6 DQw1.18(17,23,25)    - - - - - - - - - -  - - - - - - - - - -  - - - - - - F - - -  - - - Q - - - - - -   -
DR5  DQw3 (27)            - - - - - - - - - Y  - - - S - - - - - -  - - - - - - - - - -  - - - - - G - - - -   -
DRw8 DQw4 (23)            - - - - - - - - - Y  - - - S - - - - - -  - - - - - - - - - -  - - - - - G - - - -   -
DR4  DQw3 (23-26)         - - - - - - - - - Y  - - - S - - - - - -  - - S - - - - - - -  - - - - - - - - - -   -
DR7  DQw2 (28)            - - - - - - - - - Y  - - - S - - - - - -  - - F - - - - - - -  - - - - - - - - - -   -
```

```
                              41        50         60         70         80
DR1  DQw1.1               R K E T A W R P E   F S K F G G F D P Q  G A L R N M A V A K  H N L N I M I K R Y  N S T A A T N
DR2  DQw1.2               - - - - - - - - -   - - - - - - - - - -  - - - - - - - - - -  - - - - - - - - - -  - - - - - - -
DRw6 DQw1.18              K - - - - - - - -   - - - - - - - - - -  - - - - - - - - - -  - - - - - - - - - -  - - - - - - -
DR5  DQw3                 - - V - - C L - V   L R Q - R [ ] - - -  F - - T - I - L - -  - - S - L - - - - -  - - - - - - -
DRw8 DQw4                 - - V - - C L - V   L R Q - R [ ] - - -  F - - T - I - T - -  - - L - - S - L - -  - - - - - - -
DR4  DQw3                 - - V - - - Q - L   - R R - R [ ] - - -  F - - T - I - L - -  V - - L - - - - - -  - - S - - - -
DR7  DQw2                 - - V - K L - L -   H R L R [ ] - - - -  F - - T - I - L - -  L - - L - - - S - -  - - - - - - -
```

DQβ First Domain Amino Acid Sequences

Residues 1–50:

```
                                 10          20          30          40          50
DR1  DQw1.1   (1)      R D S P E D F V Y Q  F K G L C Y F T N G  T E R V R G V T R H  I Y N R E E Y V R F  D S D V G V Y R A V
DR2  DQw1.2   (6)      - - - - - - - - - F  - - - M - - - - - -  - - - - - L Y - - Y  - - - - - - - - A -  - - - - - - - - - -
DR2  DQw1.12  (6)      - - - - - - - - - L  - - A M - - - - - -  - - - - - - Y - - Y  - - - - - - - - A -  - - - - - - - - - -
DR2  DQw1.AZH (6)      - - - P - - - - - -  - - A M - - - - - -  - - - - - Y D - - Y  - - - - - - - - A -  - - - - - - - - - -
DRw6 DQw1.9   (23)     - - - - - - - - - -  - - - - - - - - - -  - - - - - - - - - -  - - - - - - - - - -  - - - - - - - - - -
DRw6 DQw1.18 (17,23)   - - - - - - - - - -  - - - M - - - - - -  - - - L - L - - - -  - - - - - - - - A -  - E F - - - - - - -
DRw6 DQw1.19  (29)     - - - - - - - - - -  - - - M - - - - - -  - - - L - L - - - -  - - - - - - - - A -  - - - - - - - - - -

DR3  DQw2     (30)     - - - - - - - - - -  - - - M - - - - - -  - - - - - L S - - S  - - - - - - - - I -  - E - - - - - - - -

DR4  DQw3.1   (23)     - - - - - - - - - -  - - A M - - - - - -  - - - - - Y Y - - Y  - - - - - - - - A -  - - - - - - - - - -
DR4  DQw3.2   (31)     - - - - - - - - - -  - - - M - - - - - -  - - - - - L Y - - Y  - - - - - - - - A -  - E - - - - - - - -
DR7  DQw3.3   (32)     - - - - - - - - - -  - - - M - - - - - -  - - - - - L Y - - Y  - - - - - - - - A -  - - - - - - - - - -

DR4  DQw4     (10)     - - - - - - F - - -  - - - M - - - - - -  - - - - - - Y - - Y  - - L - - - - - A -  - - - - - - - - - -
DRw8 DQw4     (23)     - - - - - - F - - -  - - - M - - - - - -  - - - - - - Y - - Y  - - - - - - - - A -  - - - - - - - - - -
```

Residues 51–94:

```
                                 60          70          80          90
DR1  DQw1.1            T P Q G R P V A E Y  W N S Q K E V L E G  A R A S V D R V C R  H N Y E V A Y R G I  L Q R R
DR2  DQw1.2            - - - - - - - D - -  - - - - - - - - - -  T - - - E L - T - -  - - - - - - - F - -  - - - -
DR2  DQw1.12           - - - - - - - D - -  - - - - - D I R - -  T - - - E L - T - -  - - - - - - - F - -  - - - -
DR2  DQw1.AZH          - - - - - - - S - -  - - - - - - - - - -  T - - - E L - T - -  - - - - - - - - - -  - - - -
DRw6 DQw1.9            - - - - - - - D - -  - - - - - - - - - -  - - - - E L - - - -  - - - - - - - - - -  - - - -
DRw6 DQw1.18           - - - - - - - D - -  - - - - - D I R - -  T - - - E L - T - -  - - - - - - F - - -  - - - -
DRw6 DQw1.19           - - - - - - - - - -  - - - - - - - - - -  - - - - E L - T - -  - - - - - - G - - -  - - - -

DR3  DQw2         L L  - - - - - - - A - -  - - - - - - - - - -  - K - - A - - - - -  Q L E L - - - - - -  L - T T

DR4  DQw3.1     L - -  - - - - P - - D - -  - - - - - D I R - -  T - - - E L - T - -  Q L E L - - - - - -  L - T T
DR4  DQw3.2     L - -  - - - - P - - A - -  - - - - - - - R - -  T - - - E L - T - -  Q L E L - - - - - -  L - T T
DR7  DQw3.3     L - -  - - - - P - - D - -  - - - - - - - R - -  T - - - E L - T - -  Q L E L - - - - - -  L - T T

DR4  DQw4       L - -  - - - - L - - D - -  - - - - - D I - - -  - - - - - - - T T - -  Q L E L - - - - - -  L - T T
DRw8 DQw4       L - -  - - - - L - - D - -  - - - - - D I - - -  - - - - - - - T T - -  Q L E L - - - - - -  L - T T
```

Note: DQα and DQβ chain alleles are tentatively designated by representative haplotypes on which they are commonly found. However, DQα and DQβ chains of identical structure may also be found on other haplotypes. These are summarized below. The examples given refer only to first domain amino acid sequences; silent nucleotide changes are found between some structurally identical alleles from different haplotypes.

DQα Chains   **Ref.**

DR1 DQw1.1 = DRw6 DQw1.9   33
DR2 DQw1.2 = DR2 DQw1.AZH = DRw6 DQw1.19   23, 25
DRw6 DQw1.18 = DRw8 DQw1 = DR5 DQw1   23, 25
DRw5 DQw3 = DR3 DQw2 = DRw14 DQw3 = DR2 DQw3   17, 23—25, 34, 35
DRw8 DQw4 = DR3 DQw4   36
DR4 DQw3 = DR9 DQw3[a]   23—25, 28
DR7 DQw2 = DR7 DQw3   28, 32

DQβ Chains   **Ref.**

DR1 DQw1.1 = DRw10 DQw1.1   34
DR3 DQw2 = DR7 DQw2   19, 30
DR4 DQw3.1 = DR5 DQw3.1 = DRw14 DQw3.1   17, 23
DR7 DQw3.3 = DR9 DQw3.3   23, 32
DRw8 DQw4 = DR3 DQw4   23, 36
DR4 DQw3.2 = DR2 DQw3.2   34

[a] A DR9 DQα chain has also been reported which contains a deletion of 5 codons (positions −1 to +4) (Reference 28).

REFERENCES

1. **Tonelle, C., DeMars, R., and Long, E. O.,** DOβ: a new β chain gene in HLA-D with a distinct regulation of expression, *EMBO J.,* 4, 2839, 1985.
2. **Merryman, P., Gregersen, P. K., Lee, S., Silver, J., Nunez-Roldan, A., Crapper, R., and Winchester, R. J.,** Nucleotide sequence of a DRw10 β chain cDNA clone, *J. Immunol.,* 140, 2447, 1988.
3. **Hurley, C. K., Ziff, B. L., Silver, J., Gregersen, P. K., Hartzman, R., and Johnson, A. H.,** Polymorphism of the HLA-DR1 haplotype in American blacks, *J. Immunol.,* 140, 4019, 1988.
4. **Erlich, H. A.,** personal communication.
5. **Wu, S., Saunders, T. L., and Bach, F. H.,** Polymorphism of human Ia antigens generated by reciprocal intergenic exchange between two DRβ loci, *Nature (London),* 324, 676, 1986.
6. **Lee, B. S. M., Rust, N. A., McMichael, A. J., and McDevitt, H. O.,** HLA-DR2 subtypes for an additional supertypic family of DRβ alleles, *Proc. Natl. Acad. Sci. U.S.A.,* 84, 4591, 1987.
7. **Gustafsson, K., Wiman, K., Emmoth, E., Larhammar, D., Bohme, J., Hyldig-Nielsen, J. J., Ronne, H., Peterson, P., and Rask, L.,** Mutations and selection in the generation of class II histocompatibility antigen polymorphism, *EMBO J.,* 3, 1655, 1984.
8. **Hurley, C. K., Gregersen, P. K., Gorski, J., Steiner, N., Robbins, F. M., Hartzman, R., Johnson, A. H., and Silver, J.,** The DR3(w18), DQw4 haplotype differs from DR3 (w17), DQw2 haplotypes at multiple class II loci, *Human Immunol.,* 25, 37, 1989.
9. **Spies, T., Sorrentino, R., Boss, J. M., Okada, K., and Strominger, J. L.,** Structural organization of the DR subregion of the human major histocompatibility complex, *Proc. Natl. Acad. Sci. U.S.A.,* 82, 5165, 1985.
10. **Gregersen, P. K., Shen, M., Song, Q., Merryman, P., Degar, S., Seki, T., Maccari, J., Goldberg, D., Murphy, H., Schwenzer, J., Wang, C. Y., Winchester, R. J., Nepom, G. T., and Silver, J.,** Molecular diversity of HLA-DR4 haplotypes, *Proc. Natl. Acad. Sci. U.S.A.,* 83, 2642, 1986.
11. **Cairns, J. S., Curtsinger, J. M., Dahl, C. D., Freeman, S., Alter, B. J., and Bach, F.,** Sequence polymorphism of HLA-DRβ1 alleles relating to T cell recognized determinants, *Nature (London),* 317, 166, 1985.
12. **Gregersen, P. K., Goyert, S. M., Song, Q.-L., and Silver, J.,** Microheterogeneity of HLA-DR4 haplotypes: DNA sequence analysis of LD"KT2" and LD"TAS" haplotypes, *Human Immunol.,* 19, 287, 1987.
13. **Gregersen, P. K.,** personal communication.
14. **Tieber, V. L., Abruzzini, L. F., Didier, D. K., Schwartz, B. D., and Rotwein, P.,** Complete characterization and sequence of an HLA class II β chain from the DR5 haplotype, *J. Biol. Chem.,* 261, 2738, 1986.
15. **Bell, J. I., Denney, D., Foster, Belt, T., Todd, J. A., and McDevitt, H. O.,** Allelic variation in the DR subregion of the human major histocompatibility complex, *Proc. Natl. Acad. Sci. U.S.A.,* 84, 6234, 1987.
16. **Gorski, J. and Mach, B.,** Polymorphism of human Ia antigens: gene conversion between two DRβ loci results in a HLA-D/DR specificity, *Nature (London),* 322, 67, 1986.
17. **Kao, H.-T., Gregersen, P. K., Tang, J. C., Takahashi, T., Wang, C. Y., and Silver, J.,** Molecular analysis of the HLA class II genes in two DRw6-related haplotypes, DRw13 DQw1 and DRw14 DQw3, *J. Immunol.,* 142, 1743, 1989.
18. **Gregersen, P. K., Moriuchi, T., Karr, R. W., Obata, F., Moriuchi, J., Maccari, J., Goldberg, D., Winchester, R. J., and Silver, J.,** Polymorphisms of HLA-DR β chains in DR4, −7, and −9 haplotypes: implications for the mechanisms of allelic variation, *Proc. Natl. Acad. Sci. U.S.A.,* 83, 9149, 1986.
19. **Karr, R., Gregersen, P. K., Obata, F., Goldberg, D., Maccari, J., Alber, C., and Silver, J.,** Analysis of DRβ and DQβ chain cDNA clones from a DR7 haplotype, *J. Immunol.,* 137, 2886, 1986.
20. **Navarrete, C., Seki, T., Miranda, A., Winchester, R. W., and Gregersen, P. K.,** DNA sequence analysis of the HLA-DRw12 allele, *Hum. Immunol.,* 25, 51, 1989.
21. **Gorski, J., Eckels, D. D., Thiercy, J. M., Ucla, C., and Mach, B.,** Sequence analysis of the DRw13 β chain gene: the Dw19 specificity may be encoded by the DRβIII locus, in *Immunogenetics and Histocompatibility, Vol. 2,* Dupont, B., Ed., Springer Verlag, New York, in press.
22. **Curtsinger, J. M., Hilden, J. M., Cairns, J. S., and Bach, F. H.,** Evolutionary and genetic implications of sequence variation in two nonallelic HLA-DR β chain cDNA sequences, *Proc. Natl. Acad. Sci. U.S.A.,* 84, 209, 1987.
23. **Todd, J. A., Bell, J. I., and McDevitt, H. O.,** HLA-DQβ gene contributes to susceptibility and resistance to insulin dependent diabetes mellitus, *Nature (London),* 329, 599, 1987.
24. **Saiki, R. K., Bugawan, T. L., Horn, G. T., Mullis, K. B., and Erlich, H. A.,** Analysis of enzymatically amplified β-globin and HLA-DQα DNA with allele-specific oligonucleotide probes, *Nature (London),* 324, 163, 1986.

25. **Horn, G. T., Bugawan, T. L., Long, C. M., and Erlich, H. A.,** Allelic sequence variation of the HLA-DQ loci: relationship to serology and to insulin-dependent diabetes susceptibility, *Proc. Natl. Acad. Sci. U.S.A.,* 85, 6012, 1988.

26. **Auffray, C., Lillie, J. W., Arnot, D., Grossberger, D., Kappes, D., and Strominger, J.,** Isotypic and allotypic variation of human class II histocompatibility antigen α-chain genes, *Nature (London),* 308, 372, 1984.

27. **Schiffenbauer, J. D., Didier, K., Klearman, M., Rice, K., Shuman, S., Tiber, V. L., Kittlesen, D. J., and Schwartz, B. D.,** Complete sequence of the HLA DQα and DQβ cDNA from a DR5/DQw3 cell line, *J. Immunol.,* 139, 228, 1987.

28. **Moriuchi, J., Moriuchi, T., and Silver, J.,** Nucleotide sequence of an HLA-DQα chain derived from a DRw9 cell line: genetic and evolutionary implications, *Proc. Natl. Acad. Sci. U.S.A.,* 82, 3420, 1985.

29. **Turco, E., Care, A., Compagnone-Post, P., Robinson, C., Cascino, I., and Trucco, M.,** Allelic forms of the alpha- and beta-chain genes encoding DQw1-positive heterodimers, *Immunogenetics,* 26, 282, 1987.

30. **Boss, J. M. and Strominger, J. L.,** Cloning and sequence analysis of the human major histocompatibility complex gene DC-3β, *Proc. Natl. Acad. Sci. U.S.A.,* 81, 5199, 1984.

31. **Larhammar, D., Hyldig-Nielsen, J. J., Servenius, B., Andersson, G., Rask, L., and Peterson, P.,** Exon-intron organization and complete nucleotide sequence of a human major histocompatibility antigen DCβ gene, *Proc. Natl. Acad. Sci. U.S.A.,* 84, 209, 1987.

32. **Song, Q. L., Gregersen, P. K., Karr, R. W., and Silver, J.,** Recombination between DQα and DQβ genes generates human histocompatibility leukocyte antigen class II haplotype diversity, *J. Immunol.,* 139, 2993, 1987.

33. **Todd, J.,** personal communication.

34. **Schenning, L., Larhammar, D., Bill, P., Wiman, K., Jonsson, A.-C., Rask, L., and Peterson, P.,** Both α and β chain of HLA-DC class II histocompatibility antigens display extensive polymorphism in their amino terminal domains, *EMBO J.,* 3, 447, 1984.

35. **Liu, C. P., Bach, F. H., and Wu, S.,** Molecular studies of a rare DR2/LD-5a/DQw3 HLA class II haplotype. Multiple genetic mechanisms in the generation of polymorphic HLA class II genes, *J. Immunol.,* 140, 3631, 1988.

36. **Hurley, C. K., Gregersen, P. K., Steiner, N., Bell, J., Hartzman, R., Nepom, G., Silver, J., and Johnson, A. H.,** Polymorphism of the HLA-D region in American blacks, *J. Immunol.,* 140, 885, 1988.

37. **Gorski, J.,** First domain sequence of the HLA-DRβ1 chain from two HLA-DRw14 homozygous typing cell lines: TEM (Dw9) and AMALA (Dw16), *Hum. Immunol.,* 24, 145, 1989.

INDEX

Nucleosomes, 228
Nucleotide sequence analysis
  Class II allelic polymorphisms, 157—164
  MHC Class II polymorphism, 150—154
Nucleotide sequences
  DVβ gene, 6—7
  homogenization vs. diversification, by conversion, 192—193
Nucleotide substitutions, 193

O

OBP, see Octamer binding protein
Octamer, 229, 237
Octamer binding protein (OBP), 212
Oligo typing, 206—207
Oligonucleotide hybridization, 205—207
Oligonucleotide probes, 136, 139—142, 159, 162, 170—174, 180
  allele-specific, 170—172, 176, 178
  family analysis, 176
  insulin-dependent diabetes mellitus, 174—176
  locus-specific, 170—172, 174
  methodologic aspects of use of, 173—174
  shared-sequence, 170, 172—173, 179
Oligonucleotide typing, 206
Oligonucleotides, 206, 231—232, 236—237

P

PCR, see Polymerase chain reaction
Pemphigus vulgaris (PV), 130, 135—136, 139—143, 157, 162—163
  DR4-associated susceptibility, 139—141
  DRw6 susceptibility, 139, 141—143
  nucleotide sequence analysis, 159, 162
Peptide mapping, 45
PFGE, 2, 69, 149, 226—227
  distances between adjacent subregions, 12
  gene order in HLA Class II gene region, 14
  HLA probes used for, 3
  restriction enzymes for analysis of HLA Class II gene region, 2—5
Plasma membrane, 72
Plasmids, 192, 229—231, 242
Point mutations, 67, 110, 152, 154, 189—190, 194, 240
Pokeweed mitogen, 72
Polymerase chain reaction (PCR), 130, 150
Polymorphic residues, 131, 140—141, 143—144, 152, 155—156, 160—161
Polymorphism, 108, 144, 192
  defined, 187
  DQα genes, 113—114
  DQβ genes, 114—116
  DRβ, 122
  DR4 subtypes, 121
  extent of, 188
  HLA Class II genes, 109—110
  long-range vs. short-range variations, 194—195
  mechanisms for generation of, 152

MHC Class I genes, 185—200
Polymorphism within SD, DR, and DQ specificities, 27, 29—31
Primed lymphocyte (LD) typing (PLT) test, 22—24
Probing for disease susceptibility, 169—184
Promoter elements, 228—233
Promoter sequences, 241—242
Promoters, 240
Prostaglandins, 226
Protein kinase C, 72
Protein polymorphism, 33
Pseudogenes, 86, 187, 196, 203
Pulsed field gel electrophoresis, see PFGE
PV, see *Pemphigus vulgaris*
Pyrimidine tract, 229, 232, 242

R

RA, see Rheumatoid arthritis
Reciprocal gene conversion, 33
Recombination, 108, 116—121, 149, 154, 181, 191—195, 203
Regulatory polymorphism, 60
Resistance, 134—135
Resting macrophages, 226
Resting T cells, 71—72
Restriction enzymes
  infrequently cutting, 9
  PFGE analysis of HLA Class II gene region, 2—5
Restriction fragment length polymorphism, see RFLP
RF-X, 212, 214—217
RFLP
  Class II allelic polymorphism, 157
  DNA typing, 206
  Dw/LD subtype, 30—31
  genomic DNA analysis, 45
RFLP analysis, 136
RFLP studies, 14—15
  DVβ location, 6, 8
  MHC Class II polymorphism, 150
Rheumatoid arthritis (RA), 120, 122—125, 139, 141, 156, 161, 163, 170, 172, see also Juvenile rheumatoid arthritis
  genes of interest, 180
  nucleotide sequence analysis, 159
  probing for susceptibility to, 178—180
  shared sequence epitope, 173
Rheumatoid factor, 178
RM2, 238—240
RM3, 238—240
Rubella, 136, 138

S

Schistosoma japonicum, 72
SCIDs, 215—216, 226
SD, 21, 187
Selection, 67, 193
Selection pressures, 110, 119, 196
Selfish DNA, 15
Sequence comparisons, 189